New Insights into Novel Catalysts for Treatment of Pollutants in Wastewater

New Insights into Novel Catalysts for Treatment of Pollutants in Wastewater

Editors

Hao Xu
Yanbiao Liu

Basel • Beijing • Wuhan • Barcelona • Belgrade • Novi Sad • Cluj • Manchester

Editors

Hao Xu
Department of Environmental Science and Engineering
Xi'an Jiaotong University
Xi'an
China

Yanbiao Liu
College of Environmental Science and Engineering
Donghua University
Shanghai
China

Editorial Office
MDPI
St. Alban-Anlage 66
4052 Basel, Switzerland

This is a reprint of articles from the Special Issue published online in the open access journal *Catalysts* (ISSN 2073-4344) (available at: www.mdpi.com/journal/catalysts/special_issues/catalysts_pollutants_treament_wastewater).

For citation purposes, cite each article independently as indicated on the article page online and as indicated below:

Lastname, A.A.; Lastname, B.B. Article Title. *Journal Name* **Year**, *Volume Number*, Page Range.

ISBN 978-3-0365-9866-6 (Hbk)
ISBN 978-3-0365-9865-9 (PDF)
doi.org/10.3390/books978-3-0365-9865-9

Contents

Preface . . . . **vii**

Siyuan Guo, Zhicheng Xu, Wenyu Hu, Duowen Yang, Xue Wang and Hao Xu et al.
Progress in Preparation and Application of Titanium Sub-Oxides Electrode in Electrocatalytic Degradation for Wastewater Treatment
Reprinted from: *Catalysts* **2022**, *12*, 618, doi:10.3390/catal12060618 . . . . **1**

Haochen Yan, Fuqiang Liu, Jinna Zhang and Yanbiao Liu
Facile Synthesis and Environmental Applications of Noble Metal-Based Catalytic Membrane Reactors
Reprinted from: *Catalysts* **2022**, *12*, 861, doi:10.3390/catal12080861 . . . . **23**

Haiqin Lu, Guilu Xu and Lu Gan
N Doped Activated Biochar from Pyrolyzing Wood Powder for Prompt BPA Removal via Peroxymonosulfate Activation
Reprinted from: *Catalysts* **2022**, *12*, 1449, doi:10.3390/catal12111449 . . . . **36**

Ruicheng Li, Jianhua Xiong, Yuanyuan Zhang, Shuangfei Wang, Hongxiang Zhu and Lihai Lu
Catalytic Ozonation of Norfloxacin Using Co-Mn/CeO_2 as a Multi-Component Composite Catalyst
Reprinted from: *Catalysts* **2022**, *12*, 1606, doi:10.3390/catal12121606 . . . . **48**

Fanxi Zhang, Dan Shao, Changan Yang, Hao Xu, Jin Yang and Lei Feng et al.
New Magnetically Assembled Electrode Consisting of Magnetic Activated Carbon Particles and Ti/Sb-SnO_2 for a More Flexible and Cost-Effective Electrochemical Oxidation Wastewater Treatment
Reprinted from: *Catalysts* **2022**, *13*, 7, doi:10.3390/catal13010007 . . . . **61**

Guodong Wang, Shirong Zong, Hang Ma, Banglong Wan and Qiang Tian
Removal Efficiency and Performance Optimization of Organic Pollutants in Wastewater Using New Biochar Composites
Reprinted from: *Catalysts* **2023**, *13*, 184, doi:10.3390/catal13010184 . . . . **73**

Xuanqi Kang, Jia Wu, Zhen Wei, Bo Jia, Qing Feng and Shangyuan Xu et al.
Modification of Ti/Sb-SnO_2/PbO_2 Electrode by Active Granules and Its Application in Wastewater Containing Copper Ions
Reprinted from: *Catalysts* **2023**, *13*, 515, doi:10.3390/catal13030515 . . . . **85**

Peiguo Zhou, Zongbiao Dai, Tianyu Lu, Xin Ru, Meshack Appiah Ofori and Wenjing Yang et al.
Degradation of Rhodamine B in Wastewater by Iron-Loaded Attapulgite Particle Heterogeneous Fenton Catalyst
Reprinted from: *Catalysts* **2022**, *12*, 669, doi:10.3390/catal12060669 . . . . **98**

Jiali Yu, Shihu Shu, Qiongfang Wang, Naiyun Gao and Yanping Zhu
Evaluation of Fe^{2+}/Peracetic Acid to Degrade Three Typical Refractory Pollutants of Textile Wastewater
Reprinted from: *Catalysts* **2022**, *12*, 684, doi:10.3390/catal12070684 . . . . **118**

Wei Qian, Wangtong Hu, Zhifei Jiang, Yongyi Wu, Zihuan Li and Zenghui Diao et al.
Degradation of Tetracycline Hydrochloride by a Novel CDs/g-C_3N_4/$BiPO_4$ under Visible-Light Irradiation: Reactivity and Mechanism
Reprinted from: *Catalysts* **2022**, *12*, 774, doi:10.3390/catal12070774 . . . . **132**

Yishi Qian, Kai Chen, Guodong Chai, Peng Xi, Heyun Yang and Lin Xie et al.
Performance Optimization and Toxicity Effects of the Electrochemical Oxidation of Octogen
Reprinted from: *Catalysts* **2022**, *12*, 815, doi:10.3390/catal12080815 . . . . **147**

Yizan Gao, Xiaodan Yang, Xinwei Lu, Minrui Li, Lijun Wang and Yuru Wang
Kinetics and Mechanisms of Cr(VI) Removal by nZVI: Influencing Parameters and Modification
Reprinted from: *Catalysts* **2022**, *12*, 999, doi:10.3390/catal12090999 . . . . **161**

Wenchang Zhao, Yuling Dai, Wentian Zheng and Yanbiao Liu
Peroxymonosulfate Activation by Photoelectroactive Nanohybrid Filter towards Effective Micropollutant Decontamination
Reprinted from: *Catalysts* **2022**, *12*, 416, doi:10.3390/catal12040416 . . . . **179**

Hao Xu and Yanbiao Liu
New Insights into Novel Catalysts for Treatment of Pollutants in Wastewater
Reprinted from: *Catalysts* **2023**, *13*, 840, doi:10.3390/catal13050840 . . . . **191**

Preface

This scientific work focuses its research efforts on wastewater treatment technologies, with a particular emphasis on the development and optimization of novel catalysts designed for the effective removal of both organic and inorganic pollutants from wastewater. The scope of the research encompasses a variety of catalysts' synthesis and modification, including precious metals, modified electrodes, magnetic activated carbon particles, and nanomaterials. These catalysts are utilized in different wastewater treatment processes, such as electrochemical oxidation, Fenton reactions, and photoelectrocatalysis. The purpose of these studies is to provide new insights and develop cost-effective and more efficient methods for pollutant removal. This research aims to improve the feasibility and acceptability of wastewater treatment by introducing innovative catalysts and technologies, thereby mitigating environmental pollution and safeguarding ecosystems. The list of authors who contributed to this Special Issue, arranged in alphabetical order, is as follows: Banglong Wan, Bo Jia, Changan Yang, Dan Shao, Donggi Wang, Duowen Yang, Fanxi Zhang, Fuqiang Liu, Guilu Xu, Guodong Chai, Guodong Wang, Haiqin Lu, Hang Ma, Hao Xu, Haochen Yan, Haojie Song, Heyun Yang, Hongxiang Zhu, Hui Jin, Jia Wu, Jiali Yu, Jianhua Xiong, Jiaxin Hou, Jin Yang, Jinna Zhang, Kai Chen, Lei Feng, Lihai Lu, Lijun Wang, Lin Xie, Lu Gan, Lu Qin, Meshack Appiah Ofori, Mingyu Li, Minrui Li, Naiyun Gao, Peiguo Zhou, Peng Xi, Qiang Tian, Qing Feng, Qiongfang Wang, Ruicheng Li, Shangyuan Xu, Shihu Shu, Shirong Zong, Shuangfei Wang, Siyuan Guo, Sizhe Wang, Tianyu Lu, Wangtong Hu, Wei Qian, Wei Yan, Wenchang Zhao, Wenjing Yang, Wentian Zheng, Wenyu Hu, Xiaodan Yang, Xiaohua Jia, XiaoliangLi, Xin Ru, Xing Xu, Xinwei Lu, Xuanqi Kang, Xue Wang, Yanbiao Liu, Yanping Zhu, Yishan Lin, Yishi Qian, Yizan Gao, Yong Li, Yongyi Wu, Yuanyuan Zhang, Yuling Dai, Yunhai Wang, Yuru Wang, Zenghui Diao, Zhen Wei, Zhi Long, Zhicheng Xu, Zhifei Jiang, Zihuan Li, Zongbiao Dai. Compiling these research findings into a single publication creates a centralized repository of knowledge on the latest wastewater treatment technologies for the academic community, industry professionals, and policymakers. The target audience for this scientific compilation includes environmental scientists and engineers, academic researchers and students, investors, and entrepreneurs. We extend our gratitude to the journals, their editors, and all the authors who contributed to this Special Issue, as well as to all the funding bodies and institutions that supported the associated scientific work.

Hao Xu and Yanbiao Liu
Editors

Review

Progress in Preparation and Application of Titanium Sub-Oxides Electrode in Electrocatalytic Degradation for Wastewater Treatment

Siyuan Guo [1,†], Zhicheng Xu [1,†], Wenyu Hu [1], Duowen Yang [1], Xue Wang [1], Hao Xu [1,2,3,*], Xing Xu [3], Zhi Long [4] and Wei Yan [1,2]

1 Department of Environmental Science Engineering, Xi'an Jiaotong University, Xi'an 710049, China; 18735815148@163.com (S.G.); kylezcxu@foxmail.com (Z.X.); hwy2018@stu.xjtu.edu.cn (W.H.); yangduowen@xjtu.edu.cn (D.Y.); wangxue_19@stu.xjtu.edu.cn (X.W.); yanwei@xjtu.edu.cn (W.Y.)
2 Research Institute of Xi'an Jiaotong University, Hangzhou 311200, China
3 Shandong Shenxin Energy Saving and Environmental Protection Technology Co., Ltd., Industrial Recirculating Water Treatment Engineering Technology Centre of Zaozhuang City, Tengzhou 277531, China; xuxing5957@sina.com
4 Henan Longxing Titanium Industry Technology Co., Ltd., Jiyuan 454650, China; longzhi5669@126.com
* Correspondence: xuhao@xjtu.edu.cn
† These authors contributed equally to this work.

Abstract: To achieve low-carbon and sustainable development it is imperative to explore water treatment technologies in a carbon-neutral model. Because of its advantages of high efficiency, low consumption, and no secondary pollution, electrocatalytic oxidation technology has attracted increasing attention in tackling the challenges of organic wastewater treatment. The performance of an electrocatalytic oxidation system depends mainly on the properties of electrodes materials. Compared with the instability of graphite electrodes, the high expenditure of noble metal electrodes and boron-doped diamond electrodes, and the hidden dangers of titanium-based metal oxide electrodes, a titanium sub-oxide material has been characterized as an ideal choice of anode material due to its unique crystal and electronic structure, including high conductivity, decent catalytic activity, intense physical and chemical stability, corrosion resistance, low cost, and long service life, etc. This paper systematically reviews the electrode preparation technology of Magnéli phase titanium sub-oxide and its research progress in the electrochemical advanced oxidation treatment of organic wastewater in recent years, with technical difficulties highlighted. Future research directions are further proposed in process optimization, material modification, and application expansion. It is worth noting that Magnéli phase titanium sub-oxides have played very important roles in organic degradation. There is no doubt that titanium sub-oxides will become indispensable materials in the future.

Keywords: electrocatalytic oxidation; anode materials; Magnéli phase; titanium oxide; wastewater treatment

Citation: Guo, S.; Xu, Z.; Hu, W.; Yang, D.; Wang, X.; Xu, H.; Xu, X.; Long, Z.; Yan, W. Progress in Preparation and Application of Titanium Sub-Oxides Electrode in Electrocatalytic Degradation for Wastewater Treatment. *Catalysts* **2022**, *12*, 618. https://doi.org/10.3390/catal12060618

Academic Editor: Pedro Modesto Alvarez Pena

Received: 9 May 2022
Accepted: 2 June 2022
Published: 6 June 2022

1. Electrocatalytic Oxidation Technologies

The rapid development of modern industries in various fields directly brings about growing pollution and ever-increasing refractory and toxic substances in the different liquid effluent. Traditional wastewater treatment methods have difficulty meeting the environmental requirements for such organic wastewater [1]. Advanced oxidation technology has developed rapidly in recent years as a practical approach to constructing a carbon-neutral treatment mode and realizing the green development of wastewater treatment technologies. Electrocatalytic oxidation technology is a promising way of removing organic pollutants. It is a green chemical technology that adopts electrodes with favorable catalytic performance to generate hydroxyl radicals (or other radicals and groups) with strong oxidation ability and react with the organic pollutants in the solution to decompose them into H_2O and

CO_2 [2,3]. Compared with traditional water purification technologies, it has drawn significant interest thanks to its advantages, such as strong oxidation capacity, with no chemical agents added, and clean and mild conditions [4–6].

1.1. Technique Principle

As shown in Figure 1, the principle of electrocatalytic oxidation technology can be divided into direct oxidation and indirect oxidation [7,8]. Direct oxidation refers to the process in which the pollutant is adsorbed directly on the anode surface and oxidized by losing its electron. It can be oxidized to substances easily for less toxic biochemical treatment, or can even be directly and deeply oxidized to H_2O and CO_2 to achieve the purpose of removing organic pollutants from wastewater. Kirk et al. [9] studied the anodic oxidation process of aniline and believed that aniline was oxidized through an electron transfer reaction, which was a typical direct anodic oxidation process.

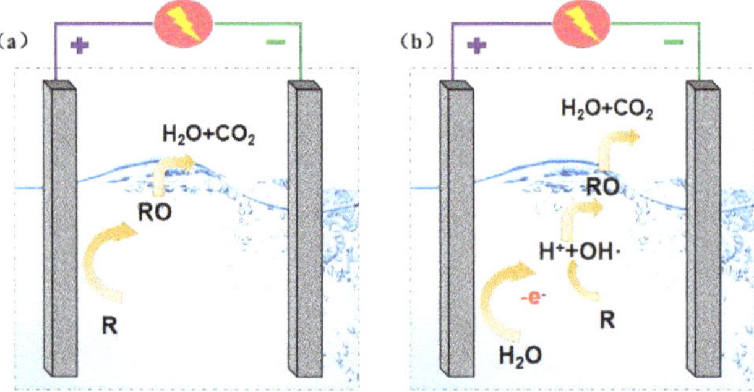

Figure 1. Schematic diagram of direct oxidation (**a**) and indirect oxidation (**b**) of electrocatalytic oxidation technology.

Indirect oxidation refers to strong oxidizing free radicals from an anode reaction (• OH [10], O_2 •, • HO_2, etc.), or products from an intermediate reaction (such as a chlorine-active substance) that are generated in the electrode surface or spread to the aqueous solution for pollutants oxidation, ultimately achieving the degradation of organic pollutants. Taking hydroxyl radicals as an example, the reaction processes are as follows:

$$H_2O \rightarrow \cdot OH + H^+ + e^- \tag{1}$$

$$R + 2 \cdot OH \rightarrow RO + H_2O \tag{2}$$

$$2 \cdot OH \rightarrow H_2O + \frac{1}{2}O_2 \tag{3}$$

In most actual electrocatalytic oxidation processes, there is no definite sharp boundary between direct oxidation and indirect oxidation. However, the above two processes are usually both included [11]. Li et al. [8] studied removing azo dyes from water by oxidation using different kinds of electrodes. They found that the oxidation process was mainly determined by direct oxidation for some active electrodes (Ru and IrO_2). In contrast, both direct oxidation and indirect oxidation of active radicals influenced the oxidation process of inactive electrodes (BDD and PbO_2).

1.2. Anode

The electrocatalytic oxidation system is a composite system, mainly including an anode, cathode, power supply, electrolytic cell, and other supporting equipment [12]. Electrode material is the key point of electrocatalytic technology because any research on electrocatalytic technology must be based on a certain electrode material with superior performance [13–15]. The electrode not only maintains the conductive process (i.e., conductivity) but also activates the reactants (i.e., catalytic activity), increases the electron transfer rate, and performs the promotion and selection functions for electrochemical reactions. In addition, to meet the requirements of practical use, electrode materials are also supposed to have satisfactory properties in stability and economic performance.

The active substances with oxidizing ability produced by the anode play an essential role in the removing of organic pollutants. At the same time, side reactions (such as oxygen evolution reactions) can also occur at the anode, which will burden the reaction energy consumption of the system. Therefore, selecting electrode materials (especially anode materials) in the electrocatalytic oxidation system is crucial. The electrode materials used in academic research and engineering practice mainly include graphite electrodes (graphite electrode and BDD electrode), noble metal electrodes (Pt electrode), and titanium-based metal oxide electrodes.

Graphite electrode refers to the electrode composed of carbon, which is usually used for dye decolorization studies due to its high adsorption and larger specific surface area [16–18]. Its strong adsorption can also incur adverse effects for the spread of the intermediate, causing the accumulation of the products on the electrode surface and covering the active site to impair the electrode, resulting in a loss of catalytic activity, which extensively limited its widespread application. Common noble metal electrodes are Pt, Ru, Au, etc. [19,20]. This electrode has good corrosion resistance and high oxygen evolution potential, while the high expenditure restrains its use in the industrial field of electrocatalytic oxidation. The BDD electrode is considered competent in electrocatalytic oxidation to organic matters with superior performance, which has good electrical conductivity, high oxygen evolution potential, strong corrosion resistance, and good stability [21]. However, its preparation process is relatively complex with high costs. The electrode surface can form more thin-polymer film in the degradation process, leading to electrode passivation, which cannot be ideally used in practical engineering [22,23]. Titanium-based metal oxide electrode uses a titanium sheet as the substrate and is covered with one or more metal oxide coatings, also known as a dimension stable anode (DSA). Common metal oxide coatings on DSA electrodes mainly include Sb-SnO_2 [24–27], IrO_2 [28], RuO_2 [29], and PbO_2 [30–34]. At present, there are many investigations on the doping modification methods of the Ti/PbO_2 electrode, mainly including the doping ion [35–37], introducing the intermediate layer [38], doping nanoparticles [39,40] and regulating the micro-morphology of electrode materials [41,42]. Compared with the instability of graphite electrodes and the high price of noble metal electrodes and boron-doped diamond electrodes, titanium-based metal oxide electrodes have attracted great attention and are widely applied in practice because of their relatively low-price and simple preparation as well as easy functionalization and modification [43].

Nonetheless, there are still many obstacles when using titanium-based metal oxide electrodes in the practical application of the electrocatalytic oxidation system, mainly focusing on the following four aspects, namely safety, catalytic activity, economy, and stability:

(1) Potential safety hazards: Due to intrinsic properties, traditional titanium-based metal oxide electrodes (including the PbO_2 electrode, Sb-SnO_2 electrode, and Ir/Ru/Ta series electrode) have an issue of surface metal element dissolution in reactions, which will last during the entire life cycle of the electrode, resulting in a significant risk of secondary pollution [44]. In addition, there is a significant scaling phenomenon on the cathode surface in the engineering practice of electrocatalytic oxidation technology. Once the scale layer thickens and results in the connection between anode and cathode, the anode surface layer will be penetrated and corroded, seriously impairing the safe and stable operation of the system [45,46];

(2) Inadequate catalytic capacity: The yield of the hydroxyl radical on the surface of the Sb-SnO_2 electrode and Ir/Ru/Ta electrode is not sufficient. Although the hydroxyl radical yield of the PbO_2 electrode is higher than that of the former two, it is mainly in the adsorption state [47] and incompetent for effective degradation, bringing its oxidation capacity even lower than that of the Sb-SnO_2 electrode [48,49];

(3) Poor economy: Although it is relatively cheap compared with the noble metal electrode and BDD electrode, the titanium-based metal oxide electrode still requires the use of noble metal salt to prepare a brush coating solution in the preparation process, resulting in high cost and vulnerability to fluctuations in noble metal market price [50,51];
(4) Stability to be improved: The above-mentioned PbO_2 electrode, Sb-SnO_2 electrode, and Ir/Ru/Ta series electrode all have a surface-active element dissolution, surface oxide layer loss, and titanium base passivation in the use process (especially with F^- ion or high concentration Cl^- ion), which leads to the irreversible inactivation of the electrode [52,53]. Thus, the stability needs further improvement.

The above discussion demonstrates many problems in the practical application of the anode materials currently used. Hence, it is urgent to develop new electrode materials to meet the requirements of electrocatalytic oxidation technology.

1.3. *Magnéli Titanium Sub-Oxides Material*

Titanium is a metal material with abundant storage (only second to iron) and excellent biosafety performance. However, it was used as the substrate material for electrodes because of the easy formation of an insulating oxide layer on its surface [54]. However, when the valence state of titanium is lower than Ti^{3+}, its properties will be completely different from those of traditional titanium oxides [55,56], which have attracted widespread attention since its discovery.

1.3.1. Crystal Structure of Titanium Sub-Oxides Material

Magnéli titanium sub-oxide is a general term for a series of non-stoichiometric titanium oxides, whose generic formula is Ti_nO_{2n-1} ($4 \leq n \leq 10$) [57], including a series of titanium oxides, such as Ti_4O_7, Ti_5O_9, and Ti_6O_{11}.

The crystal structure of Magnéli titanium sub-oxides can be regarded as a network structure composed of a two-dimensional chain of titanium dioxide with multiple titanium atoms at the center and oxygen atoms at the apex in an octahedral structure [58]. For example, as shown in Figure 2a, the Ti_4O_7 crystal is composed of three regular octahedral TiO_2 layers and one TiO layer due to the vacancy of an oxygen atom in every four layers and the apparent shear plane of the crystal structure. The TiO layer forms a shared edge perpendicular to the surface due to the vacancy of oxygen atoms, resulting in titanium atoms closer together to form a conductive band. In contrast, a TiO_2 layer on both sides of the TiO layer can wrap the conductive band of titanium sub-oxides, which determines the strong corrosion resistance of Ti_4O_7.

1.3.2. Physicochemical Properties of Titanium Sub-Oxides Material

The unique lattice structure makes titanium sub-oxide materials possess unique physical and electrochemical properties, such as high conductivity, superior chemical stability, and a wide electrochemical stability potential window [59]. The conductivity of titanium sub-oxides is in the same order of magnitude as carbon rods. As shown in Table 1, the conductivity of Ti_4O_7 material is the highest, which can reach 1500 $S{\cdot}cm^{-1}$, about twice that of graphite (727 $S{\cdot}cm^{-1}$).

Recent research shows that Ti_4O_7 is exceptionally stable in a strong corrosive solution and an alkali environment. As shown in Table 2, the mass loss of Ti_4O_7 in the fluorine-containing acidic electrolyte is only 0.29% after 350 h. In a high concentration HF/HNO_3 electrolyte, the mass loss is only 12.7%, far better than that of metal Ti. Some studies have also stated that the expected half-life of Ti_4O_7 is 50 years in 1.0 M H_2SO_4 at room temperature [60].

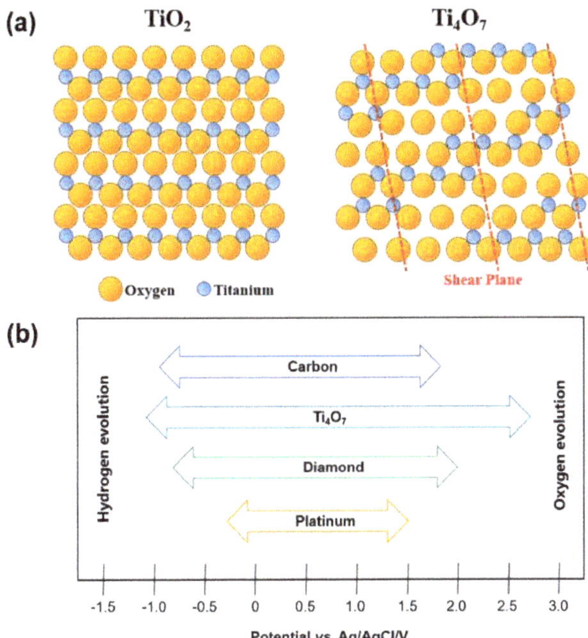

Figure 2. Comparison of crystal structures of TiO_2 and Ti_4O_7 (**a**); hydrogen and oxygen evolution potentials of some electrode materials (1.0 M H_2SO_4) (**b**) [58].

Table 1. Electrical conductivity of each phase of titanium sub-oxides at 298 K [58].

Ti_nO_{2n-1} (s)	Electrical Conductivity (σ/S cm^{-1})	Lg (σ/S cm^{-1})
Ti_3O_5	630	2.8
Ti_4O_7	1035	3.0
Ti_4O_7 *	1995	3.3
Ti_5O_9	631	2.8
Ti_6O_{11}	63	1.8
Ti_8O_{15}	25	1.4
$Ti_3O_5 + Ti_4O_7$	410	2.6
$Ti_4O_7 + Ti_5O_9$	330	2.5
$Ti_5O_9 + Ti_6O_{11}$	500	2.7

* Another preparation method was used to prepare Ti_4O_7.

Table 2. Comparison of corrosion resistance between Ti_4O_7 and Ti [58].

Sample	Electrolyte	150 h Quality Loss/%	350 h Quality Loss/%
Ti	HF	22	100
Ti_4O_7		0.017	0.29
Ti	HF/HNO_3	100	100
Ti_4O_7		0.56	12.7

In addition, Figure 2b illustrates that the difference between the oxygen evolution potential and hydrogen evolution potential of Ti_4O_7 is nearly 4 V in a 1.0 m H_2SO_4 solution, which exhibits a promising application prospect in the field of electrocatalytic oxidation [61].

Furthermore, the titanium sub-oxide material can be considered a low-cost material, and the properties of the titanium-based material make it intrinsic for electrode safety. Therefore, titanium sub-oxide is suitable as electrode material in electrocatalytic degradation, with superior properties meeting the requirements for selecting electrodes in the electrocatalytic oxidation system [62–64].

As a kind of conductive material, titanium sub-oxide material has the advantages of high conductivity, decent catalytic activity, strong physical and chemical stability, high corrosion resistance, low cost, and long service life, etc., which is considered a kind of catalytic electrode with great development potential. At present, many domestic and foreign studies have used titanium sub-oxide material to treat various types of refractory organic wastewater, highlighting the unique advantages of titanium sub-oxides in the field of electrocatalytic oxidation.

According to the classification of different electrode forms, a great deal of work was summarized on the preparation and application of titanium sub-oxide electrodes. Then, the existing technical problems and challenges were also put forward, with further prospects for the future development direction.

2. Preparation Methods of Titanium Sub-Oxides Electrode

At present, there are three types of titanium sub-oxide materials used in the electrocatalytic oxidation wastewater systems, including a coated electrode, integrated electrodes, and composite electrodes. Different kinds of electrodes have their own advantages and disadvantages. Many researchers have made many contributions in optimizing the preparation process and improving the performance of electrodes, which is conducive to broadening the application prospect of titanium sub-oxides electrodes.

2.1. Preparation of Titanium Sub-Oxides Powder

As the primary raw material for electrode preparation, titanium sub-oxides powder has been used in many processes. Current synthesis methods of titanium sub-oxides powder focus on reducing of titanium dioxide at a high temperature between 600 and 1000 °C [65–67]. In addition, there are also studies on the preparation of titanium sub-oxides powder using metallic titanium, organic titanium, inorganic titanium, and other raw materials as precursors [68–70]. Briefly, titanium dioxide is used as a prevalent raw material due to its abundant reserves and relatively low cost [71]. Table 3 lists the principles and process comparisons of the three methods of carbothermal reduction, hydrogen reduction, and metallothermic reduction using titanium dioxide as the precursor.

Table 3. Synthesis methods of titanium sub-oxides powder with TiO_2 as precursor.

Synthesis Method	Principle	Process Condition and Results	References
Carbothermal reduction	$nTiO_2(s) + C(s) = Ti_nO_{2n-1}(s) + CO(g)$	C-Ti_4O_7 was obtained at 1025 °C in N_2 gas flow.	[72]
Hydrogen reduction	$nTiO_2(s) + H_2(g) = Ti_nO_{2n-1}(s) + H_2O(g)$	Ti_4O_7 was obtained at 1050 °C in H_2 gas flow.	[73]
Metallothermic reduction	$nTiO_2(s) + Me(s) = Ti_nO_{2n-1}(s) + MeO(s)$	$Ti_4O_7$and Ti_6O_{11} were obtained by mechanical activation of Ti and TiO_2 powder and annealing at 1333–1353 K in Ar gas flow for 4 h.	[74]
	$(2n-1)\ TiO_2(s) + Ti(s) = 2Ti_nO_{2n-1}(s) (n = 1,2,3\ldots)$	With silicon as reducing agent and calcium chloride as additive, TiO_2 powder was reduced to various mixed phases of titanium sub-oxide under different experimental conditions.	[75]

2.2. Titanium Sub-Oxides-Coated Electrode

Titanium sub-oxides-coated electrode refers to the preparation of titanium sub-oxides powder by reduction and other methods, which is then coated or deposited on the anode substrate, which is adopted as a catalytic electrode. The preparation methods of coating electrodes mainly include the coating methods, magnetron sputtering methods, the electrodeposition method, and the sol–gel method, etc. As shown in Figure 3, the performance of the electrodes prepared vary depending on the preparation processes.

2.2.1. Coating Method

The coating method includes coating, spraying, and powder curing technology, among which plasma spraying is a new kind of multipurpose spraying method with relative maturity and precision. Thanks to its characteristics of ultra-high temperature, fast jet particles, compact coating density, high bond strength, inert gas as the working gas, and no easy oxidation of spraying material, etc., it is widely used in the preparation of titanium sub-oxides electrodes. Teng et al. [76] coated titanium sub-oxides powder on titanium mesh and titanium plate by plasma spraying technology, which successfully prepared a Ti/Ti_4O_7 electrode. The preparation process and characterizations are shown in Figure 3a,b. Ti_4O_7 powder was approximately spherical and uniformly and compactly covered on a Ti matrix, providing a large surface area and more active sites for the electrochemical reaction. The electrode had high conductivity and low oxygen evolution potential, presenting fair oxidation activity to sulfadiazine. Soliu et al. [77] also prepared a Ti/Ti_4O_7 electrode by plasma spraying technology, which was employed for the electrocatalytic degradation of amoxicillin. In the plasma spraying system, 20–60 μm of Ti_4O_7 particles were sprayed on the surface of the pretreated titanium plate under the conditions of a 700 A current, argon, and with hydrogen-mixed gas (19.6% H_2) as the carrier at 2 rpm.

Al, carbon cloth, and other conductive substrates are also alternatives for the matrix and the coating on a titanium base. Han et al. [78] successfully coated Ti_4O_7 ceramic particles on an Al substrate by plasma spraying. The system was operated under the condition of a 10 V spray voltage in a nitrogen atmosphere, and the accelerated life of the prepared electrode was 69.6% longer than that of the commonly used titanium-based SnO_2 electrode. The current efficiency and energy consumption were increased by 5.59% and 16.2%, respectively, with electrochemical performance greatly improved. Geng et al. [79] impregnated a layer of TiO_2 on a porous Al_2O_3 matrix by the coating reduction method. The as-prepared electrode was aged at room temperature for 2 h, dried at 100 °C for 1 h, and sintered in air at 800 °C, which was all repeated two to eight times, and it was finally reduced in the H_2 atmosphere at 1050 °C to obtain the Ti_4O_7 layer. The conductivity of the prepared electrode is more than 200 S cm^{-1}, with the particle size of the coating in the 200–300 nm range and the average pore size being 350 nm, exhibiting superior electrochemical properties.

This technology pretreated the substrate and coated the anode surface with a dense and uniform Ti_4O_7 particle coating, which has the advantages of simple operation, low equipment maintenance cost, and flexible regulation performance. However, there are still technical problems, such as the weak adhesion between the substrate and the coating, which impairs the stability of the electrode.

2.2.2. Magnetron Sputtering Method

Magnetron sputtering is a physical vapor deposition method with the advantages of easy control, a large coating area, and strong adhesion. In Ar + O_2 plasma, Wong et al. [80] deposited a series of $TiO_x(0 < x \leq 2)$ films with different O/Ti ratios on silicon substrates by reactive magnetron sputtering and discussed the effects of the oxygen flow rate and substrate temperature on the structural properties, mechanical properties, and electrical properties of the films, respectively. The results showed that the titanium, Ti(-O) solid solution, TiO, Ti_2O_3, Magnéli, rutile, and anatase TiO_2 phases appeared successively with the increase of the oxygen flow rate. The conductivity decreased with the rise of oxygen content but increased with the elevation of the deposition temperature. The hardness of the films changed in the order of TiO > Ti_2O_3 > Ti_4O_7. Titanium sub-oxides films with high performance can be obtained at the deposition temperature of 500 °C. These thin films possessed properties in terms of a small crystal diameter, the proper crystal structure of defects and grain boundaries, and good electrocatalytic oxidation performance.

Li et al. [81] also explored the magnetron sputtering method to deposit titanium sub-oxide on MGO-ZrO_2 plasma electrolytic oxidation (PEO) coating at different lining temperatures. XRD results shown in Figure 3e demonstrate that the relative content of titanium sub-oxide in the coating is relatively high at a higher temperature, while the content is almost zero at room temperature. Therefore, the temperature of the lining greatly influences the composition of the coating.

This technology is easy to control, with a stable prepared electrode morphology and a strong coating adhesion. In contrast, the composition is relatively impure, and is greatly affected by various preparation factors.

2.2.3. Electrodeposition Method

Ertekin et al. [82] successfully prepared titanium sub-oxide thin-film electrodes by the electrochemical deposition of titanium peroxide solution on indium tin oxide-coated glass with acetonitrile and hydrogen peroxide as supporting electrolytes. This study also confirmed the feasibility of the electrochemical deposition of crystalline Ti_nO_{2n-1} film, a preparation method with an electrochemical application prospect. SEM characterization in Figure 3c shows that the surface of the film electrode was porous and granular. Figure 3d XRD results confirm that different potentials and temperatures greatly influence the crystallinity of the film. At different electrochemical deposition potentials and temperatures, Ti_nO_{2n-1} was Ti_3O_5, Ti_4O_7, and Ti_5O_9, among which there was a low content of the Ti_4O_7 phase with the best performance.

This method is simple to operate with a relatively low cost. However, much the same as magnetron sputtering, the biggest problem of this technology is that the coating composition is not easy to control, which will have a significant influence on the performance of the electrode.

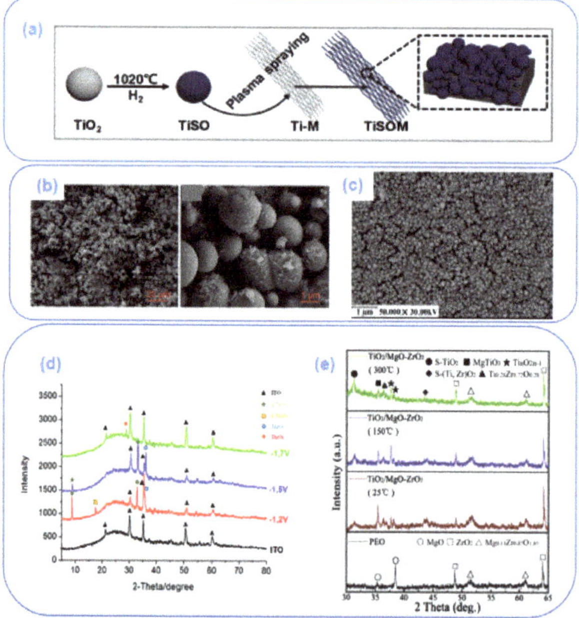

Figure 3. Processes of plasma spraying technology [76] (**a**); SEM images of coating electrode [76] (**b**); SEM images of electrodeposition electrode [82] (**c**); XRD pattern of electrodeposition electrode [82] (**d**); XRD pattern of magnetron sputtering electrode [81] (**e**).

2.2.4. Sol–Gel Method

The sol–gel method can reduce the temperature required for titanium sub-oxide electrode preparation. Sasmita et al. [60] used synthetic titanium dioxide powder dispersed in an aqueous solution of polyacrylic acid as a raw material, with the addition of monomer (MAM) and a crosslinking agent (MBAM) to the slurry in a ratio of 4:1. Next, 0.01 wt% APS was added to the mix for 6 h. The slurry was injected into a Teflon mold by adding 10 mL of TEMED to the slurry for catalytic polymerization. Polymerization took place within 5 min, and the pellets were cast. After drying at 50 °C, gel-cast electrodes were prepared by sintering for 6 h in a tubular furnace at 1050 °C.

Perica Paunovic et al. [83] used the Magnéli phase instead of a carbon-based carrier material as a catalyst carrier for the hydrogen/oxygen evolution electrocatalyst. Magnéli phase raw materials were pulverized by a ball mill at 200 rpm of acceleration in a dry ball mill for 4, 8, 12, 16, and 20 h, respectively. The sol–gel method deposited the Co (10 wt%) metal phase on the Magnéli matrix. With the extension of mechanical treatment time, the size of the Magnéli phase decreased and the catalytic activity of the corresponding electrocatalyst increased. However, due to the low surface area of the Magnéli phase, its activity is not as good as that of corresponding electrocatalysts deposited on other support materials, such as multi-walled carbon nanotubes.

This method is flexible because of the low cost required for preparation. However, the catalytic activity of the electrode could be affected to a certain extent due to the doping of complex coagulant aids.

2.3. Titanium Sub-Oxides-Integrated Electrode

The methods of preparing integrated electrodes mainly include compression reduction, powder sintering, and membrane preparation by hydrothermal reduction, etc. The preparation diagrams of different processes are shown in Figure 4.

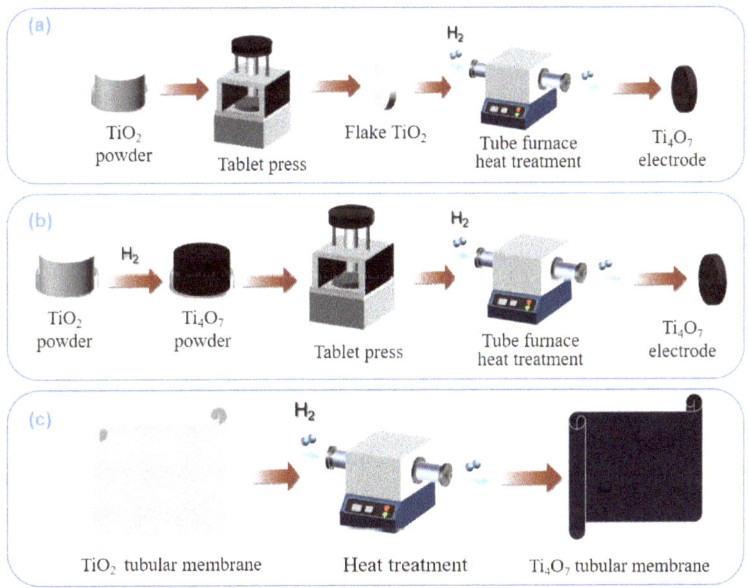

Figure 4. Preparation processes of compression reduction method (**a**); powder sintering method (**b**); membrane preparation by hydrothermal reduction method (**c**).

2.3.1. Compression Reduction Method

As shown in Figure 4a, TiO_2 powder is first pressed, molded, and then calcined in a reducing atmosphere to produce a titanium sub-oxides-integrated electrode. Regonini et al. [84] mixed titanium dioxide particles with polyethylene glycol as binder under pressure, sintered them into spheres, and then reduced them into titanium sub-oxide electrodes under the protection of Ar gas at 1300 °C. As shown in SEM images of Figure 5a, compared with titanium dioxide, the prepared electrode had conspicuous furrows to provide more active sites but did not form porous particles. The XRD results in Figure 5b also reveal different kinds of titanium sub-oxide phases such as Ti_4O_7, Ti_5O_9, and Ti_6O_{11} in the integrated electrode due to the reduction after pressing. Difficulty in controlling the reduction degree and an insufficient internal reduction degree resulted in the composite structure of different layers. The TiO_2 layer remaining in the inner layer generated the internal barrier layer capacitor (IBLC) effect, significantly weakening the electrode performance.

2.3.2. Powder Sintering Method

The powder sintering method can effectively avoid the disadvantages of the compression reduction method. The technical route is shown in Figure 4b, where titanium sub-oxides powder is prepared before compression sintering. Lin et al. [85] first synthesized Ti_4O_7 nanopowder by reducing titanium dioxide nanopowder at 950 °C in a H_2 atmosphere. Then, the prefabricated Ti_4O_7 nanopowder was mixed with polyacrylamide/polyvinyl alcohol binder to form slurry, and spray dried to small ceramic particles (40–80 mesh, water volume 5%). After the ceramic particles were loaded into the mold and fully vibrated, the prefabricated ceramic parts were pressed for 5 min by a 60 MPa static press. The macroporous Ti_4O_7 ceramic-integrated electrode was successfully prepared by drying the ceramic preform and sintering it for 11 h in a vacuum at 1350 °C. The average pore size of the electrode surface was 2.6 μm, and the porosity was 21.6%. The electrode had excellent electrochemical performance, including ultra-high oxygen evolution potential, decent electrochemical stability, high electroactive surface, and weak adsorption performance.

In addition, Lin et al. [86] studied a new spark plasma sintering (SPS) process, in which a pulse current was directly applied between powder particles to heat and sintering. Firstly, 2g performed Ti_4O_7 powder was put into a graphite mold and vacuumed to less than 1 Pa. Start sintered at 1100 °C for 40 min, then rapid cooling in the vacuum to below 60 °C with maintaining sintering pressure is 5 MPa. Finally, the Ti_4O_7-integrated electrode was prepared. The characteristics of this process can be evenly heated with a high heating rate, low sintering temperature, and time, which could avoid the electrode surface cracks to improve the stability of the electrode.

Sasmita et al. [60] also successfully adopted the powder sintering method to successfully produce a titanium sub-oxides-integrated electrode. They mixed Ti_4O_7 powder and paraffin oil, pressed and formed under the action of a hydraulic press, and sintered at 1050 °C for 6 h in a H_2 atmosphere. As shown in Figure 5c,d, the electrode surface is porous and granular, and the component is still Ti_4O_7.

The binder selection varies in different studies, but the principle is that the selected binder will volatilize entirely as far as possible in the operation process of high-temperature reduction. To sum up, a pure, stable, and uniformly integrated titanium sub-oxides electrode can be prepared by the powder sintering method; however, its application is limited to the shape and size. The cost is relatively high because the integrated titanium sub-oxides electrode with a large volume is not accessible to press and sinter.

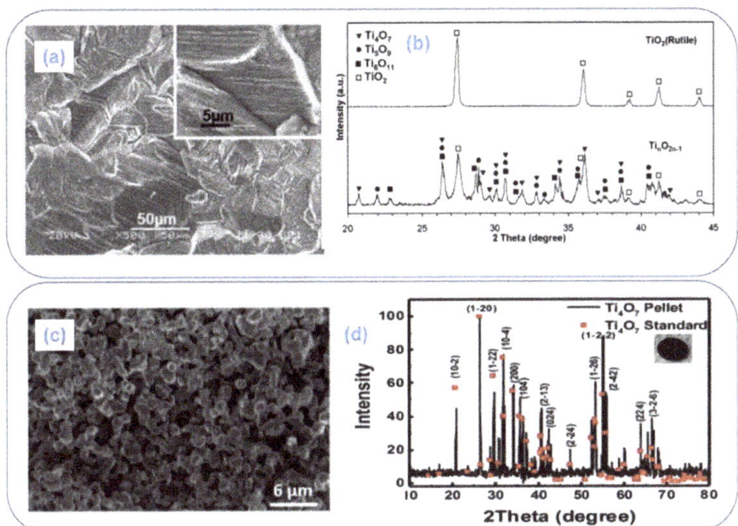

Figure 5. XRD pattern of compression reduction method [84] (**a**); SEM image of compression reduction method [84] (**b**); XRD pattern of powder sintering method [60] (**c**); SEM figure of powder sintering electrode [60] (**d**).

2.3.3. Membrane Preparation by Hydrothermal Reduction Method

A tubular titanium dioxide membrane can also be used to synthesize a titanium sub-oxides electrode, where titanium dioxide will be reduced to Ti_4O_7 after hydrothermal treatment. The specific path is shown in Figure 4c. Guo et al. [85] used a tubular titanium dioxide ultrafiltration membrane and reduced it to Ti_4O_7 by a hydrothermal method in a tubular furnace at 1050 °C in a H_2 atmosphere. With different treatment times, the main composition of the membrane also changed, and the phase change from Ti_6O_{11} to Ti_4O_7 occurred from 30 h to 50 h. Based on this, Liang et al. [86] designed a set of tubular electrode assembly reactors using titanium oxide thin film, in which stainless steel tubes (SSP) were used as cathodes, and tubular titanium oxide film was placed in the center of the tubes as anodes, making full use of advantages such as the good stability and low internal resistance of thin-film electrodes. The optimal conditions for the catalytic degradation of methylene blue are as follows: the removal rate was close to 100% when the current density was 9 mA·cm^{-2} for 90 min.

Besides, Mitsuhiro et al. [87] further studied the preparation of Ti_4O_7 film by the gradual oxidation of titanium foil. Ti_4O_7 films with a thickness of 3 μm were successfully prepared on the surface of titanium foil by annealing in air and heating to 1173 K at low-oxygen partial pressure. The surface of the electrode was characterized by four layers of several hundred nanometer-sized equiaxed particles. The thin film has excellent electrocatalytic oxidation performance and can absorb light in visible and near-infrared regions, making it have specific application prospects in photocatalysis.

2.4. Titanium Sub-Oxides Composite Electrode

The composite electrode is also a recent research hotspot for electrode performance improvement. It can be mainly divided into titanium sub-oxides modified by other kinds of electrodes and titanium sub-oxides doped with other metals.

2.4.1. Titanium Oxide-Doped Composite Electrode

By doping Magnéli phase Ti_4O_7, the electrocatalytic activity and the stability and conductivity of specific electrode materials can be increased to some extent. Zhang et al. [88]

prepared a Ru@Pt-type core–shell catalyst containing Ti_4O_7 (Ru@Pt/Ti_4O_7) by microwave pyrolysis. In this method, the mixture of the ruthenium precursor and TiO_2 was heat-treated in a H_2 atmosphere, Ru nuclei were obtained on the Ti_4O_7 carrier, and platinum shells were generated by microwave radiation. The characterization results showed that the catalytic material had a core–shell structure of Ru core and Pt shell, which significantly improved PT-Ru's durability. Zhao et al. [89] synthesized a MoS_2/Ti_4O_7 composite HER electrocatalyst by the hydrothermal method. Under the condition of 0.5 M H_2SO_4, the hydrogen evolution activity of the MoS_2/Ti_4O_7 composite catalyst was significantly improved. The results presented that, with the addition of conductive carrier Ti_4O_7, the HER activity of MoS_2 could be significantly enhanced by the formation of an interface SeO-Ti bond, thus improving the electrochemically active surface area fast charge transferability.

Wang et al. [90] compared the morphology, structure, and electrochemical properties of Ti_4O_7-modified $LiNi_{0.8}Co_{0.15}Al_{0.05}O_2$ (NCA) with the original NCA. Compared with the original one, the Ti_4O_7 modified NCA electrode exhibited better cycling performance and specific capacity thanks to its improved stability and conductivity. At the current density of 2000 $mA{\cdot}g^{-1}$, the electrode with 0.5 wt% Ti_4O_7 had a capacity of 159 $mA{\cdot}g^{-1}$, higher than the original electrode with a capacity of 94 $mA{\cdot}g^{-1}$. After 50 cycles, the capacity retention rate increased to 88% at 40 $mA{\cdot}g^{-1}$. Guo et al. [91] studied the morphology, crystal structure, electrochemical properties, electrocatalytic oxidation performance, and stability of a PbO_2 electrode modified with different doses of Ti_4O_7 by the electrochemical deposition method. The results showed that Ti_4O_7 modification could significantly improve the surface morphology of the electrode, improve the current response of the electrode, and reduce the impedance of the electrode. Under 1.0$g{\cdot}L^{-1}$ as the best deposition amount, the decolorization rate of dye wastewater can reach 99%, and the TOC removal rate can reach 65%, in 180 min. The accelerated lifetime of the electrode was 175 h, 1.65 times longer than that of the PbO_2 electrode (105.5 h). The above studies indicate that the composite electrode doped with Ti_4O_7 has a good application prospect.

2.4.2. Metals-Doped Composite Titanium Sub-Oxides Electrode

In addition, there are many types of research on the modification of titanium sub-oxides electrodes doped with other metals. For example, Linet et al. [86] doped a Ce^{3+} electrode preparation by the powder sintering method. Compared with the electrode before doping, the increased oxygen evolution potential and the reduced internal resistance of the charge transfer of the prepared electrode could be obtained, and the degradation effect of perfluorooctane sulfonic acid (PFOS) was further improved because the doping of Ce^{3+} resulted in more active sites on the electrode surface, which improved the rapid transfer of charge. Huang et al. [92] studied the effects of loaded crystalline and amorphous Pd on titanium sub-oxides electrodes prepared by plasma spraying technology. The results showed that the performance of electrodes doped with Pd was greatly improved. The effect of amorphous Pd was more obvious due to the disordered atomic arrangement of amorphous metals and the unsaturated coordination of amorphous metals. The amorphous metal formed a stronger interaction with the substrate, and this electron-metal interaction support (EMSI) enhanced electron transfer, which ultimately promoted the oxidation kinetics of the electrode. The above study demonstrated that doping Pt, Ce, Pd, and other active substances positively affects the performance of the titanium sub-oxides electrode, which means it is of further research interest.

3. Application of Titanium Sub-Oxides Electrode in Electrocatalytic Oxidation Wastewater

Compared with other commercial electrodes, including commercial graphite, stainless steel, and dimensional stability anode (DSA) electrodes, the electrochemical oxidation rate of the Ti_4O_7 anode was reported to be higher than that of other electrodes. Table 4 lists the performance comparison of several common electrodes with the Ti_4O_7 anode. The Ti_4O_7 anode has superior electrocatalytic activity, high oxygen evolution overpotential, and stable material properties that make it suitable for the electrochemical treatment of many pollutants, such as drugs, dyes, and phenolic organic matter (Table 5 lists the effects of titanium sub-oxide anode treatment on different organic matters).

Table 4. Performance comparison of several common electrodes with Ti_4O_7 anode.

Type	Oxygen Evolution Potential (OEP V vs. Ag/AgCl)	Accelerated Life * (h)	Type
Graphite	1.0	-	[93]
Pt	1.51	-	[93]
BDD	2.5	-	[94]
Ti/PbO_2	1.85	0.5	[95]
Ti/SnO_2	1.81	12	[93]
$Ti/SnO_2 + Sb_2O_3/PbO_2$	1.98	16	[96]
Ti_4O_7	2.28	31.2	[78]

* In 1 M H_2SO_4 solution, it was considered as failure when the voltage rose to 10 V under the current density of 1 A cm^{-2}.

Table 5. Titanium sub-oxides as electrode electro-oxidation to degrade organic matter.

Organic Matter	Electrode	Processing Conditions	Treatment Effect	Reference
Sulfamerazine (SMR)	TF/Ti_4O_7	Current density 10 mA·cm^{-2}, pH 2.	After 60 min, the removal rate of SMR was 48.09%.	[97]
Tetracycline (TC)	Ti_4O_7	Current density 10 mA·cm^{-2}, initial pH 4.51.	The degradation rate of TC within 3 h was 97.95%.	[77]
Methyl orange (MO)	Ti_4O_7	Current density 10 mA·cm^{-2}, initial dye concentration is 100 mg L^{-1}.	COD removal rate reaches 91.7%.	[98]
Acid red B	Ti_4O_7	Voltage 0.5 V, pH 7.0, acid red B concentration 400 mg L^{-1}.	After 7 h, the dye removal rate can reach 91.95%.	[99]
Phenol	Titanium Sub-Oxide	Initial concentration is 100 mg L^{-1}, voltage 12 V, pH 3.0.	The degradation rate of phenol within 3 h was 92.22%, COD removal rate 94.26%.	[100]
4-Chlorophenol	Titanium Sub-Oxide Membrane	Initial concentration is 20 mg L^{-1}, current density 5 mA·cm^{-2}.	After 2 h, COD met the discharge standard, and the mineralization rate of 4-chlorophenol was 64%.	[101]

3.1. *Antibiotic Wastewater*

Antibiotic pollutants have attracted attention due to high COD concentration, difficulty in biodegradation, and strong pollution. Titanium sub-oxide materials can play a strong role in the degradation of antibiotic pollutants.

Teng et al. [76] successfully prepared Ti/Ti_4O_7 electrodes using plasma spraying technology. They studied the electrochemical catalytic oxidation reaction of sulfadiazine (SDZ), a typical antibiotic, using the electrode as an electrode anode. The results showed that under the degradation conditions of 0.05 mol L^{-1} sodium sulfate, pH 6.33, and a current density of 10 mA·cm^{-2}, the SDZ removal rate reached almost 100% after 60 min. It is demonstrated that the Ti/Ti_4O_7 anode had a high oxygen evolution potential and long-term stability in the treatment of actual pharmaceutical wastewater. Moreover, the network structure of the Ti/Ti_4O_7 anode can obtain a large electrochemically active surface area and improve the mass transfer between the electrolyte solution and anode.

Soliu et al. [77] also used plasma spraying technology to prepare Ti_4O_7 electrodes and compared it with several common commercial electrodes to study the electrochemical catalytic oxidation reaction of amoxicillin. The results revealed that, under low current intensity, the mineralization efficiency of the DSA and Pt anode was low. The TOC removal rate of the DSA and Pt anode was 36% and 41%, respectively. In comparison, the TOC removal rate of the Ti_4O_7 anode can reach 69%.

Wang et al. [102] evaluated the performance and complete pathway of tetracycline (TC) electrochemical oxidation on a Ti/Ti_4O_7 anode prepared by plasma spraying. The results showed that the electrochemical oxidation of TC on the Ti/Ti_4O_7 anode followed pseudo-first-order kinetics, and the TC removal efficiency can reach 95.8% in 40 min. The anode also had high stability, and the TC removal efficiency remained above 95% after five times of repeated use.

3.2. *Dye Wastewater*

Dye wastewater is also a significant field of electrocatalytic oxidation wastewater treatment technology. Wang et al. [98] successfully prepared titanium sub-oxides electrode material by spark plasma sintering technology and conducted electrocatalytic oxidation research on azo dye methyl orange (MO). The results showed that with the increase of the current density and the decrease of the initial concentration of MO, the removal rates of MO and COD presented an upward trend, and the complete removal of MO was achieved in a relatively short time. When the current density is 10 $mA \cdot cm^{-2}$ and the initial dye concentration is 100 $mg \cdot L^{-1}$, the COD removal rate can reach 91.7%. The degradation pathways of MO on the Ti_4O_7 electrode can be divided into two types: (1) the active substance first attacked the azo bond and the large conjugated system formed by benzene ring; (2) the active substance first attacked the azo bond and the C−N bond on the benzene ring. As the degradation reaction goes on, the final products are H_2O and CO_2.

Wang Yu [99] compared the carbon cloth and the titanium sub-oxides electrochemical oxidation device on the degradation of methylene blue decolorizing effects, and the results demonstrated that the reaction rate of the titanium sub-oxides anode was 3.7 times that of carbon cloth and the titanium sub-oxides electrode of methylene blue mineralization rate can reach 83.5%, but carbon cloth was only 42.6%. Therefore, the degradation of methylene blue by titanium sub-oxides was mainly mineralization. The experimental results showed that when the pH of the electrolyte is 3.01, the current density is 1.13 $mA \cdot cm^{-2}$, and the concentration is less than 100 $mg \cdot L^{-1}$; therefore, the electrochemical oxidation device of titanium sub-oxides can be operated with a better degradation rate and mineralization rate, higher Coulomb efficiency, and a lower reaction energy consumption.

3.3. *Wastewater Containing Phenols*

The refractory organic wastewater containing phenols has the characteristics of comprehensive sources, brutal treatment, and substantial destruction. At the same time, the titanium sub-oxides electrode can also achieve the effective degradation of phenolic pollutants. For example, Tan Yang [101] studied the electrochemical oxidation process of 4-chlorophenol wastewater by anodizing the titanium sub-oxides film material. The results showed that the target substance diffused faster at the titanium sub-oxides film electrode/solution interface compared with the BDD film electrode. When the initial concentration of 4-chlorophenol was 20 $mg \cdot L^{-1}$, the electrolyte solution, membrane flux, and current density were 0.04 $mol \cdot L^{-1}$ sodium sulfate solution, 0.023 $mL \cdot cm^{-2} \cdot s^{-1}$, and 5 $mA \cdot cm^{-2}$, respectively, and the removal rate of 4-chlorophenol can reach 100%. After 150 min, both COD and chromaticity can reach the discharge standard with low energy consumption.

In addition, Ping Geng [62] synthesized a pure Magnéli Ti_4O_7 nanotube array (NTA) by reducing lithium titanate with hydrogen at 850 °C. Compared with Ti_4O_7 particles and Ti_4O_7 NTA, the chemical oxygen demand (COD) removal rate of phenol increased by 20%. The degradation coefficient, COD removal rate, and current efficiency of pure Ti_4O_7 NTA were higher than those of boron-doped diamond and other Magnéli phase nanotube arrays.

Wang et al. [101] prepared three anodes (Ti/Ti_4O_7, Ti/Ti_4O_7-PbO_2-Ce, and Ti/Ti_4O_7 nanotube array (NTA)), which were used to treat a p-nitrophenol (PNP) solution by electrocatalytic oxidation. After 30 min of treatment, the removal rate of PNP by Ti/Ti_4O_7 NTA and Ti/Ti_4O_7 was 89–92%, 10–60% higher than that of Ti/Pt, Ti/RuO_2-IrO_2, and Ti/

IrO_2-Ta_2O_5, and equivalent to BDD (95%). In addition, Ti/Ti_4O_7 NTA and Ti/Ti_4O_7 have a decent mineralization effect on the PNP solution.

Maria et al. [103] successfully prepared Ti_4O_7-porous electrodes with a continuous layered structure. When the liquid was recirculated through the layered structure of the electrode, the performance was significantly improved. After 6 h of p-bendasone treatment, the removal rate was 85% and the mineralization rate was 57%, which was much better than the commercial electrode currently developed. Ti_4O_7 is a potential anode material for electrochemical oxidation due to its high efficiency and low energy consumption.

3.4. Treatment of Mixed Pollutants

The titanium sub-oxides electrode has sound degradation effects on single pollutants and is suitable for treating wastewater containing mixed contaminants.

Wang et al. [104] studied the feasibility and effectiveness of a titanium sub-oxides anode in simultaneously removing bacterial pathogens, antibiotics, and antibiotic resistance genes in water by using an integrated electrode prepared by high-temperature sintering. The results showed that the degradation rates of tetracycline (TC) and sulfamedimethazine (SDM) in the same treatment system reached more than 95% after 3 h of treatment. Further studies found that with the extension of treatment time the activity of bacterial pathogens was basically zero after 15 min, and the resistance genes were effectively removed. This was because the EO process produces a large amount of •OH, which can effectively attack pathogens, resulting in cell damage and reduction of cytoplasm content. This study confirmed the practical electrochemical degradation ability to mix medical wastewater in a titanium sub-oxides anode system.

Dan et al. [105] used the titanium sub-oxides-coated electrode prepared by plasma spraying technology as an anode and compared it with a Ti/RuO_2-IrO_2 anode to conduct the electrochemical treatment of coking wastewater (CW). The degradation results presented that the titanium sub-oxides electrode had apparent advantages in the COD removal rate, TOC removal rate, current efficiency, and energy consumption. The COD removal rate of CW was 78.7% and the TOC removal rate of CW was 50.3%, which were higher than that of the Ti/RuO_2-IrO_2 anode. In addition, the removal effects of the two electrodes for different kinds of PAHs in mixed wastewater were compared, where the titanium sub-oxides electrode had better performance for most organic compounds.

Pei et al. [61] employed a titanium sub-oxides membrane electrode to treat printing and dyeing industrial wastewater, which proved that the membrane electrode prepared by a high-temperature reduction method was mainly composed of Ti_4O_7 and a small amount of Ti_5O_9, with good conductivity, and high oxygen precipitation potential and electrochemical stability. When the current density was 8 $mA \cdot cm^{-2}$ and the membrane flux was 2.31×10^{-3} $mL \cdot cm^{-2} \cdot s^{-1}$, electrolysis for 1.5 h can effectively treat the actual printing and dyeing industrial wastewater. The COD removal rate was up to 96.07%, the current efficiency up to 24.22%, and the energy consumption was reduced by 32.99% compared with that without the membrane flux. It has important research value and development potential in small-scale decentralized industrial wastewater treatment.

3.5. Coupling Technology of Titanium Sub-Oxides Anodic Electrocatalytic Oxidation System

The titanium sub-oxides anode electrocatalytic oxidation system still cannot achieve the best effect for treating complex practical wastewater, which has the disadvantages of high energy consumption and low electronic utilization rate [106,107]. Therefore, the coupling application of electrocatalytic oxidation and other technologies has a promising prospect for development. The coupling technologies mainly focus on photoelectron-Fenton, electro-Fenton, and ultrasonic enhancement.

Han et al. [108] used a TF/Ti_4O_7 electrode prepared by loading titanium sub-oxides powder on the surface of titanium foam (TF) material as the anode and carbon with nitrogen-modified nickel foam (NF/CN) as the cathode (the experimental device is shown in Figure 6b). When the current density was 10 $mA \cdot cm^{-2}$ and the initial pH was 2, the removal rate of sulfamethazine (SMR) was 48.09% after the electrocatalytic reaction took 60 min. When the electro-Fenton anodic oxidation system constructed with NF/CN-TF/Ti_4O_7 degraded SMR again, the removal rate of 8 h could reach 99.48%, which was far better than that of the electrocatalytic system. The results showed that the electro-Fenton mechanism was the primary way to degrade pollutants in the electrochemical system, with a contributing rate of 71.60%. The coupling of the electro-Fenton and anodic oxidation system resulted in significant growth in the production of hydroxyl radicals, thus improving the degradation capacity of the whole system.

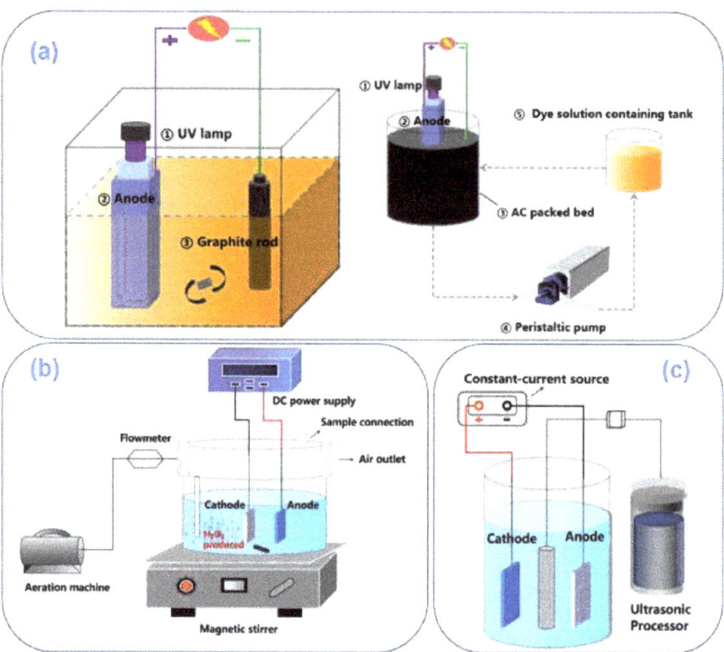

Figure 6. Photoelectro-Fenton coupling device [108] (**a**); electro-Fenton coupling device [107] (**b**); ultrasonic strengthening coupling device (**c**).

Zwane et al. [109] also used electro-Fenton (EF)-, anodic oxidation (AO)-, and electro-Fenton-combined anodic oxidation (EF/AO) to degrade tetracycline on the titanium sub-oxides (Ti_4O_7) layer (anode) deposited on carbon felt (cathode)/titanium substrate. The EF/AO coupling system had the highest pollutant removal efficiency compared with EF and AO. When the initial concentration of tetracycline was 20 ppm and 50 ppm, the removal rate of total organic carbon was 69 $\pm$ 1% and 68 $\pm$ 1%, respectively. The reason was that the EF/AO process with Ti_4O_7 as anode and CF as a cathode can generate hydroxyl radicals on the surface of the inactive Ti_4O_7 electrode and the electro-Fenton process solution on the cathode carbon blanket, which could reduce the energy consumption and improve the removal efficiency of tetracycline. It was observed from HPLC data that the optimal conditions for 20 ppm and 50 ppm tetracycline concentrations were 120 mA and 210 mA, with complete removal of tetracycline after 30 min.

Estrada et al. [110] studied an embedded activated carbon-filled bed cathode and a TiO_2/Ti_4O_7 photoelectric anode (the experimental device is shown in Figure 6a). The experimental results of the photocatalysis (PC), electrooxidation (EO), and photoelectrochemical (PEC) of the methyl orange (Mo) model dye solution showed that TiO_2/Ti_4O_7 composite film combined the advantages of the two materials so that anatase TiO_2 had better photocatalytic performance. The partially reduced Ti_4O_7 had higher conductivity, thus promoting the effective decolorization of the dye solution, and the decolorization rate was close to 70% within 60 min. Based on simulating calculations of (1) photoelectric chemicals-induced discoloration on the surface of the anode (electric catalytic oxidation and light contribution), (2) the adsorption and desorption process of dye molecules on the surface of activated carbon particles and pores, and the (3) Fenton-induced decolorization of the dye on the surface of the cathode existing in the system, it was concluded that the surface of the anode photoelectric catalytic contribution was up to 54%. Technical coupling [111] can improve the yield and utilization rate of hydroxyl radicals in the system. The Fenton process can also avoid electron quenching in the degradation process and reduce the energy consumption. In addition, it was proposed that further optimization of the area/volume ratio of the membrane to the reactor can improve the removal efficiency of target pollutants.

Yang et al. [112] used a Ti_4O_7-plate electrode as an anode prepared by plasma spraying, through coupling ultrasonic strengthening, and the electrical catalytic degradation of chloramphenicol (CAP) wastewater was conducted (experimental research device, as shown in Figure 6c). The results confirmed that under the condition of the ultrasonic, targets in the titanium sub-oxides film electrode/solution interface diffused faster, improving the efficiency of mass transfer. When the initial concentration of chloramphenicol was 20 $mg \cdot L^{-1}$, the electrolyte solution, ultrasonic power, current density, and pH were 0.05 $mol \cdot L^{-1}$ sodium sulfate solution, 150 W, 20 $mA \cdot cm^{-2}$, and 5, respectively, and the removal rate of chloramphenicol could reach 100%. The degradation mechanism was direct electrochemical oxidation induced by •OH and the indirect electrooxidation of Cl^-/ClO^- intermediates.

4. Summary and Outlook

Titanium sub-oxide materials are considered a candidate for the electrode with great potential in electrocatalytic oxidation due to their excellent electrochemical performance. At present, some achievements have been made in the research of titanium sub-oxides in electrode preparation and wastewater treatment. A variety of stable electrode preparation technology routes were developed to be applied in the treatment process of wastewater with various contaminants, further coupling with other technologies to enhance the treatment efficiency. However, there are still great challenges in the research process, where the current preparation process is still cumbersome and inefficient, the conditions for scaling-up production are not available, the stability of the electrode needs to be improved, and the degradation effect of the complex water system is still to be studied. Thus, the future development direction of titanium sub-oxide materials can be focused on the following aspects:

(1) Optimization of preparation process: The current relatively stable preparation route still has the disadvantages of high energy consumption requirements, high temperature, and low efficiency, in which high-temperature calcination is prone to agglomeration. Therefore, it is necessary to further optimize the preparation route at high temperatures to prepare relatively pure titanium sub-oxides electrodes with high activity. In addition, the medium- and low-temperature synthesis should be further explored to avoid the disadvantages resulting from high temperatures and reduce the preparation costs;

(2) Modification of electrode materials: Studies show that the performance of electrodes can be significantly promoted by doping with highly active metal elements, and its degradation mechanisms need further elucidation. By identifying the mechanisms, the titanium sub-oxide electrode can be further modified and optimized to retain its excellent electrochemical performance. The original defects are to be made up by doping and other technical means to develop high-efficiency and low-consumption electrode products;

(3) Application expansion: The performance of titanium sub-oxides is not limited to electrocatalysis. Combining an electrocatalytic oxidation system and other technical means can better develop the hidden performance of the electrode to achieve twice the

result with half the effort. In addition, the effective construction of an electrocatalytic oxidation system can also make the electrode material fulfill the maximum function;

(4) Principles exploration: Titanium sub-oxides are a kind of material with excellent stability. Therefore, it is of great importance to explore the deactivation mechanism of electrodes for prolonging electrode life. In the electrocatalytic oxidation system, the effects of different impurity ions on the electrode and the synergistic degradation principle of complex pollutants demand to be further studied.

Author Contributions: S.G.: conceptualization, investigation, writing—original draft; Z.X.: writing-reviewing and editing, investigation; W.H.: investigation; D.Y.: writing—reviewing and editing; X.W.: investigation; H.X.: conceptualization, writing—reviewing and editing, resources, funding acquisition; X.X. and Z.L.: resources, investigation; W.Y.: supervision, reviewing. All authors have read and agreed to the published version of the manuscript.

Funding: This research was funded by [the Natural Science Basic Research Plan in the Shaanxi Province of China] grant number [2021JM-012], [the Welfare Technology Research Plan of Zhejiang Province] grant number [Program No. LZY21E080003]. and [the Fundamental Research Funds for the Central Universities] grant number [Program No. xjh012020037].

Conflicts of Interest: The authors declare that they have no known competing financial interests or personal relationships that could have appeared to influence the work reported in this paper.

References

1. Moreira, F.C.; Boaventura, R.A.R.; Brillas, E.; Vilar, V.J.P. Electrochemical advanced oxidation processes: A review on their application to synthetic and real wastewaters. *Appl. Catal. B Environ.* **2017**, *202*, 217–261. [CrossRef]
2. Panizza, M.; Michaud, P.-A.; Iniesta, J.; Comninellis, C.; Cerisola, G. Electrochemical oxidation of phenol at boron-doped diamond electrode. *Electrochim. Acta* **2001**, *46*, 3573–3578.
3. Lei, J.; Xu, Z.; Xu, H.; Qiao, D.; Liao, Z.; Yan, W.; Wang, Y. Pulsed electrochemical oxidation of acid Red G and crystal violet by PbO_2 anode. *J. Environ. Chem. Eng.* **2020**, *8*, 103773.
4. Zhang, Y.; Zhang, C.; Shao, D.; Xu, H.; Rao, Y.; Tan, G.; Yan, W. Magnetically assembled electrodes based on Pt, RuO_2-IrO_2-TiO_2 and Sb-SnO_2 for electrochemical oxidation of wastewater featured by fluctuant Cl^- concentration. *J. Hazard. Mater.* **2022**, *421*, 126803. [CrossRef]
5. Kobya, M.; Hiz, H.; Senturk, E.; Aydiner, C.; Demirbas, E. Treatment of potato chips manufacturing wastewater by electrocoagulation. *Desalination* **2006**, *190*, 201–211. [CrossRef]
6. Dargahi, A.; Shokoohi, R.; Asgari, G.; Ansari, A.; Nematollahi, D.; Samarghandi, M.R. Moving-bed biofilm reactor combined with three-dimensional electrochemical pretreatment (MBBR-3DE) for 2,4-D herbicide treatment: Application for real wastewater, improvement of biodegradability. *RSC Adv.* **2021**, *11*, 9608–9620. [CrossRef]
7. Martínez-Huitle, C.A. Electrochemical oxidation of organic pollutants for the wastewater treatment: Direct and indirect processes. *Chem. Soc. Rev.* **2006**, *35*, 1324–1340. [CrossRef]
8. Li, A.; Weng, J.; Yan, X.; Li, H.; Shi, H.; Wu, X. Electrochemical oxidation of acid orange 74 using Ru, IrO_2, PbO_2, and boron doped diamond anodes: Direct and indirect oxidation. *J. Electroanal. Chem.* **2021**, *898*, 115622. [CrossRef]
9. Kirk, D.W.; Sharifian, H.; Foulkes, F.R. Anodic oxidation of aniline for waste water treatment. *J. Appl. Electrochem.* **1985**, *15*, 285–292. [CrossRef]
10. Chen, Z.; Zhang, Y.; Zhou, L.; Zhu, H.; Wan, F.; Wang, Y.; Zhang, D. Performance of nitrogen-doped graphene aerogel particle electrodes for electro-catalytic oxidation of simulated Bisphenol A wastewaters. *J. Hazard. Mater.* **2017**, *332*, 70–78. [CrossRef]
11. Chen, G. Electrochemical technologies in wastewater treatment. *Sep. Purif. Technol.* **2004**, *38*, 11–41. [CrossRef]
12. Xu, H.; Qiao, D.; Xu, Z.; Guo, H.; Chen, S.; Xu, X.; Gao, X.; Yan, W. Application of electro-catalytic oxidation technology in organic wastewater treatment. *Ind. Water Treat.* **2021**, *41*, 1–9. (In Chinese)
13. Liu, Y.; Liu, H. Comparative studies on the electrocatalytic properties of modified PbO_2 anodes. *Electrochim. Acta* **2008**, *53*, 5077–5083. [CrossRef]
14. Patel, P.S.; Bandre, N.; Saraf, A.; Ruparelia, J.P. Electro-catalytic Materials (Electrode Materials) in Electrochemical Wastewater Treatment. *Procedia Eng.* **2013**, *51*, 430–435. [CrossRef]
15. Ignasi, S. Electrochemical advanced oxidation processes: Today and tomorrow. A review. *Environ. Sci. Pollut. Res. Int.* **2014**, *21*, 8336–8367.
16. Ihara, I.; Umetsu, K.; Kanamura, K.; Watanabe, T. Electrochemical oxidation of the effluent from anaerobic digestion of dairy manure. *Bioresour. Technol.* **2006**, *97*, 1360–1364. [CrossRef] [PubMed]
17. Zhuo, Q.; Yang, B.; Deng, S.; Huang, J.; Wang, B.; Yu, G. Electrochemical Anodic Materials Used for Degradation of Organic Pollutants. *Prog. Chem.* **2012**, *24*, 628–636. (In Chinese)

18. Dargahi, A.; Hasani, K.; Mokhtari, S.A.; Vosoughi, M.; Moradi, M.; Vaziri, Y. Highly effective degradation of 2,4-Dichlorophenoxyacetic acid herbicide in a three-dimensional sono-electro-Fenton (3D/SEF) system using powder activated carbon (PAC)/Fe3O4 as magnetic particle electrode. *J. Environ. Chem. Eng.* **2021**, *9*, 105889. [CrossRef]
19. Salazar-Banda, G.R.; Santos, G.d.O.S.; Duarte Gonzaga, I.M.; Dória, A.R.; Barrios Eguiluz, K.I. Developments in electrode materials for wastewater treatment. *Curr. Opin. Electrochem.* **2021**, *26*, 100663. [CrossRef]
20. Comninellis, C.; Pulgarin, C. Anodic oxidation of phenol for waste water treatment. *J. Appl. Electrochem.* **1991**, *21*, 703–708. [CrossRef]
21. Bai, H.; He, P.; Pan, J.; Chen, J.; Chen, Y.; Dong, F.; Li, H. Boron-doped diamond electrode: Preparation, characterization and application for electrocatalytic degradation of m-dinitrobenzene. *J. Colloid Interface Sci.* **2017**, *497*, 422–428. [CrossRef] [PubMed]
22. Martínez-Huitle, C.A.; Brillas, E. Decontamination of wastewaters containing synthetic organic dyes by electrochemical methods: A general review. *Appl. Catal. B Environ.* **2009**, *87*, 105–145. [CrossRef]
23. Zhao, G.; Li, P.; Nong, F.; Li, M.; Gao, J.; Li, D. Construction and High Performance of a Novel Modified Boron-Doped Diamond Film Electrode Endowed with Superior Electrocatalysis. *J. Phys. Chem. C* **2010**, *114*, 5906–5913. [CrossRef]
24. Yang, B.; Jiang, C.; Yu, G.; Zhuo, Q.; Deng, S.; Wu, J.; Zhang, H. Highly efficient electrochemical degradation of perfluorooctanoic acid (PFOA) by F-doped Ti/SnO_2 electrode. *J. Hazard. Mater.* **2015**, *299*, 417–424. [CrossRef] [PubMed]
25. Yuan, M.; Salman, N.M.; Guo, H.; Xu, Z.; Xu, H.; Yan, W.; Liao, Z.; Wang, Y. A 2.5D Electrode System Constructed of Magnetic Sb-SnO_2 Particles and a PbO_2 Electrode and Its Electrocatalysis Application on Acid Red G Degradation. *Catalysts* **2019**, *9*, 875. [CrossRef]
26. Wu, J.; Zhu, K.; Xu, H.; Yan, W. Electrochemical oxidation of rhodamine B by PbO_2/Sb-SnO_2/TiO_2 nanotube arrays electrode. *Chin. J. Catal.* **2019**, *40*, 917–927. [CrossRef]
27. Zhang, Z.; Wang, Z.; Sun, Y.; Jiang, S.; Shi, L.; Bi, Q.; Xue, J. Preparation of a novel Ni/Sb co-doped Ti/SnO_2 electrode with carbon nanotubes as growth template by electrodeposition in a deep eutectic solvent. *J. Electroanal. Chem.* **2022**, *911*, 116225. [CrossRef]
28. Chatzisymeon, E.; Dimou, A.; Mantzavinos, D.; Katsaounis, A. Electrochemical oxidation of model compounds and olive mill wastewater over DSA electrodes: 1. The case of Ti/IrO_2 anode. *J. Hazard. Mater.* **2009**, *167*, 268–274. [CrossRef]
29. Qu, J.P.; Zhang, X.G.; Wang, Y.G.; Xie, C.X. Electrochemical reduction of CO_2 on RuO_2/TiO_2 nanotubes composite modified Pt electrode. *Electrochim. Acta* **2005**, *50*, 3576–3580. [CrossRef]
30. Samet, Y.; Elaoud, S.C.; Ammar, S.; Abdelhedi, R. Electrochemical degradation of 4-chloroguaiacol for wastewater treatment using PbO_2 anodes. *J. Hazard. Mater.* **2006**, *138*, 614–619. [CrossRef]
31. Guo, H.; Xu, Z.; Qiao, D.; Wan, D.; Xu, H.; Yan, W.; Jin, X. Fabrication and characterization of porous titanium-based PbO_2 electrode through the pulse electrodeposition method: Deposition condition optimization by orthogonal experiment. *Chemosphere* **2020**, *261*, 128157.
32. Zhang, X.; Shao, D.; Lyu, W.; Xu, H.; Yang, L.; Zhang, Y.; Wang, Z.; Liu, P.; Yan, W.; Tan, G. Design of magnetically assembled electrode (MAE) with Ti/PbO_2 and heterogeneous auxiliary electrodes (AEs): The functionality of AEs for efficient electrochemical oxidation. *Chem. Eng. J.* **2020**, *395*, 125145. [CrossRef]
33. Shao, D.; Wang, Z.; Zhang, C.; Li, W.; Xu, H.; Tan, G.; Yan, W. Embedding wasted hairs in Ti/PbO_2 anode for efficient and sustainable electrochemical oxidation of organic wastewater. *Chin. Chem. Lett.* **2022**, *33*, 1288–1292. [CrossRef]
34. Samarghandi, M.R.; Ansari, A.; Dargahi, A.; Shabanloo, A.; Nematollahi, D.; Khazaei, M.; Nasab, H.Z.; Vaziri, Y. Enhanced electrocatalytic degradation of bisphenol A by graphite/beta-PbO_2 anode in a three-dimensional electrochemical reactor. *J. Environ. Chem. Eng.* **2021**, *9*, 106072. [CrossRef]
35. Yang, B.; Wang, J.; Jiang, C.; Li, J.; Yu, G.; Deng, S.; Lu, S.; Zhang, P.; Zhu, C.; Zhuo, Q. Electrochemical mineralization of perfluorooctane sulfonate by novel F and Sb co-doped Ti/SnO_2 electrode containing Sn-Sb interlayer. *Chem. Eng. J.* **2017**, *316*, 296–304. [CrossRef]
36. Xu, Z.; Yu, Y.; Liu, H.; Niu, J. Highly efficient and stable Zr-doped nanocrystalline PbO_2 electrode for mineralization of perfluorooctanoic acid in a sequential treatment system. *Sci. Total Environ.* **2017**, *579*, 1600–1607. [CrossRef]
37. Li, L.; Huang, Z.; Fan, X.; Zhang, Z.; Dou, R.; Wen, S.; Chen, Y.; Chen, Y.; Hu, Y. Preparation and Characterization of a Pd modified Ti/SnO_2-Sb anode and its electrochemical degradation of Ni-EDTA. *Electrochim. Acta* **2017**, *231*, 354–362. [CrossRef]
38. Zhang, Q.; Guo, X.; Cao, X.; Wang, D.; Wei, J. Facile preparation of a Ti/α-PbO_2/β-PbO_2 electrode for the electrochemical degradation of 2-chlorophenol. *Chin. J. Catal.* **2015**, *36*, 975–981. [CrossRef]
39. Wu, W.; Huang, Z.-H.; Hu, Z.-T.; He, C.; Lim, T.-T. High performance duplex-structured SnO_2-Sb-CNT composite anode for bisphenol A removal. *Sep. Purif. Technol.* **2017**, *179*, 25–35. [CrossRef]
40. Xu, Z.; Liu, H.; Niu, J.; Zhou, Y.; Wang, C.; Wang, Y. Hydroxyl multi-walled carbon nanotube-modified nanocrystalline PbO_2 anode for removal of pyridine from wastewater. *J. Hazard. Mater.* **2017**, *327*, 144–152. [CrossRef]
41. Xu, L.; Liang, G.; Yin, M. A promising electrode material modified by Nb-doped TiO_2 nanotubes for electrochemical degradation of AR 73. *Chemosphere* **2017**, *173*, 425–434. [CrossRef] [PubMed]
42. Wang, Q.; Jin, T.; Hu, Z.; Zhou, L.; Zhou, M. TiO_2-NTs/SnO_2-Sb anode for efficient electrocatalytic degradation of organic pollutants: Effect of TiO_2-NTs architecture. *Sep. Purif. Technol.* **2013**, *102*, 180–186. [CrossRef]
43. Shao, D.; Zhang, Y.; Lyu, W.; Zhang, X.; Tan, G.; Xu, H.; Yan, W. A modular functionalized anode for efficient electrochemical oxidation of wastewater: Inseparable synergy between OER anode and its magnetic auxiliary electrodes. *J. Hazard. Mater.* **2020**, *390*, 122174. [CrossRef] [PubMed]

44. Xu, H.; Yan, W.; Yang, H. Surface Analysis of Ti/Sb-SnO_2/PbO_2 Electrode after Long Time Electrolysis. *Rare Met. Mater. Eng.* **2015**, *44*, 2637–2641.
45. Guo, Y.; Xu, Z.; Guo, S.; Chen, S.; Xu, H.; Xu, X.; Gao, X.; Yan, W. Selection of anode materials and optimization of operating parameters for electrochemical water descaling. *Sep. Purif. Technol.* **2021**, *261*, 118304. [CrossRef]
46. Xu, H.; Xu, Z.; Guo, Y.; Guo, S.; Xu, X.; Gao, X.; Wang, L.; Yan, W. Research and application progress of electrochemical water quality stabilization technology for recirculating cooling water in China: A short review. *J. Water Process Eng.* **2020**, *37*, 101433. [CrossRef]
47. Guo, H.; Xu, Z.; Wang, D.; Chen, S.; Qiao, D.; Wan, D.; Xu, H.; Yan, W.; Jin, X. Evaluation of diclofenac degradation effect in "active" and "non-active" anodes: A new consideration about mineralization inclination. *Chemosphere* **2022**, *286*, 131580. [CrossRef]
48. Xing, X.; Ni, J.; Zhu, X.; Jiang, Y.; Xia, J. Maximization of current efficiency for organic pollutants oxidation at BDD, Ti/SnO2-Sb/PbO2, and Ti/SnO2-Sb anodes. *Chemosphere* **2018**, *205*, 361–368. [CrossRef]
49. Zhou, C.; Wang, Y.; Chen, J.; Niu, J. Electrochemical degradation of sunscreen agent benzophenone-3 and its metabolite by Ti/SnO2-Sb/Ce-PbO2 anode: Kinetics, mechanism, toxicity and energy consumption. *Sci. Total Environ.* **2019**, *688*, 75–82. [CrossRef]
50. Kaur, R.; Kushwaha, J.P.; Singh, N. Electro-oxidation of amoxicillin trihydrate in continuous reactor by Ti/RuO2 anode. *Sci. Total Environ.* **2019**, *677*, 84–97. [CrossRef]
51. Duan, P.; Hu, X.; Ji, Z.; Yang, X.; Sun, Z. Enhanced oxidation potential of Ti/SnO2-Cu electrode for electrochemical degradation of low-concentration ceftazidime in aqueous solution: Performance and degradation pathway. *Chemosphere* **2018**, *212*, 594–603. [CrossRef] [PubMed]
52. Xia, Y.; Feng, J.; Fan, S.; Zhou, W.; Dai, Q. Fabrication of a multi-layer CNT-PbO2 anode for the degradation of isoniazid: Kinetics and mechanism. *Chemosphere* **2021**, *263*, 128069. [CrossRef] [PubMed]
53. Xu, H.; Guo, H.; Feng, J.; Wang, D.; Liao, Z.; Wang, Y.; Wei, Y. Electrochemical Oxidation Combined with Adsorption: A Novel Route for Low Concentration Organic Wastewater Treatment. *Int. J. Electrochem. Sci.* **2019**, *14*, 8110–8120.
54. Kong, D.; Lue, W.; Feng, Y.; Bi, S. Advances and Some Problems in Electrocatalysis of DSA Electrodes. *Prog. Chem.* **2009**, *21*, 1107–1117. (In Chinese)
55. Tang, C.; Zhou, D.; Zhang, Q. Synthesis and characterization of Magnéli phases: Reduction of TiO_2 in a decomposed NH_3 atmosphere. *Mater. Lett.* **2012**, *79*, 42–44. [CrossRef]
56. Geng, P.; Chen, G. Antifouling ceramic membrane electrode modified by Magnéli Ti_4O_7 for electro-microfiltration of humic acid. *Sep. Purif. Technol.* **2017**, *185*, 61–71. [CrossRef]
57. Smith, J.R.; Walsh, F.C.; Clarke, R.L. Reviews in applied electrochemistry. Number 50—Electrodes based on Magnéli phase titanium oxides: The properties and applications of Ebonex (R) materials. *J. Appl. Electrochem.* **1998**, *28*, 1021–1033. [CrossRef]
58. Walsh, F.C.; Wills, R.G.A. The continuing development of Magnéli phase titanium sub-oxides and Ebonex (R) electrodes. *Electrochim. Acta* **2010**, *55*, 6342–6351. [CrossRef]
59. Pei, S.; Teng, J.; Ren, N.; You, S. Low-Temperature Removal of Refractory Organic Pollutants by Electrochemical Oxidation: Role of Interfacial Joule Heating Effect. *Environ. Sci. Technol.* **2020**, *54*, 4573–4582. [CrossRef]
60. Nayak, S.; Chaplin, B.P. Fabrication and characterization of porous, conductive, monolithic Ti_4O_7 electrodes. *Electrochim. Acta* **2018**, *263*, 299–310. [CrossRef]
61. Pei, S.; Zhu, L.; Zhang, Z.; Teng, J.; Liu, X.; You, S. Electrochemical properties of titanium sub-oxide membrane electrode and application for electro-oxidation treatment of dyeing wastewater. *Acta Sci. Circumstantiae* **2020**, *40*, 3658–3665. (In Chinese)
62. Geng, P.; Su, J.; Miles, C.; Comninellis, C.; Chen, G. Highly-Ordered Magnéli Ti_4O_7 Nanotube Arrays as Effective Anodic Material for Electro-oxidation. *Electrochim. Acta* **2015**, *153*, 316–324. [CrossRef]
63. Lin, H.; Xiao, R.; Xie, R.; Yang, L.; Tang, C.; Wang, R.; Chen, J.; Lv, S.; Huang, Q. Defect Engineering on a Ti_4O_7 Electrode by Ce3+ Doping for the Efficient Electrooxidation of Perfluorooctanesulfonate. *Environ. Sci. Technol.* **2021**, *55*, 2597–2607. [CrossRef] [PubMed]
64. Pei, S.; Shen, C.; Zhang, C.; Ren, N.; You, S. Characterization of the Interfacial Joule Heating Effect in the Electrochemical Advanced Oxidation Process. *Environ. Sci. Technol.* **2019**, *53*, 4406–4415. [CrossRef]
65. Xu, B.; Sohn, H.Y.; Mohassab, Y.; Lan, Y. Structures, preparation and applications of titanium suboxides. *RSC Adv.* **2016**, *6*, 79706–79722. [CrossRef]
66. Malik, H.; Sarkar, S.; Mohanty, S.; Carlson, K. Modelling and synthesis of Magnéli Phases in ordered titanium oxide nanotubes with preserved morphology. *Sci. Rep.* **2020**, *10*, 8050. [CrossRef]
67. Zhang, X.; Liu, Y.; Ye, J.; Zhu, R. Fabrication and characterisation of Magnéli phase Ti_4O_7 nanoparticles. *Micro Nano Lett.* **2013**, *8*, 251–253. [CrossRef]
68. Xu, B.; Zhao, D.; Sohn, H.Y.; Mohassab, Y.; Yang, B.; Lan, Y.; Yang, J. Flash synthesis of Magnéli phase (Ti_nO_{2n-1}) nanoparticles by thermal plasma treatment of H_2TiO_3. *Ceram. Int.* **2018**, *44*, 3929–3936. [CrossRef]
69. Han, W.-Q.; Wang, X.-L. Carbon-coated Magnéli-phase TinO2n-1 nanobelts as anodes for Li-ion batteries and hybrid electrochemical cells. *Appl. Phys. Lett.* **2010**, *97*, 243104. [CrossRef]
70. Conze, S.; Veremchuk, I.; Reibold, M.; Matthey, B.; Michaelis, A.; Grin, Y.; Kinski, I. Magnéli phases Ti_4O_7 and Ti_8O_{15} and their carbon nanocomposites via the thermal decomposition-precursor route. *J. Solid State Chem.* **2015**, *229*, 235–242. [CrossRef]

71. Wang, G.; Liu, Y.; Ye, J.; Qiu, W. Synthesis, microstructural characterization, and electrochemical performance of novel rod-like Ti_4O_7 powders. *J. Alloy. Compd.* **2017**, *704*, 18–25. [CrossRef]
72. Toyoda, M.; Yano, T.; Tryba, B.; Mozia, S.; Tsumura, T.; Inagaki, M. Preparation of carbon-coated Magnéli phases Ti_nO_{2n-1} and their photocatalytic activity under visible light. *Appl. Catal. B-Environ.* **2009**, *88*, 160–164. [CrossRef]
73. You, S.; Liu, B.; Gao, Y.; Wang, Y.; Tang, C.Y.; Huang, Y.; Ren, N. Monolithic Porous Magnéli-phase Ti_4O_7 for Electro-oxidation Treatment of Industrial Wastewater. *Electrochim. Acta* **2016**, *214*, 326–335. [CrossRef]
74. Gusev, A.; Avvakumov, E.; Medvedev, A.; Masliy, A. Ceramic electrodes based on Magnéli phases of titanium oxides. *Sci. Sinter.* **2007**, *39*, 51–57. [CrossRef]
75. Hauf, C.; Kniep, R.; Pfaff, G. Preparation of various titanium suboxide powders by reduction of TiO_2 with silicon. *J. Mater. Sci.* **1999**, *34*, 1287–1292. [CrossRef]
76. Teng, J.; Liu, G.; Liang, J.; You, S. Electrochemical oxidation of sulfadiazine with titanium suboxide mesh anode. *Electrochim. Acta* **2019**, *331*, 135441. [CrossRef]
77. Ganiyu, S.O.; Oturan, N.; Raffy, S.; Cretin, M.; Esmilaire, R.; van Hullebusch, E.D.; Esposito, G.; Oturan, M.A. Sub-stoichiometric titanium oxide (Ti_4O_7) as a suitable ceramic anode for electrooxidation of organic pollutants: A case study of kinetics, mineralization and toxicity assessment of amoxicillin. *Water Res.* **2016**, *106*, 171–182. [CrossRef]
78. Han, Z.; Xu, Y.; Zhou, S.; Zhu, P. Preparation and electrochemical properties of Al-based composite coating electrode with Ti_4O_7 ceramic interlayer for electrowinning of nonferrous metals. *Electrochimica Acta* **2019**, *325*, 134940. [CrossRef]
79. Geng, P.; Chen, G. Magnéli Ti_4O_7 modified ceramic membrane for electrically-assisted filtration with antifouling property. *J. Membr. Sci.* **2016**, *498*, 302–314. [CrossRef]
80. Wong, M.-S.; Lin, Y.-J.; Pylnev, M.; Kang, W.-Z. Processing, structure and properties of reactively sputtered films of titanium dioxide and suboxides. *Thin Solid Films* **2019**, *688*, 137351. [CrossRef]
81. Li, H.; Lu, S.; Li, Y.; Qin, W.; Wu, X. Tunable thermo-optical performance promoted by temperature selective sputtering of titanium oxide on MgO-ZrO2 coating. *J. Alloy. Compd.* **2017**, *709*, 104–111. [CrossRef]
82. Ertekin, Z.; Tamer, K.; Pekmez, U. Cathodic electrochemical deposition of Magnéli phases TinO2n-1 thin films at different temperatures in acetonitrile solution. *Electrochim. Acta* **2015**, *163*, 77–81. [CrossRef]
83. Paunović, P.; Popovski, O.; Fidančevska, E.; Ranguelov, B.; Gogovska, D.S.; Dimitrov, A.T.; Jordanov, S.H. Co-Magnéli phases electrocatalysts for hydrogen/oxygen evolution. *Int. J. Hydrogen Energy* **2010**, *35*, 10073–10080. [CrossRef]
84. Regonini, D.; Adamaki, V.; Bowen, C.R.; Pennock, S.R.; Taylor, J.; Dent, A.C.E. AC electrical properties of TiO_2 and Magnéli phases, Ti_nO_{2n-1}. *Solid State Ion.* **2012**, *229*, 38–44. [CrossRef]
85. Guo, L.; Jing, Y.; Chaplin, B.P. Development and Characterization of Ultrafiltration TiO_2 Magnéli Phase Reactive Electrochemical Membranes. *Environ. Sci. Technol.* **2016**, *50*, 1428–1436. [CrossRef]
86. Liang, J.; You, S.; Yuan, Y.; Yuan, Y. A tubular electrode assembly reactor for enhanced electrochemical wastewater treatment with a Magnéli-phase titanium suboxide (M-TiSO) anode and in situ utilization. *RSC Adv.* **2021**, *11*, 24976–24984. [CrossRef]
87. Matsuda, M.; Yamada, Y.; Himeno, Y.; Shida, K.; Mitsuhara, M.; Matsuda, M. Magnéli Ti_4O_7 thin film produced by stepwise oxidation of titanium metal foil. *Scr. Mater.* **2021**, *198*, 113829. [CrossRef]
88. Zhang, L.; Kim, J.; Zhang, J.; Nan, F.; Gauquelin, N.; Botton, G.A.; He, P.; Bashyam, R.; Knights, S. Ti_4O_7 supported Ru@Pt core-shell catalyst for CO-tolerance in PEM fuel cell hydrogen oxidation reaction. *Appl. Energy* **2013**, *103*, 507–513. [CrossRef]
89. Zhao, J.; Li, W.; Wu, S.; Xu, F.; Du, J.; Li, J.; Li, K.; Ren, J.; Zhao, Y. Strong interfacial interaction significantly improving hydrogen evolution reaction performances of MoS_2/Ti_4O_7 composite catalysts. *Electrochim. Acta* **2020**, *337*, 135850. [CrossRef]
90. Wang, Y.; Zhao, D.; Zhang, K.; Li, Y.; Xu, B.; Liang, F.; Dai, Y.; Yao, Y. Enhancing the rate performance of high-capacity LINI(0.8)Co(0.15)Al(0.05)O(2) cathode materials by using Ti_4O_7 as a conductive additive. *J. Energy Storage* **2020**, *28*, 101182. [CrossRef]
91. Guo, H.; Xu, Z.; Qiao, D.; Wang, L.; Xu, H.; Yan, W. Fabrication and characterization of titanium-based lead dioxide electrode by electrochemical deposition with Ti_4O_7 particles. *Water Environ. Res.* **2021**, *93*, 42–50. [CrossRef] [PubMed]
92. Huang, D.; Wang, K.; Niu, J.; Chu, C.; Weon, S.; Zhu, Q.; Lu, J.; Stavitski, E.; Kim, J.-H. Amorphous Pd-Loaded Ti_4O_7 Electrode for Direct Anodic Destruction of Perfluorooctanoic Acid. *Environ. Sci. Technol.* **2020**, *54*, 10954–10963. [CrossRef] [PubMed]
93. Ding, H.; Feng, Y.; Liu, J. Comparison of electrocatalytic performance of different anodes with cyclic voltammetry and Tafel curves. *Chin. J. Catal.* **2007**, *28*, 646–650. (In Chinese)
94. Zhang, Y.; Zhang, R.; Ma, J.; Liu, L. Advanced Electrochemical Oxidation Process with BDD Electrodes for Organic Wastewater Treatment. *China Water Wastewater* **2006**, *22*, 15–18. (In Chinese)
95. Wei, Z.; Kang, X.; Xu, S.; Zhou, X.; Jia, B.; Feng, Q. Electrochemical oxidation of Rhodamine B with cerium and sodium dodecyl benzene sulfonate co-modified Ti/PbO_2 electrodes: Preparation, characterization, optimization, application. *Chin. J. Chem. Eng.* **2021**, *32*, 191–202. [CrossRef]
96. Qiao, D.; Xu, Z.; Guo, H.; Wang, X.; Wan, D.; Li, X.; Xu, H.; Yan, W. Non-traditional power supply mode: Investigation of electrodeposition towards a better understanding of PbO_2 electrode for electrochemical wastewater treatment. *Mater. Chem. Phys.* **2022**, *284*, 126066. [CrossRef]
97. Han, J. Degradation of Sulfamerazine by NF/CNTF/Ti_4O_7 Electrochemical System and its Mechanism. Master's Thesis, Harbin Institute of Technology, Harbin, China, 2019. [CrossRef]

98. Wang, G.; Liu, Y.; Ye, J.; Lin, Z.; Yang, X. Electrochemical oxidation of methyl orange by a Magnéli phase Ti_4O_7 anode. *Chemosphere* **2019**, *241*, 125084. [CrossRef] [PubMed]
99. Wang, Y. Anodic Oxidation for Degradation of Dyeing Wastewater in Eleochemical Systems with Titanium Sub-Oxide Anode. *Harbin Inst. Technol.* 2016. Available online: https://kns.cnki.net/kcms/detail/detail.aspx?dbcode=CMFD&dbname=CMFD201701&filename=1016913685.nh&uniplatform=NZKPT&v=dl01S0dpi5oBa4sLKxvjDs9DEAi3BmuNaVpj_rsz9T2dfFtl53U0KdpUdR80Ru5s (accessed on 8 May 2022).
100. Wang, J. Preparation of Titanium Sub-Oxide Electrode and Research on Degradation Efficiency of Phenol Wastewater by Three-dimensional Electrode System. *Harbin Inst. Technol.* **2020**. [CrossRef]
101. Tan, Y. Magnéli-phase Ti_4O_7 Conductive Membrane for Effective Electrochemical Degradation of 4-Chlorophenol Wastewater. *Harbin Inst. Technol.* 2018. Available online: https://kns.cnki.net/kcms/detail/detail.aspx?dbcode=CMFD&dbname=CMFD201901&filename=1018894521.nh&uniplatform=NZKPT&v=NX9XUuHDAefm1QlL7hWOo6L9Dfcwokx8EQsfUQ5NPnf6Wqe-A6luvnTUcpYkTcg4 (accessed on 8 May 2022).
102. Wang, J.; Zhi, D.; Zhou, H.; He, X.; Zhang, D. Evaluating tetracycline degradation pathway and intermediate toxicity during the electrochemical oxidation over a Ti/Ti_4O_7 anode. *Water Res.* **2018**, *137*, 324–334. [CrossRef]
103. Wang, H.; Li, Z.; Zhang, F.; Wang, Y.; Zhang, X.; Wang, J.; He, X. Comparison of Ti/Ti_4O_7, Ti/Ti_4O_7-PbO_2-Ce, and Ti/Ti_4O_7 nanotube array anodes for electro-oxidation of p-nitrophenol and real wastewater. *Sep. Purif. Technol.* **2021**, *266*, 118600. [CrossRef]
104. Barni, M.F.S.; Doumic, L.I.; Procaccini, R.A.; Ayude, M.A.; Romeo, H.E. Layered platforms of Ti_4O_7 as flow-through anodes for intensifying the electro-oxidation of bentazon. *J. Environ. Manag.* **2020**, *263*, 110403. [CrossRef] [PubMed]
105. Wang, B.; Shi, H.; Habteselassie, M.Y.; Deng, X.; Teng, Y.; Wang, Y.; Huang, Q. Simultaneous removal of multidrug-resistant Salmonella enterica serotype typhimurium, antibiotics and antibiotic resistance genes from water by electrooxidation on a Magnéli phase Ti_4O_7 anode. *Chem. Eng. J.* **2021**, *407*, 127134. [CrossRef]
106. Zhi, D.; Zhang, J.; Wang, J.; Luo, L.; Zhou, Y.; Zhou, Y. Electrochemical treatments of coking wastewater and coal gasification wastewater with Ti/Ti_4O_7 and Ti/RuO2-IrO2 anodes. *J. Environ. Manag.* **2020**, *265*, 110571. [CrossRef] [PubMed]
107. Guo, Y.; Xu, Z.; Guo, S.; Liu, J.; Xu, H.; Xu, X.; Gao, X.; Yan, W. Practical optimization of scale removal in circulating cooling water: Electrochemical descaling-filtration crystallization coupled system. *Sep. Purif. Technol.* **2022**, *284*, 120268.
108. Liang, S.; Lin, H.; Yan, X.; Huang, Q. Electro-oxidation of tetracycline by a Magnéli phase Ti_4O_7 porous anode: Kinetics, products, and toxicity. *Chem. Eng. J.* **2018**, *332*, 628–636. [CrossRef]
109. Zwane, B.N.; Orimolade, B.O.; Koiki, B.A.; Mabuba, N.; Gomri, C.; Petit, E.; Bonniol, V.; Lesage, G.; Rivallin, M.; Cretin, M.; et al. Combined Electro-Fenton and Anodic Oxidation Processes at a Sub-Stoichiometric Titanium Oxide (Ti_4O_7) Ceramic Electrode for the Degradation of Tetracycline in Water. *Water* **2021**, *13*, 2772. [CrossRef]
110. Becerril-Estrada, V.; Robles, I.; Martinez-Sanchez, C.; Godinez, L.A. Study of TiO_2/Ti_4O_7 photo-anodes inserted in an activated carbon packed bed cathode: Towards the development of 3D-type photo-electro-Fenton reactors for water treatment. *Electrochim. Acta* **2020**, *340*, 135972–135981. [CrossRef]
111. Safajou, H.; Khojasteh, H.; Salavati-Niasari, M.; Mortazavi-Derazkola, S. Enhanced photocatalytic degradation of dyes over graphene/Pd/TiO_2 nanocomposites: TiO_2 nanowires versus TiO_2 nanoparticles. *J. Colloid Interface Sci.* **2017**, *498*, 423–432. [CrossRef]
112. Yang, Z. Ultrasound Enhanced Electrochemical Oxidation of Chloramphenicol Wastewater with Titanium Sub-Oxide Anode. *Harbin Inst. Technol.* **2019**. [CrossRef]

 MDPI

Article

Facile Synthesis and Environmental Applications of Noble Metal-Based Catalytic Membrane Reactors

Haochen Yan [1], Fuqiang Liu [1], Jinna Zhang [2,*] and Yanbiao Liu [1,3,*]

[1] College of Environmental Science and Engineering, Textile Pollution Controlling Engineering Center of the Ministry of Ecology and Environment, Donghua University, Shanghai 201620, China
[2] State Key Laboratory of Urban Water Resource and Environment, School of Environment, Harbin Institute of Technology, Harbin 150090, China
[3] Shanghai Institute of Pollution Control and Ecological Security, Shanghai 200092, China
* Correspondence: jnzhang@hit.edu.cn (J.Z.); yanbiaoliu@dhu.edu.cn (Y.L.); Tel.: +86-21-6779-8752 (Y.L.)

Abstract: Noble metal nanoparticle-loaded catalytic membrane reactors (CMRs) have emerged as a promising method for water decontamination. In this study, we proposed a convenient and green strategy to prepare gold nanoparticle (Au NPs)-loaded CMRs. First, the redox-active substrate membrane (CNT-MoS_2) composed of carbon nanotube (CNT) and molybdenum disulfide (MoS_2) was prepared by an impregnation method. Water-diluted Au(III) precursor ($HAuCl_4$) was then spontaneously adsorbed on the CNT-MoS_2 membrane only through filtration and reduced into Au(0) nanoparticles in situ, which involved a "adsorption–reduction" process between Au(III) and MoS_2. The constructed CNT-MoS_2@Au membrane demonstrated excellent catalytic activity and stability, where a complete 4-nitrophenol transformation can be obtained within a hydraulic residence time of <3.0 s. In addition, thanks to the electroactivity of CNT networks, the as-designed CMR could also be applied to the electrocatalytic reduction of bromate (>90%) at an applied voltage of −1 V. More importantly, by changing the precursors, one could further obtain the other noble metal-based CMR (e.g., CNT-MoS_2@Pd) with superior (electro)catalytic activity. This study provided new insights for the rational design of high-performance CMRs toward various environmental applications.

Keywords: catalytic membrane reactor; noble metal catalyst; carbon nanotubes; molybdenum disulfide; water purification

Citation: Yan, H.; Liu, F.; Zhang, J.; Liu, Y. Facile Synthesis and Environmental Applications of Noble Metal-Based Catalytic Membrane Reactors. *Catalysts* **2022**, *12*, 861. https://doi.org/10.3390/catal12080861

Academic Editor: Hideyuki Katsumata

Received: 7 July 2022
Accepted: 3 August 2022
Published: 5 August 2022

1. Introduction

Recently, catalytic membrane reactors (CMRs) have attracted considerable attention for various environmental applications [1–3]. In a typical CMR, catalysts could be attached on or embedded in varying substrate materials [4]. This strategy allows the catalytic reaction and separation process to occur simultaneously in a single operational unit without the necessity for a catalyst post-separation [5,6]. In comparison to conventional batch systems, such a flow-through design features enhanced mass transport by convection, operability, and scalability. However, it is still a grand challenge to rationally design a high-performance CMR with excellent catalytic reactivity and separation efficiency.

Gold nanoparticles (Au NPs) have been regarded as promising catalysts due to their excellent catalytic performance, which have been applied in many important industrial sectors, such as CO oxidation, hydrogenation reaction, and alcohols oxidation [7–9]. The specific properties of Au NPs are highly dependent on their nanoscale size and high surface area [10]. Although these nanoparticles exhibit a better catalytic performance, the direct use of these nano-catalysts can hardly be achieved due to particle aggregations and the post-separation of catalyst from solution [11–13]. These restrictions significantly increase the cost, thus hindering wide industrial applications. To overcome these limitations, significant efforts have been devoted to constructing catalytic membrane reactors (CMRs) by integrating the noble metal nanocatalysts with supporting materials [14,15]. Researchers have developed various strategies to immobilize these nanoparticles onto carriers, including coprecipitation, impregnation, and/or in situ growth [16]. However, these typical methods usually involve toxic hazardous reducing and/or stabilizing chemical reagents [17–19]. In addition, the harsh flow conditions during catalysis may wash out these catalysts from the host surface, which would ultimately lead to an evident performance decay over continuous running cycles. It is, therefore, highly desirable to develop rapid, robust, and environmentally-friendly CMRs preparation protocols.

Alternatively, the emerging two-dimensional transition metal dichalcogenides, molybdenum disulfide (MoS_2), may have the potential to address the mentioned problems due to their advantages of low toxic and sulfur-rich properties [20,21]. Studies have demonstrated that MoS_2, with abundant active sulfur sites, have a high affinity for special metal ions (e.g., Pb(II), Ag(I) and Au(III)) via Lewis soft−soft interactions [22–24]. On the other hand, it has been found that the adsorbed noble metal ions onto MoS_2 can spontaneously capture the electrons released by Mo(IV) to achieve their synchronous reduction, avoiding the need for additional reductants (e.g., $NaBH_4$) [25–27]. For example, our recently work showed that the MoS_2 nanoflowers could achieve the effective recovery of gold from complex wastewater involving a two-step adsorption–reduction process [28–30].

Furthermore, the substrate membrane used in the CMR also plays a unique role to construct a robust system. Various substrates with high stability and water flux have been reported, such as alumina, polymeric, and graphene oxide [31–33]. Among these support materials, one-dimensional carbon nanotubes (CNT) may be considered an ideal substrate due to its porosity, high mechanical strength, large specific surface area, rich surface chemistry, and electroactivity [34–37]. Moreover, CNT can easily be assembled into 3D porous and conductive networks by the vacuum filtration route. Therefore, this easy-modification material could provide a promising platform for improving the stability and dispersion of MoS_2.

In this study, we adopted a rapid and green approach to synthesize the Au-immobilized CMR (CNT-MoS_2@Au). The MoS_2-modified CNT (CNT-MoS_2) membrane was fabricated by a simple one-step impregnation method. The MoS_2 nanoflowers on the membrane surface allowed the adsorption of $[AuCl_4]^-$ and the in situ reduction of the adsorbed anions to Au NPs free of spiking any reducing agents. Morphological, compositional, and structural characterization were performed to collectively verify the successful synthesis of Au NPs on the CNT-MoS_2-supporting membrane. The catalytic performance of the as-prepared CNT-MoS_2@Au membrane was evaluated using the hydrogenation of 4-nitrophenol (4-NP) as a model reaction, owing to its well-established characterization protocol and operational simplicity. The effects of key operational parameters on the performance of membrane reactors were systematically investigated. The electrocatalytic reduction of bromate (BrO_3^-) was also employed as another model reaction to demonstrate the excellent electrocatalytic characteristics of the CNT-MoS_2@Au membrane. Finally, the non-specificity of CNT-MoS_2 membrane for loading different noble metals was emphasized by regulating the precursors.

2. Results and Discussion

2.1. Facile Synthesis of CNT-MoS_2@Au Catalytic Membrane

Figure 1 illustrated a facile strategy for the construction of a CNT-MoS_2@Au catalytic membrane. The MoS_2 was prepared through a hydrothermal process and the CNT membrane was synthesized via a vacuum filtration route [38]. By impregnating the CNT membrane in 15 mL of well-dispersed MoS_2 ethanol solution (1 mg/mL) for 90 min, the CNT-MoS_2 membrane was obtained after a further heat treatment (140 °C for 1 h). To endow catalytic activity for the CNT-MoS_2 membrane, Au NPs were loaded onto the membrane by passing through the noble metal salt solution (20 mL of 0.1 mM $HAuCl_4$) at a flow rate of 3 mL/min and pH 4.8 for 90 min. Consequently, a CNT-MoS_2@Au catalytic membrane reactor was constructed. Notably, when compared to conventional synesis routes, the filtration approach takes less time to achieve the uniform introduction of Au NPs (e.g., 90 min vs. 48 h) [39]. As filtration proceeded, the color of the Au(III) solution visibly changed from light yellow to colorless, whereas the color of the membrane changed from black to brown-gold (Figure S1), suggesting that Au had been successfully deposited onto the surface of the CNT-MoS_2 membrane. A CNT-alone membrane was prepared as a control. After passing through 20 mL of 0.1 mM $HAuCl_4$ for 90 min, no visible color change occurred, demonstrating that a negligible Au(III) was retained by the CNT-only membrane. This observation also highlighted the essential role played by MoS_2 for anchoring Au(III).

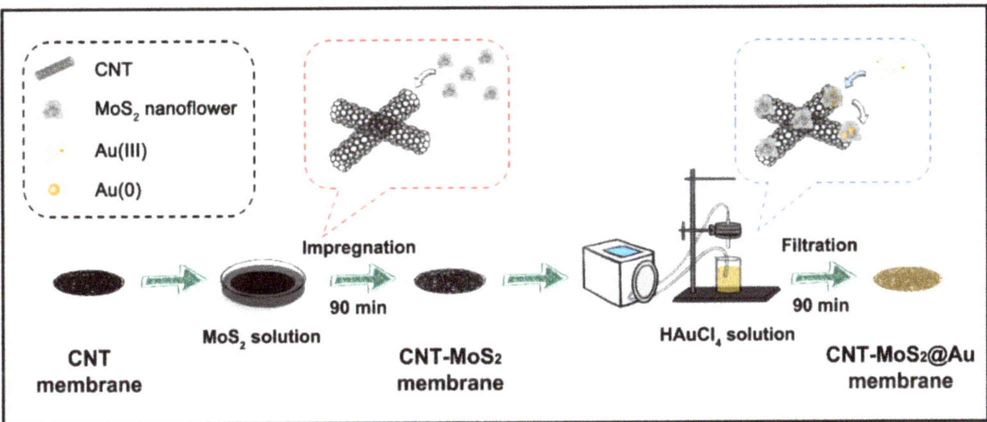

Figure 1. Schematic illustration for the synthesis of the CNT-MoS_2@Au catalytic membrane.

2.2. Characterization of the CNT-MoS_2@Au Catalytic Membrane

Figure 2a showed the FESEM image of the CNT-MoS_2@Au membrane. Typically, MoS_2 presented as nanoflowers composed of ultrathin lamellar structures with an average diameter of 400 nm (Figure S2) [38]. Numerous spherical nanoparticles were observed after the Au loading, while those MoS_2 nanoflowers were disappeared. Similar observations were reported previously [38]. These bright spherical nanoparticles were later proven to be metallic Au by using XPS and XRD characterization. The disappearance of the MoS_2 nanoflowers may be attributed to the high relative atomic mass of Au, leading to a collapse of the MoS_2 nanoflower structure during the operation. We used FESEM to monitor the evolution of the catalytic membrane surface during the Au loading process. Figure S3a showed that MoS_2 retained the nanoflower-like structure during the first 30 min and those Au NPs were observed to deposit on the edge of MoS_2 lamellar structures. The energy dispersive spectroscopy (EDS) was applied to illustrate the elemental distribution for Mo, S and Au (Figure S3b). The high-degree morphological consistence among these elements confirmed the successful loading of Au onto MoS_2 (Figures S2, S3a and 2a). The XPS spectra showed that the superficial elemental atomic ratio of the CNT-MoS_2@Au membrane was 67.83% C, 8.21% O, 4.74% S, 0.06% Mo, and 19.17% Au (Figure S4). Among them, the high-resolution Au 4f scan over a small energy window indicated two sets of Au 4f spin−orbit coupling doublets was observed (Figure 2b), with the $4f_{5/2}$ and $4f_{7/2}$ centered at 87.6 and 83.9 eV, which were associated with characteristic of Au(0) [40], suggesting an adsorption–reduction of Au(III) occurred on the membrane's surface. XRD analysis also indicated that four distinct diffraction peaks corresponding to metallic Au(0) located at a 2θ of 38.2° (111), 44.6° (200), 64.7° (220), and 77.6° (311) were identified (Figure S5) [41].

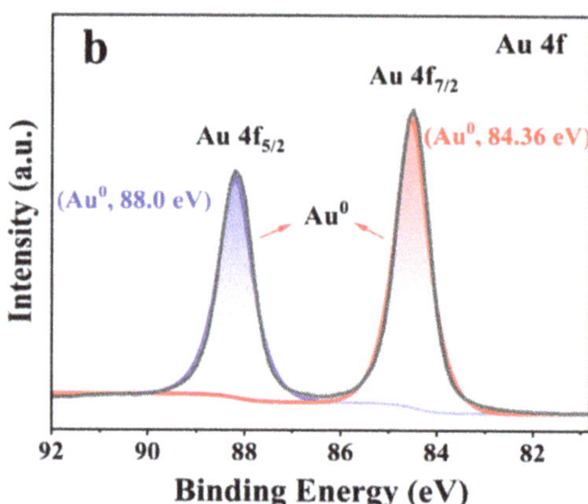

Figure 2. Preparation and characterization of the CNT-MoS_2@Au catalytic membrane. (**a**) FESEM image and (**b**) high resolution Au 4f XPS spectrum of CNT-MoS_2@Au.

2.3. Hydrogenation Reaction of 4-NP

The hydrogenation reaction of 4-NP was employed as a model catalytic reaction to compare the catalytic performance before and after the Au loading. In a control experiment, the characteristic peak of 4-NP (λ_{max} = 400 nm) occurred no distinctive change after the solution (0.1 mM 4-NP and 30 mM $NaBH_4$) passed through the CNT-MoS_2 membrane (Figure S6). This indicated that the unfunctionalized CNT-MoS_2 membrane made no effort toward the 4-NP reduction. In comparison, the 4-NP characteristic absorption peak quickly decreased once the mixture solution passing through the CNT-MoS_2@Au and CNT-MoS_2@Pd membrane. The simultaneous appearance of an alternative absorption peak at 300 nm corresponded to the presence of 4-aminophenol (4-AP). This confirmed that over 95% of 4-NP was reduced by an effective hydrogenation process just after a single-pass through the catalytic membrane in the presence of $NaBH_4$. Notably, the ultrahigh catalytic activity of the membrane was demonstrated by an extremely short hydraulic retention time of <2 s [42]. Since the $NaBH_4$ concentration significantly exceeds that of 4-NP, the reaction kinetics can be considered as pseudo-first-order (Equation (1)) [43].

$$\ln(C_t/C_0) = -kt \tag{1}$$

To optimize the catalytic performance, the effects of Au(III) loading time, flow rate, and initial 4-NP concentration on the reduction of 4-NP were investigated. First, the effect of loading time of Au(III) (30, 60, 90, and 120 min) on the catalytic performance of CNT-MoS_2@Au membrane was explored, since the loading time was directly correlated to the introduced Au catalysts. CNT-MoS_2@Au membranes were characterized by XPS to investigate the alternation of Au loading time. The XPS survey pattern at different loading times (30 and 90 min) suggested that the atomic ratio of Au increased with the loading time (Figure S4). Results indicated that increasing the loading time from 30 to 90 min led to an improvement in the reduction efficiency of 4-NP (0.1 mM) from 69.6%s to 92.5% at 2 mL/min under pH 8.0 (Figure 3a). This phenomenon can be ascribed to the better accessibility of 4-NP with Au NPs at an increased Au loading. However, the catalytic efficiency failed to increase significantly once the loading time further extended to 120 min (92.6%), which may be associated with the agglomeration of Au NPs (Figure S7) and burying certain available active sites with a longer loading times. Therefore, the loading time was fixed at 90 min for the subsequent investigations.

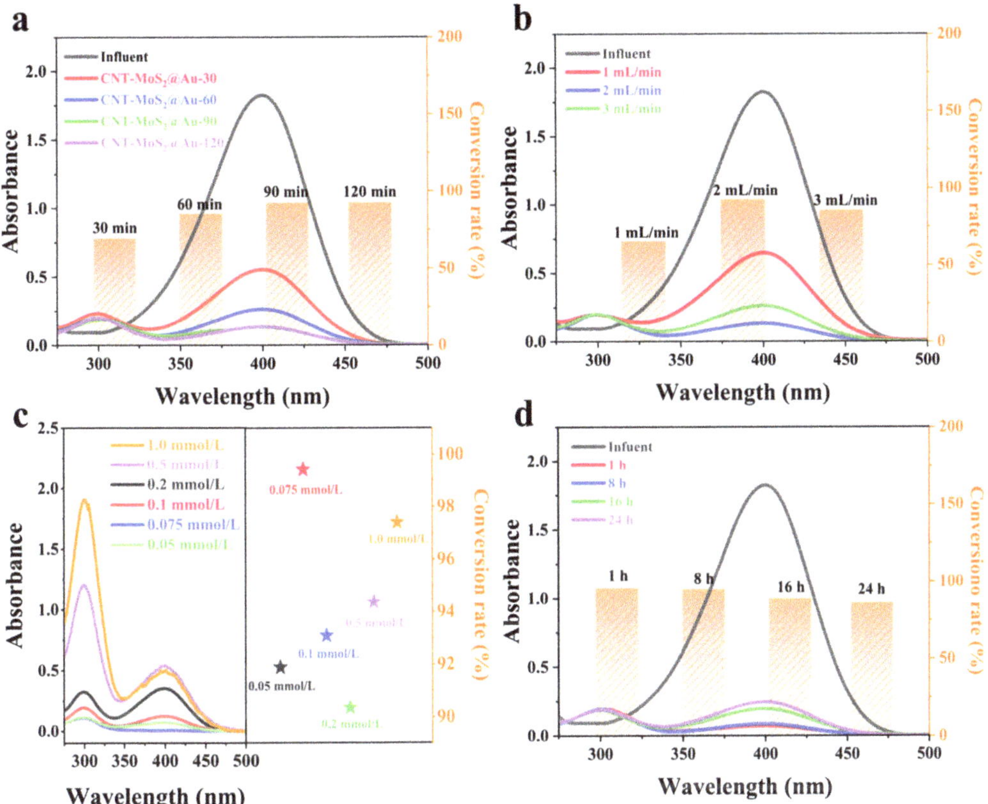

Figure 3. Effects of (**a**) Au loading time (30 to 120 min), (**b**) flow rate (1 to 3 mL/min), and (**c**) initial 4-NP concentration (0.05 to 1.0 mM) on the conversion of 4-NP using the CNT-MoS_2@Au catalytic membrane. (**d**) Operating stability in the 4-NP reduction using the CNT-MoS_2@Au catalytic membrane (24 h).

The catalytic performance of CNT-MoS_2@Au membrane was also affected by the flow rate. As shown in Figure 3b, increasing the flow rate from 1.0 to 2.0 mL/min significantly enhanced the reduction efficiency of 4-NP (0.1 mM) from 65.4% to 92.5% at pH 8.0. Nevertheless, the further increase of the flow rate to 3.0 mL/min would lead to an incomplete conversion of 4-NP (85.3%). This could be derived from the emergence of a "trade-off" effect between the 4-NP residence time within the CNT-MoS_2@Au membrane and the reaction kinetics. A lower flow rate might result in a limited mass transport within the flow-through system, whereas an excessive flow rate would bring short residence time for 4-NP and thus caused inadequate contact with the Au NPs [14].

Figure 3c showed the influence of the initial 4-NP concentration on the catalytic reduction of 4-NP. Results indicated that the CNT-MoS_2@Au membrane could effectively reduce 4-NP (>90%), regardless of the initial concentrations (from 0.05 to 1.0 mM). This suggested that the catalytic membrane had an excellent resistance to varying concentrations of 4-NP, possibly due to the effective loading of Au NPs, which provides abundant active sites for the transformation of 4-NP.

The turnover frequency (TOF), which indicates the products that are able to be generated in a catalytic reaction by the per molar amount of catalyst, was employed to quantitatively evaluate the catalytic performance. The TOF of the CNT-MoS_2@Au membrane under flow catalysis was calculated by Equation (2):

$$TOF = C_0 \times Conversion \times v / N_{Au} \quad (2)$$

where C_0 is the initial concentration of 4-NP (mol), N_{Au} is the molarity of Au in CNT-MoS_2@Au membrane (mol), and v is the flow rate (mL/min). Table S1 showed that the TOF value of CNT-MoS_2@Au CMR reached 609 h^{-1} with a flow rate of 2 mL/min, which represented a much better catalytic performance than some reported reaction systems.

To further gain insight into the operating stability of the CNT-MoS_2@Au catalytic membrane, we performed a long-time (24 h) catalytic reduction experiment for 4-NP (0.1 mM) transformation in the single-pass mode at a flow rate of 2 mL/min under pH 8.0. Results exhibited that the conversion efficiency of 4-NP could maintain >86% after the continuous operation (Figure 3d). Importantly, high catalytic activity (~95%) could be recovered just by washing the membrane with copious water for a few minutes. In addition, the catalytic performance (e.g., 95.2% after 480 min) for 4-NP was comparable or even better than that of other reported CMR, such as the β-lactoglobulin fibrils membrane (>97% after 240 min) [14] and microporous polymer monoliths (~90% after ~40 min) [44]. This suggested a promising potential application of the CNT-MoS_2@Au membrane to create CMR for industrial applications.

2.4. Electrocatalytic Reduction of BrO_3^-

Because of the excellent electrochemical properties of the CNT networks, the CNT-MoS_2@Au membrane was presumed to have excellent electrocatalytic capacity. To confirm this speculation, the electrochemical reduction of BrO_3^- was performed without dosing additional reducing agents. First, the control experiment was conducted to explore the electroreduction of BrO_3^- (0.1 mM) by the CNT-MoS_2 membrane at a flow rate of 2 mL/min under pH 4.0 (Figure S8). Negligible change in BrO_3^- concentration was observed and no Br^- was detected within 90 min regardless of the exertion of an electric field or not. This result revealed that the CNT-MoS_2 membrane was unable to electrochemically reduce BrO_3^- in the absence of Au NPs. For the CNT-MoS_2@Au membrane at 0 V, the result was similar with that of CNT-MoS_2 membrane. However, effective BrO_3^- reduction (90.1%) was achieved when a potential of −1.0 V was exerted on the CNT-MoS_2@Au membrane. The IC spectra exhibited a quick decrease of the BrO_3^- accompanied with the formation of Br^- during the electrocatalytic treatment (Figure 4c). As displayed in the inset chart, the total concentration of BrO_3^- and Br^- was slightly less than the initial total BrO_3^- concentration, which could be attributed to the formation of byproducts. This positive result demonstrated that the as-prepared CNT-MoS_2@Au membrane possess good electrocatalytic activity under electric field.

To obtain optimal electrocatalytic performance, the effect of applied potential on BrO_3^- reduction was explored. As shown in Figure 4b, all obtained data fitted the pseudo-first-order kinetic model well ($R^2 > 0.9$). Decreasing the potential from 0 to −1 V substantially improved the BrO_3^- reduction efficiency from ~0 to 90.1%. Nevertheless, further decreasing the potential to −2 V significantly inhibited the reduction of BrO_3^- (58.4%), which could be ascribed to the BrO_3^- reduction being highly potential-dependent and the occurrence of other side reactions (e.g., hydrogen evolution reaction). Appling a proper electric field (e.g., −1 V) to the CNT-MoS_2@Au membrane could significantly enhance near-surface transport by electromigration [45]. Meanwhile, electrostatic interactions between the negatively-charged BrO_3^- and positively-charged membrane surface was beneficial to the electrocatalytic reaction. However, other competitive side reactions, such as hydrogen evolution or electro-corrosion, would deteriorate the catalytic performance when the applied potential exceeded the optimal value [46–48].

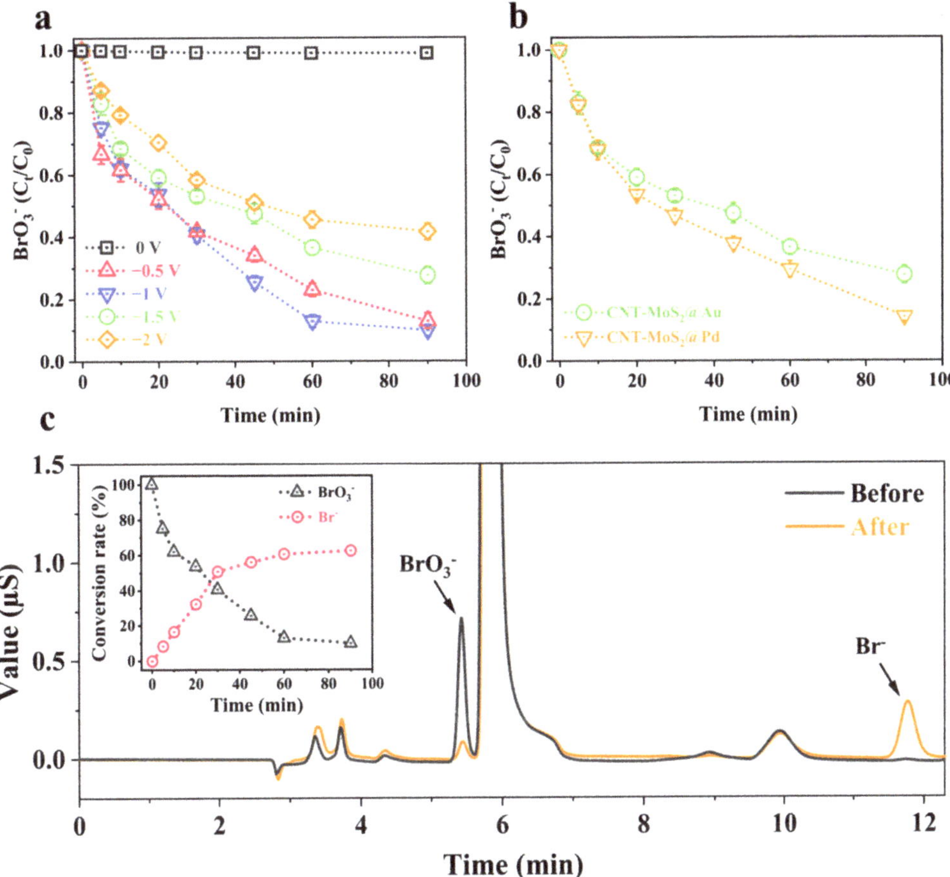

Figure 4. (**a**) Effect of applied potential (0 to −1.5 V) on $BrO_3{}^-$ reduction using the CNT-MoS_2@Au catalytic membrane. (**b**) Comparison of effects on $BrO_3{}^-$ reduction using different catalytic membrane. (**c**) IC spectra of the $BrO_3{}^-$ solution before and after the electrocatalytic treatment. Inset is a comparation of the conversion efficiency towards $BrO_3{}^-$ and Br^- within 90 min.

To better understand the different electrocatalytic activity of the CNT-MoS_2 membrane before and after Au NPs loading, CV and EIS measurements were carried out to determine the changes in electrochemical properties. As shown in Figure 5a, the redox peaks and current response of CNT-MoS_2@Au were higher and sharper than that of CNT-MoS_2. These results confirmed that CNT-MoS_2@Au exhibited better electrochemical activity and more rapid electron transfer process. In addition, it has been reported that the potential difference (Δ_{Ep}) between the oxidation peak and reduction peak is inversely correlated to the charge transfer kinetics. The Δ_{Ep} of the CNT-MoS_2@Au was 0.72-fold higher than that of CNT-MoS_2 (0.206 V vs. 0.279 V), demonstrating the charge transfer rate of CNT-MoS_2@Au was 1.37-fold than that of CNT-MoS_2 [49,50]. The EIS spectra also showed that the charge transfer resistance of CNT-MoS_2@Au (24.5 Ω) was similar with that of CNT-MoS_2 (32.5 Ω) (Figure 5b) and consistent with the CV results. This evidence suggested that the loading of Au NPs onto the CNT-MoS_2 membrane can improve the electrochemical capacitance and electron transfer kinetics, thus providing an ideal platform toward the electrocatalytic reduction of $BrO_3{}^-$.

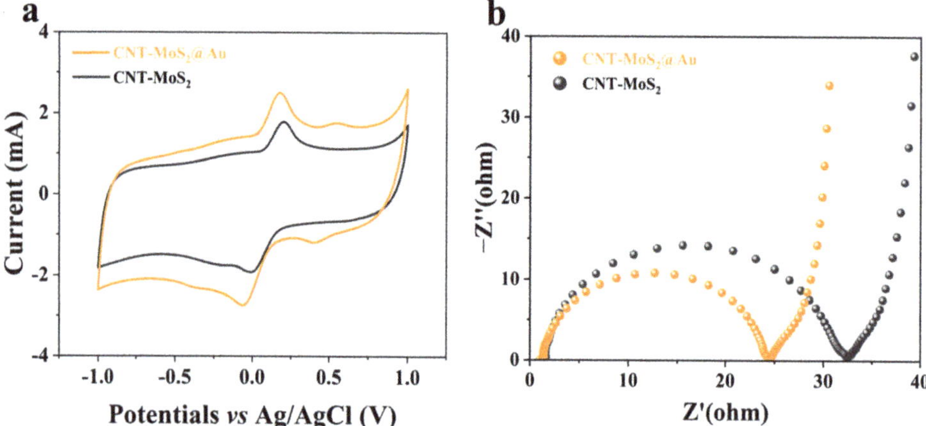

Figure 5. (**a**) CV and (**b**) EIS spectra of the CNT-MoS_2 membrane and CNT-MoS_2@Au catalytic membrane. The electrolyte contains 50 mM Na_2SO_4 and 5 mM $K_3[Fe(CN)_6]$. The CV scan rate was 5 mV/s. The amplitude and scan range in EIS were 5 mV and 10^5–10^{-2} Hz, respectively.

2.5. Generality of the Electrocatalytic Membranes

According to our previous report, besides Au, MoS_2 also exhibited a high affinity towards other noble metals (e.g., Pd). Following the similar synthesis routes, we prepared a CNT-MoS_2@Pd membrane by using different Pd-based precursors to replace $HAuCl_4$. As displayed in Figure 4b, the designed Pd-loaded catalytic membranes also demonstrated excellent performance (85.8%) toward the electrocatalytic reduction of BrO_3^- under similar reaction conditions (loading 90 min, flow rate = 2 mL/min, pH = 4, applied potential = −1.5 V). According previous reports, most reported synthesis protocols for Pd-loaded catalysts usually involve complicated processes or time-consuming steps, such as microemulsion and calcination [51,52]. It is of note that our adopted approach is highly desirable (e.g., short fabrication time and comparable or even higher catalytic activity). It not only recovers noble metal ions from water but can also directly serve as high-performance (electro)catalytic membranes for environmental applications. In other words, our proposed method can be applied to prepare multiple noble metal-loaded CMRs.

2.6. Working Mechanism of the Au-Immobilized CMR

Based on the previous discussion, it was inferred that a redox reaction occurred between Au(III) and MoS_2 following Equation (3). Au(III) was first adsorbed onto the surface of MoS_2 nanoflowers under an action of chemical chelation between Au and S, then Au(III) was able to achieve an in situ reduction to metallic AuNPs through capture the electron transfer released by the Mo(IV) [38]. This demonstrated that by impregnating the CNT-MoS_2 membrane with the Au(III) precursor (i.e., $HAuCl_4$), the Au NPs were successfully immobilized on the membrane through the adsorption–reduction process. Similarly, this simple protocol was adaptable to the loading of other noble metal catalysts.

$$6Au^{3+} + MoS_2 + 12H_2O \rightarrow 6Au^0 + MoO_4^{2-} + 2SO_4^{2-} + 24H^+ \quad (3)$$

For the 4-NP hydrogenation reaction, 4-NP molecules were firstly attached to the surface of Au NPs together with $NaBH_4$. $NaBH_4$ then transferred electrons to Au to generate molecular hydrogen, which contributed to the reduction of the nitro group of 4-NP to amino group. Finally, the generated 4-AP diffused away from the catalysts surface to free up the active catalytic sites for further reduction process [53]. Detailed steps involved are described by Equation (4) to Equation (8). Similar processes occurred in the electrochemical reduction as well. In the electrocatalytic system, molecular hydrogen was dissociated in the process of water electrolysis and adsorbed into the Au NPs forming Au hydride (Au NPs also served as the cathode) [54]. These Au hydrides participated in the removal of oxygen atoms from bromate by hydrogenation [46]. This process can be expressed by Equations (9) and (10). Overall, the immobilized Au NPs offered excellent catalytic capability in the heterogeneous catalytic system.

$$4-\text{NP}+\text{Au} \rightarrow 4-\text{NP}\cdot\text{Au} \tag{4}$$

$$\text{BH}_4{}^{-}+\text{Au} \rightarrow \text{BH}_4{}^{-}\cdot\text{Au} \tag{5}$$

$$2\text{H}_2\text{O}+\text{BH}_4{}^{-}\cdot\text{Au} \rightarrow 4\text{H}_2\cdot\text{Au}+\text{BO}_2{}^{-} \tag{6}$$

$$4-\text{NP}\cdot\text{Au}+3\text{H}_2\cdot\text{Au} \rightarrow 4-\text{AP}\cdot\text{Au}+2\text{H}_2\text{O} \tag{7}$$

$$4-\text{AP}\cdot\text{Au} \overset{desorption}{\rightarrow} 4-\text{AP}+\text{Au} \tag{8}$$

$$2\text{H}_2\text{O}+2e^{-}+\text{Au} \rightarrow \text{H}_2\cdot\text{Au}+2\text{OH}^{-} \tag{9}$$

$$3\text{H}_2\cdot\text{Au}+\text{BrO}_3{}^{-} \rightarrow \text{Br}^{-}+3\text{H}_2\text{O} \tag{10}$$

3. Materials and Methods

3.1. Chemicals and Materials

All chemicals were of analytical grade and used without further purification. Multiwalled carbon nanotubes were provided by TimesNano Co., Ltd. (Chengdu, China). N-methyl-2-pyrrolidinone (NMP, ≥99.5%), ethanol (C_2H_5OH, ≥96%), thiourea (CH_4N_2S, ≥99%), sodium molybdate ($Na_2MoO_4 \cdot 2H_2O$, ≥99%), hydrochloric acid (HCl, 36.0~38.0%), sodium bromate ($NaBrO_3$, ≥99%), sodium sulfate (Na_2SO_4, ≥99%), potassium ferricyanide ($K_3[Fe(CN)_6]$, ≥99%), palladium chloride ($PdCl_2$, ≥99%), and sodium borohydride ($NaBH_4$, ≥98%) were obtained from Sinopharm Chemical Reagent Co., Ltd. (Beijing, China). Au(III) chloride trihydrate ($HAuCl_4 \cdot 3H_2O$, ≥49%) and 4-nitrophenol (4-NP, ≥99%) were provided by Sigma-Aldrich (St. Louis, MO, USA). Ultrapure water produced from a Milli-Q Direct 8 purification system (Millipore, Burlington, MA, USA) was used for all experiments.

3.2. Fabrication of Catalytic Membrane Reactors

The CNT-MoS_2 membrane was fabricated according to a reported protocol [38]. The CNT-MoS_2@Au catalytic membrane was prepared by a simple filtration route with an effective membrane area of 7.1 cm^2 (Figure S9). In brief, the $HAuCl_4$ solution (20 mL, 1 mM) was continuously passed through the CNT-MoS_2 membrane and then returned at a flow rate of 3.0 mL/min. The Au loading on the CNT-MoS_2 membrane can be adjusted by changing the filtration time (30 to 120 min). To demonstrate the non-specificity of the CNT-MoS_2 membrane, the CNT-MoS_2@Pd catalytic membrane was also prepared by a similar procedure with $PdCl_2$ solution as precursor (20 mL, 1 mM). All propulsion was provided by a peristaltic pump (Ismatec ISM833C, Glattbruch-Zurich, Switzerland).

3.3. Catalytic Filtration Experiments

All catalytic filtration experiments were performed on a operated in a Whatman polycarbonate filtration casing (Whatman, Dassel, Germany) [36]. The hydrogenation of 4-NP was applied to evaluate the catalytic performance of the CNT-MoS_2@Au membrane. Initially, to eliminate the effect of the physical adsorption on the 4-NP removal, 100 mL of 0.1 mM 4-NP solution was first passed through the CNT-MoS_2@Au membrane to achieve adsorption saturation at 2.0 mL/min. The hydrogenation reaction was induced by passing through the membrane with 0.1 mM 4-NP solution together with 30 mM freshly prepared $NaBH_4$ at pH 8.0. The effects of Au(III) loading time (30 to 120 min), flow rate (1 to 3 mL/min), and initial 4-NP concentration (0.05 to 1.0 mM) on the catalytic performance were investigated systematically. The flow rate was controlled by a peristaltic pump and the solution pH was adjusted by 1 M NaOH and/or HCl. Effluent samples were collected at specific time intervals and characterized by a UV–vis spectrophotometer.

To demonstrate the excellent electrocatalytic reactivity of the CNT-MoS_2@Au membrane, the electrocatalytic reduction of bromate (BrO_3^-) was employed as another model reaction in a Whatman polycarbonate filtration casing with electrochemistry modifications (Figure S10) [55]. Applied potential was controlled by a CHI 660E electrochemical workstation (Chenhua Co., Ltd., Shanghai, China) in a typical three-electrode system with a CNT-MoS_2@Au working electrode, a saturated Ag/AgCl reference electrode and a Ti sheet counter electrode. In a typical electrocatalytic experiment, 20 mL of 0.1 mM $NaBrO_3$ solution with 4 mM Na_2SO_4 were passed through the membrane at 2.0 mL/min and pH 4.0 in the recirculated filtration mode. The impact of applied potential (−2 to 0 V) on the BrO_3^- reduction kinetics was optimized. Furthermore, the electrocatalytic reduction experiment was conducted by applying the CNT-MoS_2@Pd membrane to similar conditions. Effluent samples were collected with a 2 mL centrifuge tube and immediately filtered through a 0.22-μm cellulose acetate membrane. The concentration of BrO_3^- was determined by ion chromatography (IC). The conversion efficiency (%) was obtained by Equation (11):

$$\text{Conversion efficiency } (\%) = 100 \times (C_0 - C)/C_0 \quad (11)$$

where C_0 and C are the substrate concentration before or after passing through the noble metal-loaded CNT-MoS_2 membrane.

3.4. Characterization

The morphology of the CNT-MoS_2@Au membranes were characterized by field emission scanning electron microscopy (FESEM, S-4800, Hitachi, Tokyo, Japan) and energy dispersive spectroscopy (EDS, JEM-2100F). X-ray diffraction (XRD) patterns of the samples were acquired by a Rigaku D/max-2550/PC X-ray diffractometer (Rigaku, Tokyo, Japan) with Au radiation within the range of 5 to 90°. X-ray photoelectron spectroscopy (XPS) was conducted at a Thermo Fisher Scientific Escalab 250Xi (Waltham, MA, USA) under high vacuum. The concentrations of 4-NP were determined by an UV-2600 Shimadzu ultraviolet–visible spectrophotometry (Shimadzu, Kyoto, Japan) at λ_{max} of 400 nm. The concentrations of BrO_3^- were determined by Thermo Scientific Dionex Aquion IC using an IonPac AS11-HC column and 20 mM KOH eluent. The electrochemical activity of the catalytic membranes was probed by cyclic voltammetry (CV) and electrochemical impedance spectroscopy (EIS). Cyclic Voltammetry (CV) was acquired at a scan rate of 5 mV/s in a three-electrode system. EIS analysis was performed over a frequency range of 10^5 to 10^{-2} Hz at an amplitude of 5 mV.

4. Conclusions

In summary, we have successfully developed a facile and green approach to prepare varying noble metal-based CMRs for water purification. The as-synthesized redox-active CNT-MoS_2 membrane served as a robust platform, which enabled a rapid uptake of various noble metal (e.g., Au(III) and Pd(II)) precursors only by passing through the porous and conductive networks. The in situ adsorption–reduction of noble metal ions can be achieved spontaneously. In addition, the constructed CNT-MoS_2@Au membrane reactors demonstrated excellent (electro)catalytic activity and thus achieved the effective reduction of 4-NP (>99%) and bromate (>90%) within a hydraulic residence time of <2 s, which was comparable to or even better than several state-of-the-art reports. Overall, this study would significantly improve the sustainability and cost-effectiveness of (electro)catalytic reduction technologies toward water decontamination.

Supplementary Materials: The following supporting information can be downloaded at: https://www.mdpi.com/article/10.3390/catal12080861/s1. Figure S1: Electronic photos of (a) the CNT-MoS_2 membrane and (b) the CNT-MoS_2@Au membrane; Figure S2: FESEM image of CNT-MoS_2 membrane; Figure S3: (a) FESEM and (b) FESEM-EDS mapping images of CNT-MoS_2@Au-30 membrane (Au loading time, 30 min); Figure S4: Comparison of the XPS survey spectrums of CNT-MoS_2@Au catalytic membrane with different Au loading time (30 and 90 min); Figure S5: The XRD pattern of the

CNT-MoS_2@Au catalytic membrane; Figure S6: Degradation efficiency of 4-NP by using CNT-MoS_2 membrane, CNT-MoS_2@Au membrane and CNT-MoS_2@Pd membrane. Experimental conditions: $[4\text{-NP}]_0$ = 0.1 mM, flow rate = 2.0 mL/min and pH_0 = 8.0; Figure S7: FESEM image of CNT-MoS_2@Au-120 membrane (Au loading time, 120 min); Figure S8: Degradation efficiency of bromate by using CNT-MoS_2 membrane and CNT-MoS_2@Au membrane before and after applying potential. Experimental conditions: $[BrO_3^-]_0$ = 0.1 mM, flow rate = 2.0 mL/min and pH_0 = 4.0; Figure S9: Schematic illustration of the flow-through electrocatalytic filtration system; Figure S10: Schematic diagram of electrochemistry-modified Whatman polycarbonate filtration casing; Table S1: Comparison of the 4-NP reduction performance of proposed system with reported catalytic membrane reactor systems; Table S2: The k values related to bromate reduction under different applied potential according to pseudo-first order kinetic model. References [56–60] are cited in the Supplementary Materials.

Author Contributions: Conceptualization, Y.L.; Data curation, H.Y. and F.L.; Funding acquisition, J.Z.; Investigation, H.Y. and F.L.; Methodology, H.Y. and F.L.; Project administration, Y.L. and J.Z.; Resources, Y.L.; Supervision, Y.L. and J.Z.; Writing—original draft, H.Y. and F.L.; Writing—review and editing, Y.L. and J.Z. All authors have read and agreed to the published version of the manuscript.

Funding: This work was supported by the National Natural Science Foundation of China (No. 52070055) and Heilongjiang Touyan Innovation Team Program (HIT-SE-01).

Data Availability Statement: All data supporting this study are available in the Supplementary Information accompanying this paper.

Conflicts of Interest: The authors declare no conflict of interest.

References

1. Sun, M.; Wang, X.; Winter, L.R.; Zhao, Y.; Ma, W.; Hedtke, T.; Kim, J.-H.; Elimelech, M. Electrified membranes for water treatment applications. *ACS EST Eng.* **2021**, *1*, 725–752. [CrossRef]
2. Liu, Y.; Liu, X.; Yang, S.; Li, F.; Shen, C.; Huang, M.; Li, J.; Nasaruddin, R.R.; Xie, J. Rational design of high-performance continuous-flow microreactors based on gold nanoclusters and graphene for catalysis. *ACS Sustain. Chem. Eng.* **2018**, *6*, 15425–15433. [CrossRef]
3. Peng, F.; Xu, J.; Xu, H.; Bao, H. Electrostatic interaction-controlled formation of pickering emulsion for continuous flow catalysis. *ACS Appl. Mater. Interfaces* **2021**, *13*, 1872–1882. [CrossRef]
4. Zeng, Z.; Wen, M.; Yu, B.; Ye, G.; Huo, X.; Lu, Y.; Chen, J. Polydopamine induced in-situ formation of metallic nanoparticles in confined microchannels of porous membrane as flexible catalytic reactor. *ACS Appl. Mater. Interfaces* **2018**, *10*, 14735–14743. [CrossRef] [PubMed]
5. Dolatkhah, A.; Jani, P.; Wilson, L.D. Redox-responsive polymer template as an advanced multifunctional catalyst support for silver nanoparticles. *Langmuir* **2018**, *34*, 10560–10568. [CrossRef]
6. Yu, Y.; Huo, H.; Zhang, Q.; Chen, Y.; Wang, S.; Liu, X.; Chen, C.; Min, D. Nano silver decorating three-dimensional porous wood used as a catalyst for enhancing azo dyes hydrogenation in wastewater. *Ind. Crops Prod.* **2022**, *175*, 114268. [CrossRef]
7. Jia, Z.; Ben Amar, M.; Yang, D.; Brinza, O.; Kanaev, A.; Duten, X.; Vega-González, A. Plasma catalysis application of gold nanoparticles for acetaldehyde decomposition. *Chem. Eng. J.* **2018**, *347*, 913–922. [CrossRef]
8. Li, Y.; Lan, J.Y.; Liu, J.; Yu, J.; Luo, Z.; Wang, W.; Sun, L. Synthesis of gold nanoparticles on rice husk silica for catalysis applications. *Ind. Eng. Chem. Res.* **2015**, *54*, 5656–5663. [CrossRef]
9. Chen, J.; Yan, D.; Xu, Z.; Chen, X.; Chen, X.; Xu, W.; Jia, H.; Chen, J. A novel redox precipitation to synthesize Au-doped α-MnO_2 with high dispersion toward low-temperature oxidation of formaldehyde. *Environ. Sci. Technol.* **2018**, *52*, 4728–4737. [CrossRef]
10. Mistry, H.; Reske, R.; Zeng, Z.; Zhao, Z.-J.; Greeley, J.; Strasser, P.; Cuenya, B.R. Exceptional size-dependent activity enhancement in the electroreduction of CO_2 over Au nanoparticles. *J. Am. Chem. Soc.* **2014**, *136*, 16473–16476. [CrossRef]
11. Smirnov, E.; Peljo, P.; Scanlon, M.D.; Girault, H.H. Interfacial redox catalysis on gold nanofilms at soft interfaces. *ACS Nano* **2015**, *9*, 6565–6575. [CrossRef]
12. Kim, S.H. Nanoporous gold: Preparation and applications to catalysis and sensors. *Curr. Appl. Phys.* **2018**, *18*, 810–818. [CrossRef]
13. Priecel, P.; Adekunle Salami, H.; Padilla, R.H.; Zhong, Z.; Lopez-Sanchez, J.A. Anisotropic gold nanoparticles: Preparation and applications in catalysis. *Chin. J. Catal.* **2016**, *37*, 1619–1650. [CrossRef]
14. Huang, R.; Zhu, H.; Su, R.; Qi, W.; He, Z. Catalytic membrane reactor immobilized with alloy nanoparticle-loaded protein fibrils for continuous reduction of 4-nitrophenol. *Environ. Sci. Technol.* **2016**, *50*, 11263–11273. [CrossRef]
15. Tong, J.; Matsumura, Y.; Suda, H.; Haraya, K. Experimental study of steam reforming of methane in a thin (6 μM) Pd-based membrane reactor. *Ind. Eng. Chem. Res.* **2005**, *44*, 1454–1465. [CrossRef]
16. Nieto-Sandoval, J.; Gomez-Herrero, E.; Munoz, M.; de Pedro, Z.M.; Casas, J.A. Palladium-based catalytic membrane reactor for the continuous flow hydrodechlorination of chlorinated micropollutants. *Appl. Catal. B* **2021**, *293*, 120235. [CrossRef]

17. Narayanan, K.B.; Park, H.H.; Han, S.S. Synthesis and characterization of biomatrixed-gold nanoparticles by the mushroom flammulina velutipes and its heterogeneous catalytic potential. *Chemosphere* **2015**, *141*, 169–175. [CrossRef]
18. Tian, F.; Zhou, J.; Fu, R.; Cui, Y.; Zhao, Q.; Jiao, B.; He, Y. Multicolor colorimetric detection of ochratoxin a via structure-switching aptamer and enzyme-induced metallization of gold nanorods. *Food Chem.* **2020**, *320*, 126607. [CrossRef]
19. Yong, K.-T.; Sahoo, Y.; Swihart, M.T.; Prasad, P.N. Synthesis and plasmonic properties of silver and gold nanoshells on polystyrene cores of different size and of gold–silver core–shell nanostructures. *Colloids Surf. Physicochem. Eng. Asp.* **2006**, *290*, 89–105. [CrossRef]
20. Fausey, C.L.; Zucker, I.; Lee, D.E.; Shaulsky, E.; Zimmerman, J.B.; Elimelech, M. Tunable molybdenum disulfide-enabled fiber mats for high-efficiency removal of mercury from water. *ACS Appl. Mater. Interfaces* **2020**, *12*, 18446–18456. [CrossRef]
21. Wang, Z.; Mi, B. Environmental applications of 2D molybdenum disulfide (MoS_2) nanosheets. *Environ. Sci. Technol.* **2017**, *51*, 8229–8244. [CrossRef] [PubMed]
22. Wei, J.; He, P.; Wu, J.; Chen, N.; Xu, T.; Shi, E.; Pan, C.; Zhao, X.; Zhang, Y. Conversion of 2H MoS_2 to 1 T MoS_2 via lithium ion doping: Effective removal of elemental mercury. *Chem. Eng. J.* **2022**, *428*, 131014. [CrossRef]
23. Zhao, H.; Yang, G.; Gao, X.; Pang, C.H.; Kingman, S.W.; Wu, T. Hg^0 capture over $CoMoS/\gamma\text{-}Al_2O_3$ with MoS_2 nanosheets at low temperatures. *Environ. Sci. Technol.* **2016**, *50*, 1056–1064. [CrossRef] [PubMed]
24. Voiry, D.; Goswami, A.; Kappera, R.; Silva, C.d.C.C.e.; Kaplan, D.; Fujita, T.; Chen, M.; Asefa, T.; Chhowalla, M. Covalent functionalization of monolayered transition metal dichalcogenides by phase engineering. *Nat. Chem.* **2015**, *7*, 45–49. [CrossRef]
25. Wang, Z.; Sim, A.; Urban, J.J.; Mi, B. Removal and recovery of heavy metal ions by two-dimensional MoS_2 nanosheets: Performance and mechanisms. *Environ. Sci. Technol.* **2018**, *52*, 9741–9748. [CrossRef]
26. Wang, Z.; Tu, Q.; Sim, A.; Yu, J.; Duan, Y.; Poon, S.; Liu, B.; Han, Q.; Urban, J.J.; Sedlak, D.; et al. Superselective removal of lead from water by two-dimensional MoS_2 nanosheets and layer-stacked membranes. *Environ. Sci. Technol.* **2020**, *54*, 12602–12611. [CrossRef]
27. Aghagoli, M.J.; Shemirani, F. Hybrid nanosheets composed of molybdenum disulfide and reduced graphene oxide for enhanced solid phase extraction of Pb(II) and Ni(II). *Microchim. Acta* **2017**, *184*, 237–244. [CrossRef]
28. Wang, W.; Zeng, X.; Warner, J.H.; Guo, Z.; Hu, Y.; Zeng, Y.; Lu, J.; Jin, W.; Wang, S.; Lu, J.; et al. Photoresponse-bias modulation of a high-performance MoS_2 photodetector with a unique vertically stacked 2H-MoS_2/1T@2H-MoS_2 structure. *ACS Appl. Mater. Interfaces* **2020**, *12*, 33325–33335. [CrossRef]
29. Liu, B.; Han, Q.; Li, L.; Zheng, S.; Shu, Y.; Pedersen, J.A.; Wang, Z. Synergistic effect of metal cations and visible light on 2D MoS_2 nanosheet aggregation. *Environ. Sci. Technol.* **2021**, *55*, 16379–16389. [CrossRef]
30. Jia, F.; Wang, Q.; Wu, J.; Li, Y.; Song, S. Two-dimensional molybdenum fisulfide as a duperb sdsorbent for removing Hg^{2+} from water. *ACS Sustain. Chem. Eng.* **2017**, *5*, 7410–7419. [CrossRef]
31. Urbano, F.J.; Marinas, J.M. Hydrogenolysis of organohalogen compounds over palladium supported catalysts. *J. Mol. Catal. A Chem.* **2001**, *173*, 329–345. [CrossRef]
32. Alonso, F.; Beletskaya, I.P.; Yus, M. Metal-mediated reductive hydrodehalogenation of organic halides. *Chem. Rev.* **2002**, *102*, 4009–4092. [CrossRef]
33. Ma, J.; Wei, W.; Qin, G.; Xiao, T.; Tang, W.; Zhao, S.; Jiang, L.; Liu, S. Electrochemical reduction of nitrate in a catalytic carbon membrane nano-reactor. *Water Res.* **2022**, *208*, 117862. [CrossRef]
34. Ren, Y.; Liu, Y.; Liu, F.; Li, F.; Shen, C.; Wu, Z. Extremely efficient electro-Fenton-like Sb(III) detoxification using nanoscale Ti-Ce binary oxide: An effective design to boost catalytic activity via non-radical pathway. *Chin. Chem. Lett.* **2021**, *32*, 2519–2523. [CrossRef]
35. Liu, X.Q.; Wei, W.; Xu, J.; Wang, D.B.; Song, L.; Ni, B.J. Photochemical decomposition of perfluorochemicals in contaminated water. *Water Res.* **2020**, *186*, 116311. [CrossRef]
36. Liu, F.; Liu, Y.; Yao, Q.; Wang, Y.; Fang, X.; Shen, C.; Li, F.; Huang, M.; Wang, Z.; Sand, W.; et al. Supported atomically-precise gold nanoclusters for enhanced flow-through electro-Fenton. *Environ. Sci. Technol.* **2020**, *54*, 5913–5921. [CrossRef]
37. Li, Z.M.; Zhang, P.Y.; Shao, T.; Wang, J.L.; Jin, L.; Li, X.Y. Different nanostructured In_2O_3 for photocatalytic decomposition of perfluorooctanoic acid (PFOA). *J. Hazard. Mater.* **2013**, *260*, 40–46. [CrossRef]
38. Liu, F.; You, S.; Wang, Z.; Liu, Y. Redox-active nanohybrid filter for selective recovery of gold from water. *ACS EST Eng.* **2021**, *1*, 1342–1350. [CrossRef]
39. Zhong, Y.; Li, T.; Lin, H.; Zhang, L.; Xiong, Z.; Fang, Q.; Zhang, G.; Liu, F. Meso-/macro-porous microspheres confining Au nanoparticles based on PDLA/PLLA stereo-complex membrane for continuous flowing catalysis and separation. *Chem. Eng. J.* **2018**, *344*, 299–310. [CrossRef]
40. Liu, Y.; Zheng, Y.; Du, B.; Nasaruddin, R.R.; Chen, T.; Xie, J. Golden carbon nanotube membrane for continuous flow catalysis. *Ind. Eng. Chem. Res.* **2017**, *56*, 2999–3007. [CrossRef]
41. Wu, C.; Zhu, X.; Wang, Z.; Yang, J.; Li, Y.; Gu, J. Specific recovery and in situ reduction of precious metals from waste to create MOF composites with immobilized nanoclusters. *Ind. Eng. Chem. Res.* **2017**, *56*, 13975–13982. [CrossRef]
42. Zheng, W.; Liu, Y.; Liu, W.; Ji, H.; Li, F.; Shen, C.; Fang, X.; Li, X.; Duan, X. A novel electrocatalytic filtration system with carbon nanotube supported nanoscale zerovalent copper toward ultrafast oxidation of organic pollutants. *Water Res.* **2021**, *194*, 116961. [CrossRef] [PubMed]

43. Zhang, Q.; Li, M.; Luo, B.; Luo, Y.; Jiang, H.; Chen, C.; Wang, S.; Min, D. In situ growth gold nanoparticles in three-dimensional sugarcane membrane for flow catalytical and antibacterial application. *J. Hazard. Mater.* **2021**, *402*, 123445. [CrossRef]
44. Liang, M.; Su, R.; Huang, R.; Qi, W.; Yu, Y.; Wang, L.; He, Z. Facile in situ synthesis of silver nanoparticles on procyanidin-grafted eggshell membrane and their catalytic properties. *ACS Appl. Mater. Interfaces* **2014**, *6*, 4638–4649. [CrossRef] [PubMed]
45. Liu, F.; Liu, Y.; Shen, C.; Li, F.; Yang, B.; Huang, M.; Ma, C.; Yang, M.; Wang, Z.; Sand, W. One-step phosphite removal by an electroactive CNT filter functionalized with TiO_2/CeO_x nanocomposites. *Sci. Total Environ.* **2020**, *710*, 135514. [CrossRef] [PubMed]
46. Mao, R.; Zhao, X.; Qu, J. Electrochemical reduction of bromate by a Pd modified carbon fiber rlectrode: Kinetics and mechanism. *Electrochim. Acta* **2014**, *132*, 151–157. [CrossRef]
47. Gao, P.; Martin, C.R. Voltage charging rnhances ionic conductivity in gold nanotube membranes. *ACS Nano* **2014**, *8*, 8266–8272. [CrossRef] [PubMed]
48. Soares, O.S.G.P.; Ramalho, P.S.F.; Fernandes, A.; Órfão, J.J.M.; Pereira, M.F.R. Catalytic bromate reduction in water: Influence of carbon support. *J. Environ. Chem. Eng.* **2019**, *7*, 103015. [CrossRef]
49. Wang, J.; Wu, C.; Hu, N.; Zhou, J.; Du, L.; Wang, P. Microfabricated electrochemical cell-based biosensors for analysis of living cells in vitro. *Biosensors* **2012**, *2*, 127. [CrossRef]
50. Zhang, Y.; Sun, J.; Hou, B.; Hu, Y. Performance improvement of air-cathode single-chamber microbial fuel cell using a mesoporous carbon modified anode. *J. Power Sources* **2011**, *196*, 7458–7464. [CrossRef]
51. Perez-Coronado, A.M.; Soares, O.S.G.P.; Calvo, L.; Rodriguez, J.J.; Gilarranz, M.A.; Pereira, M.F.R. Catalytic reduction of bromate over catalysts based on Pd nanoparticles synthesized via water-in-oil microemulsion. *Appl. Catal. B* **2018**, *237*, 206–213. [CrossRef]
52. Yao, F.; Yang, Q.; Yan, M.; Li, X.; Chen, F.; Zhong, Y.; Yin, H.; Chen, S.; Fu, J.; Wang, D.; et al. Synergistic adsorption and electrocatalytic reduction of bromate by Pd/N-doped loofah sponge-derived biochar electrode. *J. Hazard. Mater.* **2020**, *386*, 121651. [CrossRef]
53. Madhushree, R.; UC, J.R.J.; Pinheiro, D.; KR, S.D. The catalytic reduction of 4-nitrophenol using MoS_2/ZnO nanocomposite. *Appl. Surf. Sci. Adv.* **2022**, *10*, 100265. [CrossRef]
54. Zhang, Z.-M.; Cheng, R.; Nan, J.; Chen, X.-Q.; Huang, C.; Cao, D.; Bai, C.-H.; Han, J.-L.; Liang, B.; Li, Z.-L.; et al. Effective electrocatalytic hydrodechlorination of 2,4,6-trichlorophenol by a novel $Pd/MnO_2/Ni$ foam cathode. *Chin. Chem. Lett.* **2022**, *33*, 3823–3828. [CrossRef]
55. Kishimoto, N.; Matsuda, N. Bromate ion removal by electrochemical reduction using an activated carbon felt electrode. *Environ. Sci. Technol.* **2009**, *43*, 2054–2059. [CrossRef]
56. Koga, H.; Namba, N.; Takahashi, T.; Nogi, M.; Nishina, Y. Renewable wood pulp paper reactor with hierarchical micro/nanopores for continuous-flow nanocatalysis. *Chem. Sus. Chem* **2017**, *10*, 2560–2565. [CrossRef]
57. Gopiraman, M.; Saravanamoorthy, S.; Baskar, R.; Ilangovan, A.; Ill-Min, C. Green synthesis of Ag@Au bimetallic regenerated cellulose nanofibers for catalytic applications. *New J. Chem.* **2019**, *43*, 17090–17103. [CrossRef]
58. Yu, X.-F.; Mao, L.-B.; Ge, J.; Yu, Z.-L.; Liu, J.-W.; Yu, S.-H. Three-dimensional melamine sponge loaded with Au/ceria nanowires for continuous reduction of p-nitrophenol in a consecutive flow system. *Sci. Bull.* **2016**, *61*, 700–705. [CrossRef]
59. Massaro, M.; Colletti, C.G.; Fiore, B.; La Parola, V.; Lazzara, G.; Guernelli, S.; Zaccheroni, N.; Riela, S. Gold nanoparticles stabilized by modified halloysite nanotubes for catalytic applications. *Appl. Organomet. Chem.* **2019**, *33*, e4665. [CrossRef]
60. Liu, X.; Li, Y.; Xing, Z.; Zhao, X.; Liu, N.; Chen, F. Monolithic carbon foam-supported Au nanoparticles with excellent catalytic performance in a fixed-bed system. *New J. Chem.* **2017**, *41*, 15027–15032. [CrossRef]

 catalysts

Article

N Doped Activated Biochar from Pyrolyzing Wood Powder for Prompt BPA Removal via Peroxymonosulfate Activation

Haiqin Lu [1], Guilu Xu [1] and Lu Gan [1,2,*]

[1] College of Materials Science and Engineering, Nanjing Forestry University, Nanjing 210037, China
[2] Jiangsu Co-Innovation Center of Efficient Processing and Utilization of Forest Resources, International Innovation Center for Forest Chemicals and Materials, College of Materials Science and Engineering, Nanjing Forestry University, Nanjing 210037, China
* Correspondence: ganlu@njfu.edu.cn

Abstract: In the present study, nitrogen doped biochar (N-PPB) and nitrogen doped activated biochar (AN-PPB) were prepared and used for removing bisphenol A (BPA) in water through activating peroxymonosulfate. It was found from the results that N-PPB exhibited superior catalytic performance over pristine biochar since nitrogen could brought about abundant active sites to the biochar structure. The non-radical singlet oxygen ($^1O_2$) was determined to be the dominant active species responsible for BPA degradation. Having non-radical pathway in the N-PPB/PMS system, the BPA degradation was barely influenced by many external environmental factors including solution pH value, temperature, foreign organic, and inorganic matters. Furthermore, AN-PPB had richer porosity than N-PPB, which showed even faster BPA removal efficiency than N-PPB through an adsorptive/catalytic synergy. The finding of this study introduces a novel way of designing hieratical structured biochar catalysts for effective organic pollutant removal in water.

Keywords: nitrogen dope; biochar; peroxymonosulfate; KOH activation; bisphenol A degradation

Citation: Lu, H.; Xu, G.; Gan, L. N Doped Activated Biochar from Pyrolyzing Wood Powder for Prompt BPA Removal via Peroxymonosulfate Activation. *Catalysts* **2022**, *12*, 1449. https://doi.org/10.3390/catal12111449

Academic Editors: Hao Xu and Yanbiao Liu

Received: 28 October 2022
Accepted: 14 November 2022
Published: 16 November 2022

1. Introduction

In recent years, advanced oxidation processes (AOPs) based on persulfate activations have been widely studied for catalytic degradation of organic contaminants [1,2]. Amongst various approaches for activating persulfate, transitional metal ions or metal compounds exhibit superior efficiency [3]. Nevertheless, it is intractable to overcome the metal ion leaching issue which can cause serious environmental pollution problems [4]. Recent studies have shown that carbonaceous materials can also effectively activate persulfate [5]. Thereinto, wood-derived biochar produced from pyrolyzing biomass precursor in low-oxygen or oxygen-free atmosphere has received tremendous research interest due to the advantages of abundant existence, low cost, low biotoxicity, and good stability, as well as numerous active oxygen-containing functional groups [6,7].

However, pristine biochar has low the degradation efficiency and poor recycling performance due to its limited active sites within skeletons for persulfate activation [8]. Nitrogen doping is considered to be an effective method to improve the catalytic capability of biochar through introducing diverse N-containing functional groups (pyridine N, pyrrole N, graphitic N) on the surface of biochar [9], which can not only change the electron density of local carbon atoms and increase electron mobility, but also increase edge defects of biochar [10]. At the same time, N atoms can also adjust the surface chemical properties through changing the charge/spin distribution of biochar, improve the adsorption capacity of persulfate, and add extra active sites [11], which are all in favor of catalytic degradations of organic pollutants [12]. On the other hand, biochar can be converted to activated carbon by an activation agent, through which biochar can have more surface oxygen functional groups, higher defects degree of internal carbon structure, and richer porosity [13]. Furthermore, the resulted activated biochar can be endowed with immensely increase the specific surface area along with hierarchical pore structures, which promote the adsorption capability of the biochar [14]. Therefore, it is potential to obtain N-doped activated biochar which have high organic pollutant removal efficiency in water through adsorption/degradation synergy through combining N doping and activating modification treatments to wood-based biochar.

Thus, in this study, N-doped biochar catalysts with different pyrolysis temperatures (450 °C, 600 °C, 750 °C, 900 °C) were prepared using industrial poplar powder as precursor and urea as nitrogen source. After comparing the catalytic performance through degrading bisphenol A (BPA) via activating peroxymonosulfate (PMS), the biochar pyrolyzed at 750 °C with optimal performance was then selected to prepare N-doped activated biochar using KOH as activating agent. The performance of N-doped activated biochar/PMS system towards BPA degradation was studied systematically afterwards. Furthermore, the recyclability and actual water adaptability of the reaction system were also investigated. The BPA degradation mechanisms in the reaction system were also analyzed.

2. Results and Discussion

The surface morphology structure difference between pristine poplar powder and N-PPB investigated in terms of SEM with the results shown in Figure 1. It could be observed from Figure 1a that pristine poplar powder displayed a smooth strip shape. After pyrolysis treatment, and nitrogen doping, N-PPB750 exhibited a rougher structure with richer porosity (Figure 1b), which was considerably distinguished from pristine poplar powder. At the same time, the particles of N-PPB750 became smaller compared with that of pristine poplar powder, which was attributed to the decomposition of lignocellulose component in the poplar powder during pyrolysis process [15]. The pyrolysis treatment also promoted the exfoliation of the carbon layers, which lead to the formation of abundant pores. The EDS elemental mapping diagrams shown in Figure 1d–f demonstrated that several main elements including C, N, and O were uniformly distributed N-PPB750. The contents of three major elements shown in Table 1 illustrated that a quite high content of N element (17.98 wt%) imported by urea precursor was anchored in the biochar structures.

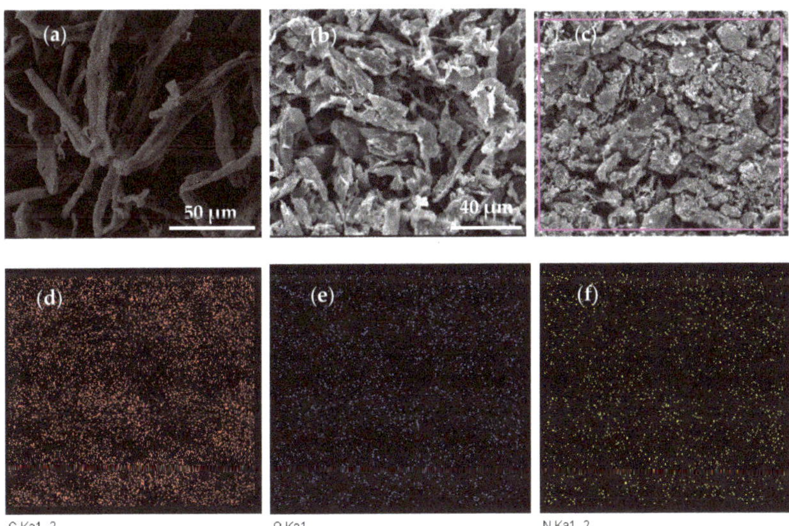

Figure 1. SEM images of (**a**) PPB-750, (**b**) N-PPB750, (**c**) selected area of N-PPB750 (red box) for EDS elemental mapping and EDS elemental mapping of (**d**) C, (**e**) O, (**f**) N.

Table 1. Elemental ratio of N-PPB750 (C, O, N).

Element	Weight%	Atomic%
C K	50.52	56.40
N K	17.98	17.21
O K	31.50	26.39
Totals	100.00	

The structures of N-PPB samples were then identified through XRD analysis with the results shown in Figure 2a. It could be seen that the pristine poplar powder had three wide diffraction peaks at 2θ = 15°, 22°, and 35°, which were affiliated to the Cellulose I structure of typical wood based biomass materials [16]. After pyrolysis, these three peaks were replaced by two broad peaks located at ~26° and 42° in N-PPB samples, which corresponded to (002) and (110) planes of classic carbon materials with graphitic structures [17]. These results further indicated the formation of fine graphitic structure in the prepared N-PPB samples. Additionally, many sharp small peaks in the range of 22–40° were also observed in all tested samples, which were related to the mineral salts such as calcium carbonate and silica existed in natural poplar powder [18]. These peaks became weaker with the elevation of pyrolysis temperature, which might be because some of the mineral salts were evaporated at higher temperature.

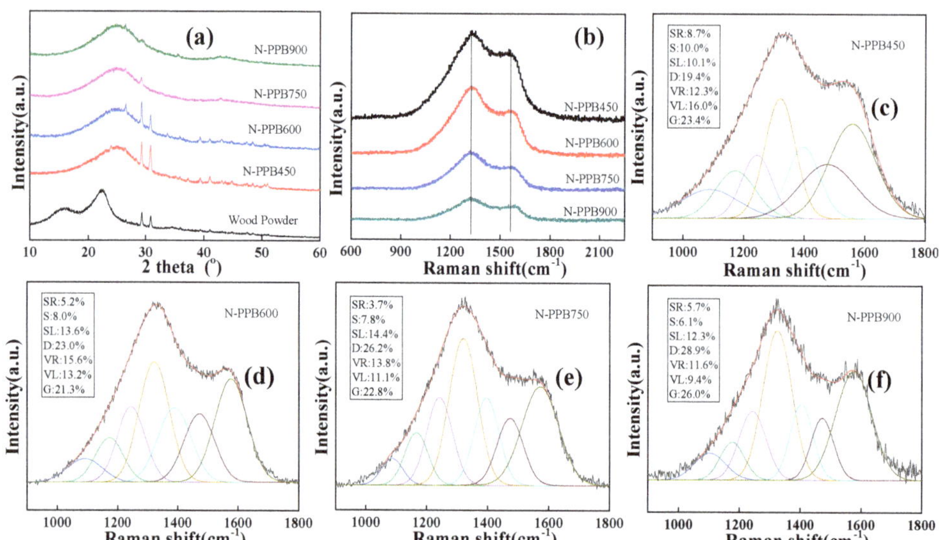

Figure 2. (**a**) XRD patterns, (**b**) Raman spectra, and (**c–f**) deconvoluted Raman spectra of N-PPB450, N-PPB600, N-PPB750, N-PPB900.

Figure 2b–f shows the Raman spectra of the N-PPB samples. As shown in Figure 2b, two characteristic peaks located at 1360 cm^{-1} and 1580 cm^{-1} could be observed for all the N doped biochar samples, which were ascribed to the characteristic D band and G band of a graphitic carbonaceous material [19]. This meant poplar wood could be converted to carbonaceous biochar when the pyrolysis temperature was higher than 450 °C. After the Raman spectra was deconvoluted, it was seen that the spectrum of each N-PPB sample were consisted of seven characteristic peaks which were ascribed to were divided into the SR band representing C-H on aromatic rings (1086 cm^{-1}), the S band representing $C_{aromatic}$-C_{alkyl} (1170 cm^{-1}), the SL band representing C_{aryl}-O-C_{alkyl} (1245 cm^{-1}), the D band representing the defective and disordered arrangement of carbon atoms (1320 cm^{-1}), the VR band representing to amorphous carbon and deterioration of crystallinity (1390 cm^{-1}), the VL band representing the semicircle ring breathing (1468 cm^{-1}), and the G band representing in-plane stretching vibration of C atomic sp^2 hybridization (1576 cm^{-1}), respectively [20]. The area ratio between D band and G band (A_D/A_G) was generally used to characterize the defect degree of the graphitic structure in a carbonaceous material [21]. It was seen that the A_D/A_G value increased from 0.83 to 1.15 with the increase in the pyrolysis temperature, indicating that higher pyrolysis temperature brought about more defect to the biochar. In the meantime, the area ratio between D band and (VR+VL) bands (A_D/A_V) represented the defect density of boundary edges in the carbon. It was seen that all the biochar samples had an A_D/A_V value of lower than 3.5, indicating that the prepared N-PPB samples were considered to contain a higher density defect of boundary edges with more unsaturated carbon atoms and a lower degree of graphitization [22].

XPS analysis was then conducted to further explore the surface element composition of the prepared N-PPB. As illustrated in Figure 3a, the main elements of C, N, and O were all observed in the spectra of the biochar samples, in which the concentration of N decreased significantly with the increase in the pyrolysis temperature. Figure 3b–d reveals the deconvoluted C1s, O1s, and N1s spectra. It was observed from Figure 3b that C1s spectrum was mainly composed of four peaks located at 284.5 eV, 285.2 eV, 286.2 eV, and 288.8 eV, which represented the C=C/C-C, C–O, C=O and COOH bonding, respectively [23,24]. Similarly, the O1s spectrum could be divided into C=O (530.9 eV), C-O (532.2 eV), and COOH (533.6 eV) (Figure 3c), and N1s spectra could be deconvoluted into four peaks corresponding to pyridinic—N (398.3 eV), pyrrolic—N (399.5 eV), graphite—N (400.4 eV), and −NOx (401.4 eV) (Figure 3d) [25]. The detailed concentrations of respective bonding were listed in Table 2. As shown, the concentration of C=C/C-C decreased with the increment of pyrolysis temperature from 450 °C to 900 °C, which was in agreement with the Raman results. Correspondingly, the proportion of C=O bonding increased with the increase in the pyrolysis temperature, which could be resulted from the conversion of C=C/C-C bonding. As for N1s bonding, the conversion of pyrrolic N or pyridine N to graphitic N could be obviously seen.

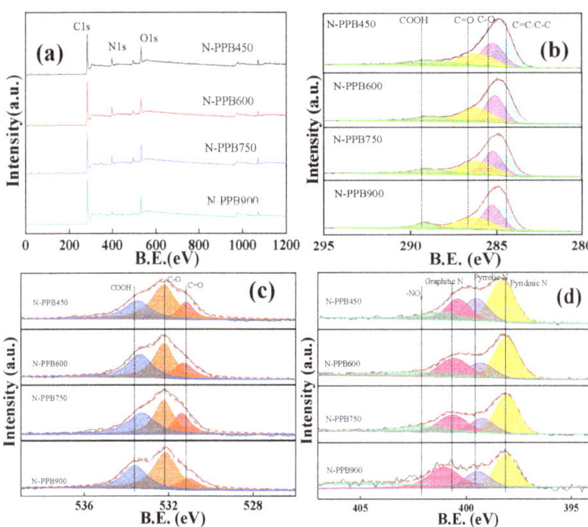

Figure 3. (**a**) Wide-scan XPS spectra, high resolution (**b**) C 1s, (**c**) N 1s, (**d**) O 1s spectra of N-PPB catalysts.

Table 2. Bonding ratio of N-PPB samples based on deconvoluted C1s and N1s XPS peak.

	C1s				N1s, %			
	C=C/C-C	C-O	C=O	–COOH	Pyridinic N	Pyroolic N	Graphitic N	$-NO_x$
N-PPB450	33.2%	32.0%	23.7%	11.1%	47.9%	22.8%	20.2%	9.1%
N-PPB600	30.7%	34.2%	26.8%	8.3%	46.6%	17.2%	21.4%	14.8%
N-PPB750	27.2%	34.5%	28.1%	10.2%	45.1%	16.8%	22.6%	15.5%
N-PPB900	31.7%	33.4%	26.4%	8.5%	31.0%	17.6%	27.5%	23.9%

The catalytic performance of the prepared N-PPB samples was investigated through degrading BPA in water with the results shown Figure 4a. As illustrated, the PPB samples showed negligible adsorption activity towards BPA since the biochar was not activated. Moreover, nitrogen doping could significantly enhance the PMS activation capability of the biochar, and all the N-PPB samples (0.5 g/L) could completely degrade 0.02 mM BPA in the solution. Within the same time span, PPB without N doping could only degrade ~10% of BPA. This indicated that N doping significantly enriched the active sites on PPB surface, which accelerated the PMS activation efficiency of the biochar sample. It was also seen that the activation capability of the N-PPB sample increased with the pyrolysis temperature, and when pyrolysis temperature increased to 600 °C or higher, the catalytic performance of the corresponding PPB did not increase much. The TOC results obtained from N-PPB600/PMS system indicated that when the reaction time was extended to 30 min, ~58% of the TOC could be removed. This result implied that the BPA could be degraded into small molecules by the prepared N-PPB catalysts and finally mineralized into inorganic molecules.

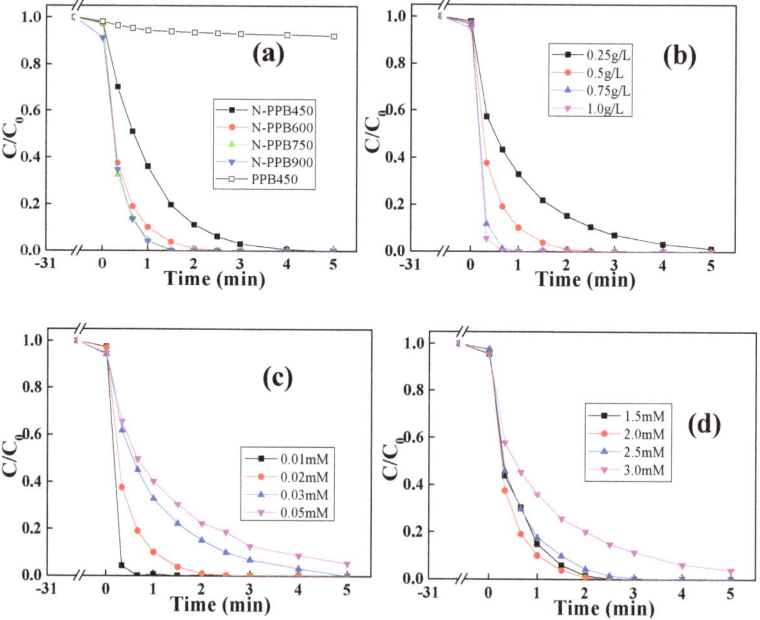

Figure 4. (**a**) Catalytic performance of N-PPB samples ($[catalyst]_0$ = 0.5 g/L, $[BPA]_0$ = 0.02 mM, $[PMS]_0$ = 1.5 mM), impact of (**b**) catalyst dosage ([BPA]0 = 0.02 mM, $[PMS]_0$ = 1.5 mM), (**c**) BPA concentration ($[catalyst]_0$ = 0.5 g/L, $[PMS]_0$ = 1.5 mM), and (**d**) PMS concentration on the performance of N-PPB600 ($[catalyst]_0$ = 0.5 g/L, $[BPA]_0$ = 0.02 mM).

N-PPB600 was selected for the following test. The impact of reaction parameters, including BPA concentration, catalyst dosage, and PMS concentration on the performance of

N-PPB600, was then investigated with the results shown in Figure 4b–d. It was revealed that the catalytic performance of N-PPB600 had a positive correlation with the catalyst dosage, and had a negative correlation with the BPA concentration. It was easy to understand that increased catalyst dosage could increase the number of active sites for PMS activation, which led to the increase in active species for BPA degradation [26]. Conversely, when the concentrations of both N-PPB600 and PMS were fixed, a constant number of active species was insufficient to degrade increased amount of BPA molecules, which resulted in the performance decline of N-PPB600. While when the concentration of PMS was increased in the reaction solution, the performance of N-PPB600 increased when PMS concentration increased from 1.5 mM to 2.0 mM, and declined when further increasing the PMS concentration. This might because an excessive amount of PMS would also consume the generated active species, which inhibited the BPA degradation [27].

The impact of solution parameters on the performance of N-PPB600 was investigated afterwards. Figure 5a shows the influence of solution pH on the BPA degradation efficiency. Overall, the catalytic performance of N-PPB600 was not much influenced by the fluctuation of solution pH value, indicating promising performance stability of the catalyst. It was further observed from Figure 5b that the activity of N-PPB600 boosted with the increase in the system temperature, indicating the endothermic nature of the PMS activation process [28].

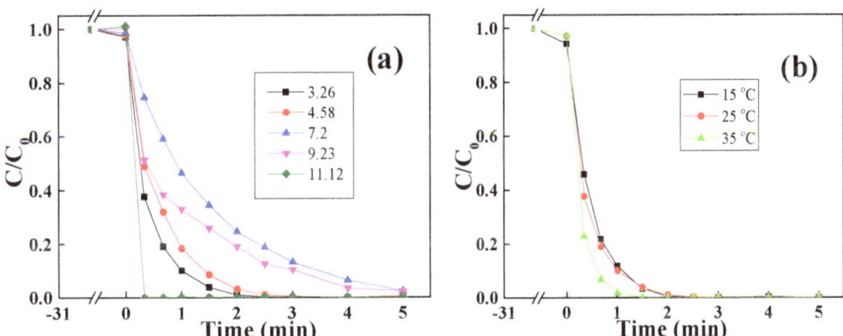

Figure 5. Impact of (**a**) pH value and (**b**) solution temperature on the performance of N-PPB600 ($[catalyst]_0$ = 0.5 g/L, $[BPA]_0$ = 0.02 mM, $[PMS]_0$ = 1.5 mM).

Figure 6 shows the impact of foreign matters including inorganic anions, natural organic matters and complex water matrices on the performance of the biochar catalyst. As illustrated in Figure 6a–d, all inorganic anions could influence the activity of N-PPB600. Specifically, $H_2PO_4^-$, HCO_3^-, and NO_3^- inhibited the BPA degradation efficiency since these anions would occupy the catalyst surface and hinder the contact between the active sites in the catalyst and PMS, resulting lower active species generation rate [29]. Unlike the other three anions, the dosage of Cl^- sped up the BPA degradation rate, and the increased amount of Cl^- would accelerate the BPA degradation rate more intensely. This was because Cl^- could react with PMS and active species and generate more chlorine based species [30], which, as a result, boosted the removal efficiency of BPA in the reaction system. It was also seen from Figure 6e–f that both humic acid and natural water had an impediment effect on the catalytic activity of N-PPB600, which was ascribed to the similar reason to the anions as discussed above. It has to be noted that even the concentration of the anions and humic acid increased to as high as 50 mM; 0.02 mM of BPA could still be effectively degraded with high removal rate, implying promising water matrix adaptability of the prepared catalyst.

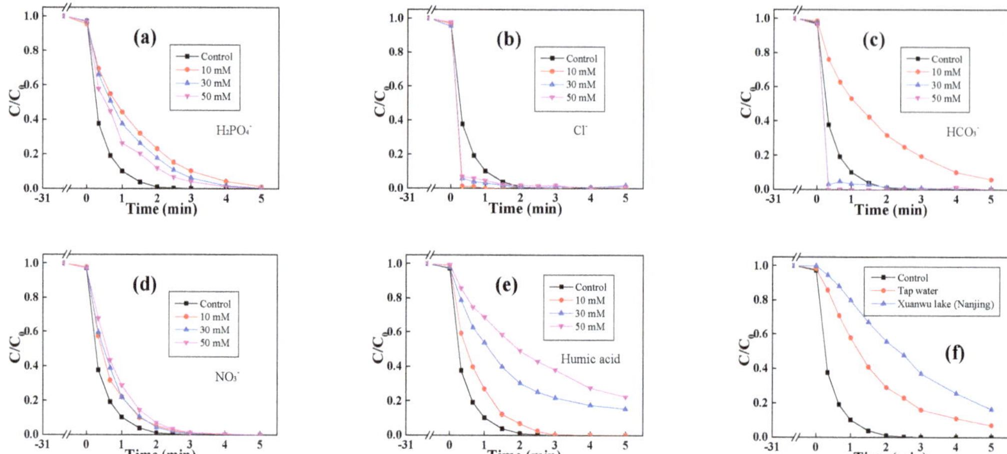

Figure 6. Impact of (**a**) $H_2PO_4^-$, (**b**) Cl^-, (**c**) HCO_3^-, (**d**) NO_3^-, (**e**) humic acid, and (**f**) natural water on the performance of N-PPB600 ([catalyst]$_0$ = 0.5 g/L, [BPA]$_0$ = 0.02 mM, [PMS]$_0$ = 1.5 mM).

The active species generated in the reaction system which dominated the BPA degradation was investigated through scavenging test with the results shown in Figure 7a. It was observed that both methanol (MeOH) and tert-butyl alcohol (TBA) did not show any inhibition effect to BPA degradation, indicating that no radical species was generated in the reaction system. When the trapping agent was changed to L-histidine and para-benzoquinone (p-BQ), the BPA degradation could be completely inhibited. Since these two agents were both non-radical scavenger which could effectively trap singlet oxygen ($^1O_2$) in the solution [31], this meant the N-PPB600/PMS system was a $^1O_2$ dominated system for BPA degradation. This could also explain why N-PPB600 could kept its catalytic activity at various water matrices since $^1O_2$ was not much influenced by the variation of external environment.

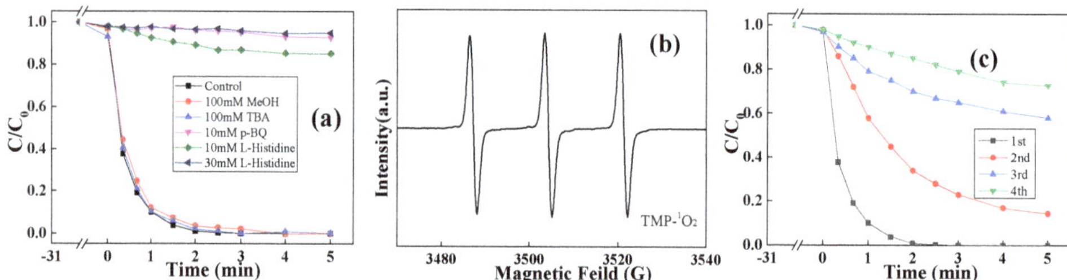

Figure 7. (**a**) BPA degradation in N-PPB600/PMS system at the existence of various scavengers, (**b**) ESR spectra test, (**c**) recyclability test for N-PPB600 ([catalyst]$_0$ = 0.5 g/L, [BPA]$_0$ = 0.02 mM, [PMS]$_0$ = 1.5 mM).

The ESR results shown in Figure 7b also illustrated that when using TMP as the trapping agent, a clear triplet peak which was ascribed to the typical TMP-$^1O_2$ adduct could be observed, indicating that abundant $^1O_2$ was existed in the N-PPB600/PMS system. Many previous studies indicated that the electron-rich Lewis basic sites in biochar could transfer the electrons to PMS and caused the generation of non-radical $^1O_2$ [32]. Since nitrogen doping could introduce abundant electron-rich nitrogen containing groups to the biochar including pyridinic N and pyrrolic N [33], the N-PPB600 could exhibited higher PMS activating efficiency than the pristine biochar.

The catalytic stability of N-PPB600 was investigated afterwards. It was seen from Figure 7c that the catalytic capability of the biochar gradually reduced after four consecutive runs, which indicated that the active sites on the catalyst surface was continuously consumed when activating PMS. Thus, the XPS spectra of the recycled N-PPB600 was investigated to illustrate the PMS activation mechanism with the results shown in Figure 8. As shown in Figure 8a, the N element was significantly consumed after consecutive BPA degradation cycles, indicating that N doping played a critical role of improving the catalytic properties of the biochar. From the deconvoluted C1s, N1s, and O1s spectra shown in Figure 8b–d, it was seen that the concentrations of C=O, pyridinic N and graphitic N were reduced after the reaction cycles, implying that these bonding participated the PMS activation and were consumed gradually, which were in accordance with many previous studies [34].

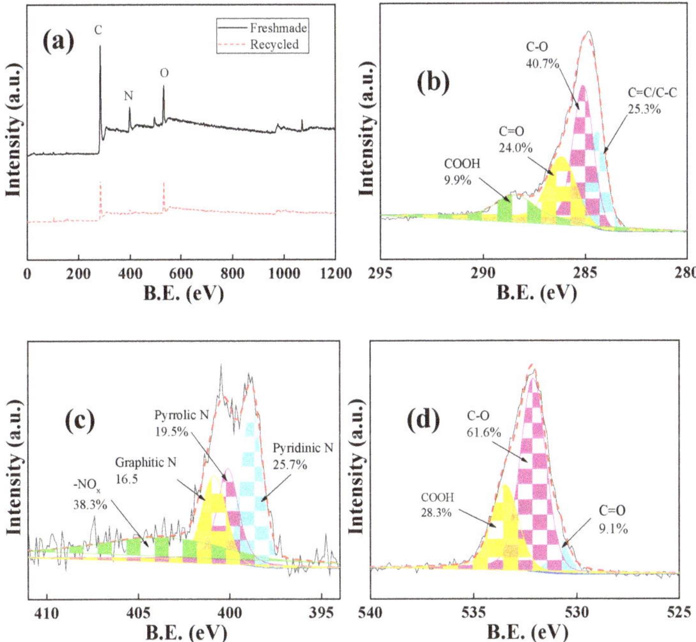

Figure 8. (**a**) Wide-scan XPS spectra, (**b**) C 1s, (**c**) N 1s, (**d**) O 1s spectra of used N-PPB600.

To improve the BPA degradation efficiency of the biochar, AN-PPB600 was prepared to enhance the adsorption capability of the biochar. It was seen that AN-PPB600 had much higher specific surface area and pore volume than N-PPB600 after KOH activation (Table 3), which indicated that AN-PPB600 should have better adsorption capability than N-PPB600. As illustrated in Figure 9, ~80% of BPA could be adsorbed by AN-PPB600 at the first 30 min, while no more than 10% of BPA was adsorbed by N-PP600. When PMS was dosed, 0.5 g/L of AN-PPB600 could completely remove BPA with high concentration of 0.1 mM through adsorptive/catalytic synergy within 3 min. Conversely, N-PPB600 could only degrade ~15% of BPA within the same time span. This indicated that through designing hieratical structured biochar with both high adsorptive and catalytic capability, organic pollutants in water could be promptly removed through the synergetic process.

Table 3. Specific surface area and pore volume of N-PPB600 and AN-PPB600.

Sample	Specific Surface Area ($m^2 \cdot g^{-1}$)	Pore Volume (cm^3 g^{-1})
N-PPB600	141	0.203
AN-PPB600	855	1.683

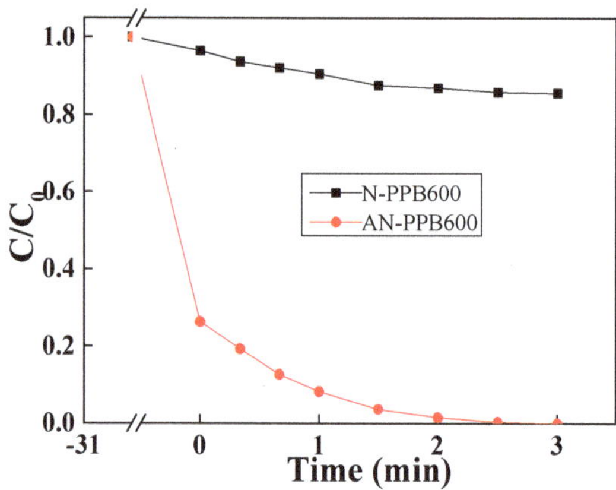

Figure 9. BPA degradation in N-PPB600/PMS and AN-PPB600/PMS systems ([catalyst]$_0$ = 0.5 g/L, [BPA]$_0$ = 0.02 mM, [PMS]$_0$ = 1.5 mM).

3. Experimental

3.1. Reagents and Chemicals

Poplar powder (300 mesh) was obtained from Yixing wood powder factory (Linyi, China). PMS (Oxone, $KHSO_5 \cdot 0.5KHSO_4 \cdot 0.5K_2SO_4$) and BPA (99%) were purchased from Sigma-Aldrich (St. Louis, MO, USA). Urea (CH_4N_2O, 99%) was provided from Nanjing Chemical Reagent Co., Ltd. (Nanjing, China). 2,2,6,6-tetramethyl-4-piperidine (TMP, 98%) was acquired from Aladdin (Shanghai, China). All the other chemicals and reagents were of analytical grade and used without further purification. Double distilled water (H_2O) was utilized throughout the whole experiments.

3.2. Preparation of the Catalysts

Typically, 3.0 g urea was first dissolved into 20 mL H_2O, after which 1.0 g poplar powder was added. The mixture was magnetically stirred 70 °C until all the H_2O was evaporated. Afterwards, the obtained powder was put into a tubular furnace, heated to desired temperature (450 °C, 600 °C, 750 °C and 900 °C) under N_2 atmosphere at 5 °C/min, pyrolyzed at respective temperature for 3 h. After cooling to room temperature, the obtained solid was further washed with H_2O for several times until the pH value of H_2O reached ~7. After being dried in an oven at 60 °C for 24 h, N-doped biochar derived from poplar powder was obtained, and was designated to be N-PPBx, where x represented the pyrolysis temperature. For comparison, activated N-doped biochar was synthesized via a similar route except that 2.0 g KOH was also dissolved in urea solution in the first step, which was named as AN-PPBx.

3.3. Characterizations

The morphology of biochar was investigated by scanning electron microscope (SEM, JEOL SEM 6490, Tokyo, Japan). The crystalline structures were studied by X-ray diffraction diffractometer (XRD, Rigaku Smartlab, Tokyo, Japan). Raman spectra of the prepared samples was investigated by a Raman spectrometer equipped with an Ar laser (532 nm, 180 mW) (Horiba Jobin Yvon HR800Tokyo, Japan). X-ray photoelectron spectroscopy (XPS,

Kratos Ultra DLD, Warwick, UK) was used to detect surface chemistry and element composition. Electron spin resonance spectroscopy (ESR, Bruker EMX-10/12, Bremen, Germany) was used to identify generated reactive oxygen species (ROS) in the biochar/PMS system. Electrochemical measurements were carried out with a CHI760E electrochemical workstation. The concentration of BPA was analyzed by high performance liquid chromatography (HPLC, Dionex Ultimate 3000, Sunnyvale, CA USA) equipped with C18 column (5 μm particle size, 250 × 4.6 mm) and UV detector at 278 nm. Total organic carbon (TOC) was measured by a MutiN/C 2100 analyzer (Jena, Germany).

3.4. Degradation Process

All the catalytic degradation experiments were conducted in a glass beaker containing 100 mL BPA solution with continuous mechanical stirring. In brief, a certain amount of biochar sample was first dispersed in H_2O and sonicated for 30 min to obtain a uniform dispersion. Afterwards, BPA solution with pre-determined concentration was obtained through injecting a certain amount of BPA stock solution into the dispersion. Before PMS was dosed to initiate BPA degradation, the mixture was stirred for 30 min to achieve the adsorption/desorption equilibrium. During the BPA degradation process, 1 mL of solution sample was periodically extracted from the reaction system, filtered with a 0.22 μm polytetrafluoroethylene membrane, and quenched with 0.5 mL methanol. The initial pH value of the solution was measured after PMS and catalyst were added, and was adjusted by NaOH (1 M) and H_2SO_4 (1 M). In a recycling test, the used biochar was separated from the reaction system through centrifuging, which was then rinsed 3 times with clean H_2O and dried at 60 °C for 12 h before being dosed to initiate next BPA degradation cycle. The ROS produced during oxidative degradation of pollutants were identified by scavenging experiments using methanol (MeOH), isobutanol (TBA), p-benzoquinone (p-BQ), and L-histdine (His) as scavengers. All experiments were conducted in triplicates, and the data were the mean values with standard deviations.

4. Conclusions

To conclude, N-PPB catalysts were prepared and used for the removal of BPA in water through activating PMS. The results shows that nitrogen doping could significantly enhance the catalytic performance of pristine biochar, in which 0.02 mM of BPA could be completely degraded within 5 min by 0.5 g/L of biochar catalyst. The scavenging test showed that non-radical $^1O_2$ was the main active species which dominated the BPA degradation. Due to this reason, the N-PPB/PMS system showed promising water matrix adaptability. To enhance the BPA removal efficiency, KOH was introduced to activate N-PPB, and the results illustrated that AN-PPB could promptly remove high concentration BPA in water through adsorptive/catalytic synergy. This study provides a new thinking of preparing biochar catalyst with heteroatom doing and hieratical structure design, which could remove organic pollutants in water with high efficiency.

Author Contributions: Investigation, writing—original draft preparation, data curation, H.L.; visualization, formal analysis, G.X.; writing—review and editing, supervision, funding acquisition, L.G. All authors have read and agreed to the published version of the manuscript.

Funding: This research was funded by the Natural Science Foundation of Jiangsu Province, China (BK20201385).

Data Availability Statement: Data are available upon request.

Acknowledgments: The Advanced Analysis and Testing Center of Nanjing Forestry University is acknowledged.

Conflicts of Interest: The authors declare no conflict of interest.

References

1. Shi, J.; Dai, B.; Fang, X.; Xu, L.; Wu, Y.; Lu, H.; Cui, J.; Han, S.; Gan, L. Waste preserved wood derived biochar catalyst for promoted peroxymonosulfate activation towards bisphenol A degradation with low metal ion release: The insight into the mechanisms. *Sci. Total Environ.* **2022**, *813*, 152673. [CrossRef] [PubMed]
2. Lv, Y.; Zong, L.; Liu, Z.; Du, J.; Wang, F.; Zhang, Y.; Ling, C.; Liu, F. Sequential separation of Cu(II)/Ni(II)/Fe(II) from strong-acidic pickling wastewater with a two-stage process based on a bi-pyridine chelating resin. *Chin. Chem. Lett.* **2021**, *32*, 2792–2796. [CrossRef]
3. Sethi, Y.A.; Panmand, R.P.; Ambalkar, A.; Kulkarni, A.K.; Patil, D.R.; Gunjal, A.R.; Gosavi, S.W.; Kulkarni, M.V.; Kale, B.B. *In situ* preparation of CdS decorated $ZnWO_4$ nanorods as a photocatalyst for direct conversion of sunlight into fuel and RhB degradation. *Sustain. Energy Fuels* **2019**, *3*, 793–800. [CrossRef]
4. Pan, Y.; Bu, Z.; Sang, C.; Guo, H.; Zhou, M.; Zhang, Y.; Tian, Y.; Cai, J.; Wang, W. EDTA enhanced pre-magnetized Fe_0/H_2O_2 process for removing sulfamethazine at neutral pH. *Sep. Purif. Technol.* **2020**, *250*, 117281. [CrossRef]
5. Lu, H.; Gan, L. Catalytic Degradation of Bisphenol A in Water by Poplar Wood Powder Waste Derived Biochar via Peroxymonosulfate Activation. *Catalysts* **2022**, *12*, 1164. [CrossRef]
6. Zhao, Z.; Zhang, X.; Lin, Q.; Zhu, N.; Gui, C.; Yong, Q. Development and investigation of a two-component adhesive composed of soybean flour and sugar solution for plywood manufacturing. *Wood Mater. Sci. Eng.* **2022**, 1–9. [CrossRef]
7. Zuo, S.; Zhang, W.; Wang, Y.; Xia, H. Low-Cost Preparation of High-Surface-Area Nitrogen-Containing Activated Carbons from Biomass-Based Chars by Ammonia Activation. *Ind. Eng. Chem. Res.* **2020**, *59*, 7527–7537. [CrossRef]
8. Bhatia, S.K.; Gurav, R.; Choi, T.-R.; Kim, H.J.; Yang, S.-Y.; Song, H.-S.; Park, J.Y.; Park, Y.-L.; Han, Y.-H.; Choi, Y.-K.; et al. Conversion of waste cooking oil into biodiesel using heterogenous catalyst derived from cork biochar. *Bioresour. Technol.* **2020**, *302*, 122872. [CrossRef]
9. You, Y.; Zhao, Z.; Song, Y.; Li, J.; Li, J.; Cheng, X. Synthesis of magnetized nitrogen-doped biochar and its high efficiency for elimination of ciprofloxacin hydrochloride by activation of peroxymonosulfate. *Sep. Purif. Technol.* **2020**, *258*, 117977. [CrossRef]
10. Baştürk, E.; Alver, A. Modeling azo dye removal by sono-fenton processes using response surface methodology and artificial neural network approaches. *J. Environ. Manag.* **2019**, *248*, 109300. [CrossRef]
11. Ye, S.; Zeng, G.; Tan, X.; Wu, H.; Liang, J.; Song, B.; Tang, N.; Zhang, P.; Yang, Y.; Chen, Q.; et al. Nitrogen-doped biochar fiber with graphitization from Boehmeria nivea for promoted peroxymonosulfate activation and non-radical degradation pathways with enhancing electron transfer. *Appl. Catal. B Environ.* **2020**, *269*, 118850. [CrossRef]
12. Huang, B.-C.; Jiang, J.; Huang, G.-X.; Yu, H.-Q. Sludge biochar-based catalysts for improved pollutant degradation by activating peroxymonosulfate. *J. Mater. Chem. A* **2018**, *6*, 8978–8985. [CrossRef]
13. Liu, X.; Zuo, S.; Cui, N.; Wang, S. Investigation of ammonia/steam activation for the scalable production of high-surface area nitrogen-containing activated carbons. *Carbon* **2022**, *191*, 581–592. [CrossRef]
14. Gao, H.; Zuo, S.; Wang, S.; Xu, F.; Yang, M.; Hu, X. Graphitic crystallite nanomaterials enable the simple and ultrafast synthesis of resorcinol-formaldehyde carbon aerogel monoliths. *Carbon* **2022**, *194*, 220–229. [CrossRef]
15. Gan, L.; Zhong, Q.; Geng, A.; Wang, L.; Song, C.; Han, S.; Cui, J.; Xu, L. Cellulose derived carbon nanofiber: A promising biochar support to enhance the catalytic performance of $CoFe_2O_4$ in activating peroxymonosulfate for recycled dimethyl phthalate degradation. *Sci. Total Environ.* **2019**, *694*, 133705. [CrossRef]
16. Geng, A.; Meng, L.; Han, J.; Zhong, Q.; Li, M.; Han, S.; Mei, C.; Xu, L.; Tan, L.; Gan, L. Highly efficient visible-light photocatalyst based on cellulose derived carbon nanofiber/BiOBr composites. *Cellulose* **2018**, *25*, 4133–4144. [CrossRef]
17. Yu, Y.; Liu, Y.; Wu, X.; Weng, Z.; Hou, Y.; Wu, L. Enhanced visible light photocatalytic degradation of metoprolol by Ag–Bi2WO6–graphene composite. *Sep. Purif. Technol.* **2015**, *142*, 1–7. [CrossRef]
18. Jia, H.; Zhao, S.; Zhu, K.; Huang, D.; Wu, L.; Guo, X. Activate persulfate for catalytic degradation of adsorbed anthracene on coking residues: Role of persistent free radicals. *Chem. Eng. J.* **2018**, *351*, 631–640. [CrossRef]
19. Hu, W.; Tong, W.; Li, Y.; Xie, Y.; Chen, Y.; Wen, Z.; Feng, S.; Wang, X.; Li, P.; Wang, Y.; et al. Hydrothermal route-enabled synthesis of sludge-derived carbon with oxygen functional groups for bisphenol A degradation through activation of peroxymonosulfate. *J. Hazard. Mater.* **2019**, *388*, 121801. [CrossRef]
20. Cheng, X.; Guo, H.; Zhang, Y.; Wu, X.; Liu, Y. Non-photochemical production of singlet oxygen via activation of persulfate by carbon nanotubes. *Water Res.* **2017**, *113*, 80–88. [CrossRef]
21. Hou, P.; Xing, G.; Tian, L.; Zhang, G.; Wang, H.; Yu, C.; Li, Y.; Wu, Z. Hollow carbon spheres/graphene hybrid aerogels as high-performance adsorbents for organic pollution. *Sep. Purif. Technol.* **2018**, *213*, 524–532. [CrossRef]
22. Liang, J.; Xu, X.; Zaman, W.Q.; Hu, X.; Zhao, L.; Qiu, H.; Cao, X. Different mechanisms between biochar and activated carbon for the persulfate catalytic degradation of sulfamethoxazole: Roles of radicals in solution or solid phase. *Chem. Eng. J.* **2019**, *375*, 121908. [CrossRef]
23. Pereira, A.T.; Henriques, P.C.; Costa, P.C.; Martins, M.C.L.; Magalhães, F.D.; Gonçalves, I.C. Graphene oxide-reinforced poly(2-hydroxyethyl methacrylate) hydrogels with extreme stiffness and high-strength. *Compos. Sci. Technol.* **2019**, *184*, 107819. [CrossRef]
24. Pan, Y.; Wang, Q.; Zhou, M.; Cai, J.; Tian, Y.; Zhang, Y. Kinetic and mechanism study of UV/pre-magnetized-Fe0/oxalate for removing sulfamethazine. *J. Hazard. Mater.* **2020**, *398*, 122931. [CrossRef] [PubMed]

25. He, J.; Zhou, Q.; Chen, S.; Tian, M.; Zhang, C.; Sun, W. Interfacial microstructures and adsorption mechanisms of benzohydroxamic acid on Pb2+-activated cassiterite (110) surface. *Appl. Surf. Sci.* **2020**, *541*, 148506. [CrossRef]
26. Liu, C.; Liu, L.; Tian, X.; Wang, Y.; Li, R.; Zhang, Y.; Song, Z.; Xu, B.; Chu, W.; Qi, F.; et al. Coupling metal–organic frameworks and g-C3N4 to derive Fe@N-doped graphene-like carbon for peroxymonosulfate activation: Upgrading framework stability and performance. *Appl. Catal. B-Environ.* **2019**, *255*, 117763. [CrossRef]
27. Wang, J.; Wang, S. Preparation, modification and environmental application of biochar: A review. *J. Clean. Prod.* **2019**, *227*, 1002–1022. [CrossRef]
28. Tao, W.; Duan, W.; Liu, C.; Zhu, D.; Si, X.; Zhu, R.; Oleszczuk, P.; Pan, B. Formation of persistent free radicals in biochar derived from rice straw based on a detailed analysis of pyrolysis kinetics. *Sci. Total Environ.* **2020**, *715*, 136575. [CrossRef]
29. Wu, J.; Grant, P.; Li, X.; Noble, A.; Aggarwal, V.K. Catalyst-Free Deaminative Functionalizations of Primary Amines by Photoinduced Single-Electron Transfer. *Angew. Chem. Int. Ed.* **2019**, *58*, 5697–5701. [CrossRef]
30. Duan, X.; Sun, H.; Wang, S. Metal-Free Carbocatalysis in Advanced Oxidation Reactions. *Acc. Chem. Res.* **2018**, *51*, 678–687. [CrossRef]
31. Liu, J.; Huang, S.; Chen, K.; Wang, T.; Mei, M.; Li, J. Preparation of biochar from food waste digestate: Pyrolysis behavior and product properties. *Bioresour. Technol.* **2020**, *302*, 122841. [CrossRef]
32. Han, R.; Fang, Y.; Sun, P.; Xie, K.; Zhai, Z.; Liu, H.; Liu, H. N-Doped Biochar as a New Metal-Free Activator of Peroxymonosulfate for Singlet Oxygen-Dominated Catalytic Degradation of Acid Orange 7. *Nanomaterials* **2021**, *11*, 2288. [CrossRef]
33. Hu, Y.; Chen, D.; Zhang, R.; Ding, Y.; Ren, Z.; Fu, M.; Cao, X.; Zeng, G. Singlet oxygen-dominated activation of peroxymonosulfate by passion fruit shell derived biochar for catalytic degradation of tetracycline through a non-radical oxidation pathway. *J. Hazard. Mater.* **2021**, *419*, 126495. [CrossRef]
34. Lata, H.; Garg, V.; Gupta, R. Adsorptive removal of basic dye by chemically activated Parthenium biomass: Equilibrium and kinetic modeling. *Desalination* **2008**, *219*, 250–261. [CrossRef]

Article

Catalytic Ozonation of Norfloxacin Using Co-Mn/CeO_2 as a Multi-Component Composite Catalyst

Ruicheng Li [1,2], Jianhua Xiong [1,2,*], Yuanyuan Zhang [1,2,3,*], Shuangfei Wang [2], Hongxiang Zhu [2] and Lihai Lu [4]

1 School of Resources, Environment and Materials, Guangxi University, Nanning 530004, China
2 Guangxi Key Laboratory of Clean Pulp and Papermaking and Pollution Control, College of Light Industry and Food Engineering, Guangxi University, Nanning 530004, China
3 School of Marine Sciences, Guangxi University, Nanning 530004, China
4 Guangxi Bossco Environmental Protection Technology Co., Ltd., Nanning 530007, China
* Correspondence: happybear99@126.com (J.X.); jiedeng05@sina.com (Y.Z.)

Abstract: In this study, a Co-Mn/CeO_2 composite was prepared through a facile sol-gel method and used as an efficient catalyst for the ozonation of norfloxacin (NOR). The Co-Mn/CeO_2 composite was characterized via XRD, SEM, BET and XPS analysis. The catalytic ozonation of NOR by Co-Mn/CeO_2 under different conditions was systematically investigated, including the effect of the initial solution's pH, Co-Mn/CeO_2 composite dose, O_3 dose and NOR concentration on degradation kinetics. Only about 3.33% of total organic carbon (TOC) and 72.17% of NOR could be removed within 150 min by single ozonation under the conditions of 60 mg/L of NOR and 200 mL/min of O_3 at pH= 7 and room temperature, whereas in the presence of 0.60 g/L of the Co-Mn/CeO_2 composite under the same conditions, 87.24% NOR removal was obtained through the catalytic ozonation process. The results showed that catalytic ozonation with the Co-Mn/CeO_2 composite could effectively enhance the degradation and mineralization of NOR compared to a single ozonation system alone. The catalytic performance of CeO_2 was significantly improved by the modification with Mn and Co. Co-Mn/CeO_2 represents a promising way to prepare efficient catalysts for the catalytic ozonation of organic polluted water. The removal efficiency of NOR in five cycles indicates that Co-Mn/CeO_2 is stable and recyclable for catalytic ozonation in water treatment.

Keywords: antibiotics; catalytic ozonation; emerging contaminant; Co-Mn/CeO_2; norfloxacin

Citation: Li, R.; Xiong, J.; Zhang, Y.; Wang, S.; Zhu, H.; Lu, L. Catalytic Ozonation of Norfloxacin Using Co-Mn/CeO_2 as a Multi-Component Composite Catalyst. *Catalysts* **2022**, *12*, 1606. https://doi.org/10.3390/catal12121606

Academic Editor: Luigi Rizzo

Received: 27 October 2022
Accepted: 5 December 2022
Published: 8 December 2022

1. Introduction

Recently, the wide applications of antibiotics have become a serious threat to the environment and public health worldwide due to their resistance to degradation and induction of resistance genes [1,2]. Norfloxacin (NOR), a typical fluoroquinolone (FQ) antibiotic, has been widely used and found in wastewater treatment plants from different routes. For instance, the concentrations of NOR detected from domestic and hospital effluents range from ng/L to μg/L [3,4]. It was found that NOR concentrations could even reach up to mg L^{-1} in pharmaceutical effluents [5]. Nevertheless, antibiotics might have a potential adverse influence on aquatic wildlife and humans even at trace levels [6]. Therefore, it is crucial to remove NOR efficiently from the aquatic environment.

Advanced oxidation processes (AOPs), which include several techniques, can generate highly oxidative species to mineralize antibiotics [7,8]. Among them, ozonation has been widely applied for the oxidative degradation of pollutants. Ozone (O_3) as a powerful oxidizing agent could degrade many organic pollutants including fluoroquinolone antibiotics. However, because of its selective oxidization of organic matters, the mineralization efficiencies of some antibiotics were relatively low when a single ozonation system was used [9,10]. A heterogeneous catalytic ozonation process could effectively improve the degradation of organic pollutants and has attracted significant attention in recent years. The heterogeneous catalysts for ozonation, such as metal oxides (MnO_2, Al_2O_3, Fe_3O_4, Co_3O_4

and CuO), metal-containing composites and carbon materials have been developed and applied in catalytic ozonation systems for the removal of various organic pollutants [11–16]. Among them, MnO_2 has been investigated and reported as a promising catalyst for O_3 due to its high efficiency and stability. For example, Nawaz et al. investigated the degradation of 4-nitrophenol (4-NP) through a heterogeneous catalytic ozonation process by using MnO_2 as the catalyst. Under the same reaction conditions, the degradation efficiency of MnO_2-catalyzed catalytic ozonation was 60.5% higher than that of ozonation alone [13]. The catalysis may be partly attributed to the role of oxygen vacancies (OVs) on MnO_2. As reported by He et al., oxygen vacancies facilitate the adsorption of O_3 onto the catalyst surface because oxygen vacancies increase the ratio of Mn^{3+}/Mn^{4+}, and then alter the charge distribution [17]. Meanwhile, Co_3O_4 also exhibited high catalytic activity for the catalytic ozonation of various refractory organic compounds. For example, Alvarez et al. investigated the degradation of pyruvic acid through a heterogeneous catalytic ozonation process by using Co_3O_4/Al_2O_3 composites as the catalyst. Under the same reaction conditions, the degradation efficiency of Co_3O_4-/Al_2O_3-catalyzed catalytic ozonation was 38% higher than that of ozonation alone. The rate of pyruvic acid disappearance is improved by the presence of cobalt, which is likely due to its catalytic effect on oxidation reactions [18].

Cerium oxides (CeO_2) has been widely applied in many research areas such as CO oxidation, VOC combustion and the water-gas shift reaction due to a low redox potential and abundant OVs [19–23]. In recent years, CeO_2, as an active component or support, has been widely investigated as an ozonation catalyst to enhance the removal of recalcitrant compounds [24–26]. For example, Li et al. found that ceria could accelerate MCM-48 to strengthen the degradation efficiency of clofibric acid (CA) by O_3 [27]. Akhtar et al. found that the presence of Fe_2O_3/CeO_2 could accelerate activated carbon to enhance the removal efficiency of sulfamethoxazole by O_3 [26]. Chen et al. reported that the introduction of a ceria catalyst can significantly enhance the catalytic ozonation of 4-chlorophenol, which could be attributed to the concentration and location of OVs [25]. However, few studies have reported on the combination of CeO_2 and Co-Mn for organic pollutant elimination via catalytic ozonation.

In this work, the Co-Mn/CeO_2 composite was fabricated by using the sol-gel method. The physical properties of the catalyst, the heterogeneous catalytic ozonation activities of Co-Mn/CeO_2 for the degradation of NOR, the performance of various operating conditions and the stability of the catalyst were evaluated.

2. Results and Discussion

2.1. Physical Properties of Catalysts

The crystal phases and crystallinities of CeO_2 and Co-Mn/CeO_2 catalysts were studied by using XRD. As shown in Figure 1a, the CeO_2 particles depicted the typical XRD patterns of pure fluorite cubic structures of CeO_2 (JCPDS 34-0349) with characteristic peaks at 2θ values of 28.6°, 33.1°, 47.5°, 56.4°, 59.1°, 69.5°, 76.8° and 79.1°, which were attributed to the (111), (200), (220), (311), (222), (400), (331) and (420) crystal planes, respectively [28–30]. The Co-Mn/CeO_2 composite samples did not show any obvious XRD diffraction for manganese oxides or cobalt oxides in Figure 1. Moreover, the XRD patterns of the Co-Mn/CeO_2 samples are quite broad compared to pristine CeO_2, which could be attributed to the formation of effective Mn-Ce, Co-Ce, and Mn- and Co-codoped solid solutions [23,31–33]. As the width and strength of XRD peaks have a close relationship with the crystallinity and crystal size of the corresponding crystal phase, the low crystallinity and small crystal size of metal oxide species can afford a large number of active sites for improved catalysis and provide a material basis for a high catalytic performance [32,34].

As shown in Figure 2a, it can be seen that the CeO_2 was in the form of irregular particles, which were evenly distributed, and there were many pores between the particles. The SEM image indicated that Co-Mn/CeO_2 was in the form of irregular flakes that were highly dispersed, and the surface was covered with a certain agglomeration and fluffy accumulation, as well as many pores with different sizes (Figure 2b). The above results indicate that Co-Mn/CeO_2 has a hierarchal micro–meso–macro porous structure. These pores were formed due to the gasification of free water and the decomposition of nitrates which acted as pore-fabricating agents in the sol-gel combustion preparation process [35]. Furthermore, in order to confirm the composition of Co-Mn/CeO_2, EDS mapping was performed, and the elemental mappings are shown in Figure 2c. The results showed a uniform dispersion of Mn, Co, Ce and O elements in the Co-Mn/CeO_2 catalysts which were consistent with the XRD and SEM results. The loose, porous structure may provide more active sites for reactant molecules, thereby promoting the performance of the catalysts [34].

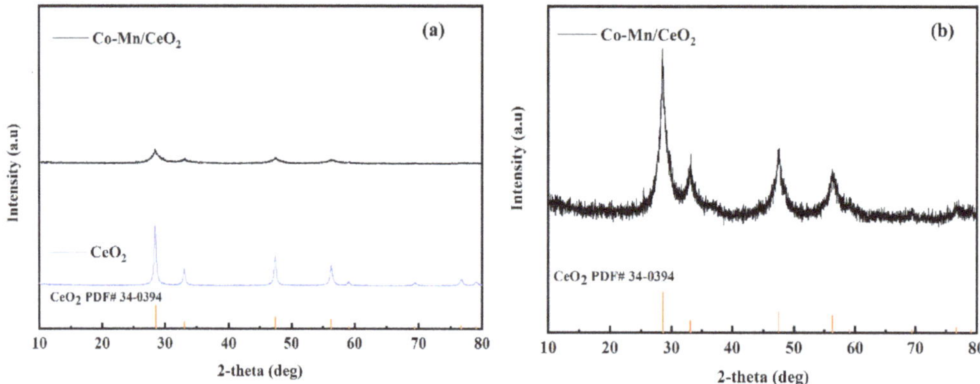

Figure 1. XRD pattern of CeO_2 and Co-Mn/CeO_2 (**a**); XRD patterns of pristine Co-Mn/CeO_2 (**b**).

Figure 2. SEM images of CeO_2 (**a**), Co-Mn/CeO_2 (**b**), and EDS mapping of Co-Mn/CeO_2 (**c_1–c_4**).

As shown in Figure 3, the N_2 adsorption–desorption isotherms of CeO_2 and Co-Mn/CeO_2 were type IV, showing that the two materials contained microporous and mesoporous structures. The data on the surface area, pore diameter and pore volume are summarized in Table 1. The BET surface area significantly increased from 34.80 m^2/g (CeO_2) to 92.43 m^2/g (Co-Mn/CeO_2). Compared with CeO_2, Co-Mn/CeO_2 has the smaller average pore size and the larger pore volume, indicating that Co-Mn/CeO_2 has more pores.

The larger specific surface area and pore number of Co-Mn/CeO_2 may be attributed to the formation of OVs and surface defects, which can provide more active sites to enhance the catalytic performance [11,36,37].

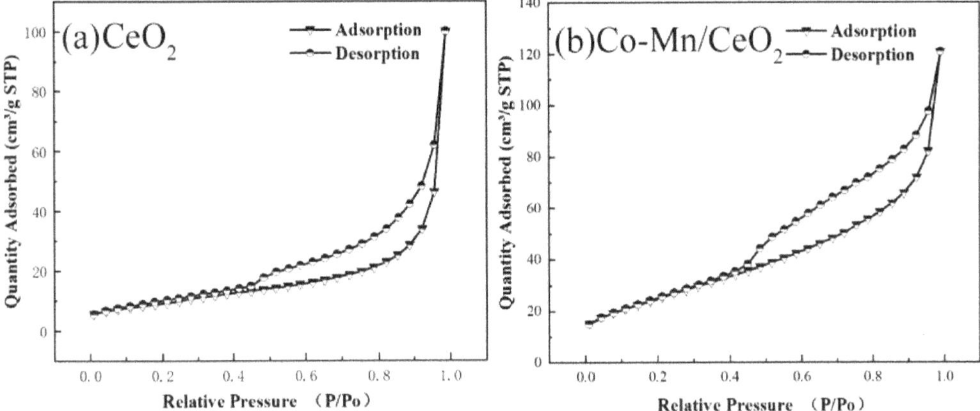

Figure 3. N_2 adsorption–desorption isotherms of CeO_2 (**a**); Co-Mn/CeO_2 (**b**).

Table 1. Surface area, average pore width and total pore volume of catalysts.

Catalyst	BET Surface Area (m^2/g)	Adsorption Average Pore Width (nm)	Total Pore Volume of Pore (cm^3/g)
CeO_2	34.804	17.8224	0.1551
Co-Mn/CeO_2	92.425	8.10891	0.1874

The element composition and chemical environment of CeO_2 and Co-Mn/CeO_2 were further identified by using XPS. As illustrated in Figure 4a, in addition to the characteristic peaks of Co2p and Mn2p, the Ce3d and O1s peaks were observed clearly in both XPS survey spectrums of CeO_2 and Co-Mn/CeO_2, which were consistent with the EDS results. For Co-Mn/CeO_2, there were two major peaks at Co2$p_{3/2}$ and Co2$p_{1/2}$, and the fitted peaks at 780.1 eV and 795.2 eV could be attributed to Co^{3+}, whereas the peaks at 781.5 eV and 796.4 eV could be ascribed to Co^{2+}. Thus, it is concluded that Co existed in the oxidation states of Co^{2+} and Co^{3+} [38]. As shown in Figure 4c, the Mn 2p XPS spectrum demonstrates two peaks centered at 642.4 eV and 653.5 eV, which can be attributed to Mn 2$p_{3/2}$ and Mn 2$p_{1/2}$ states, respectively [39]. The Mn 2$p_{3/2}$ peak of Co-Mn/CeO_2 could be fitted by two main peaks centered at 642.3 eV and 644.1 eV with a ratio of 0.86, corresponding to the chemical states of Mn^{3+} and Mn^{4+}, respectively. The results show that the content of the Mn^{4+} species was higher than that of the Mn^{3+} species.

As shown in the Ce3d spectra of Figure 4b, the relative abundance of Ce 3d in Co-Mn/CeO_2 was smaller, suggesting that some Ce^{4+} in CeO_2 may be replaced by cobalt ions or manganese ions which could result in the creation of OVs. These principle binding energies were labeled as u and v, which were attributed to the two pairs of Ce spin-orbital doublets, 3$d_{3/2}$ (higher BE) and 3$d_{5/2}$ (lower BE), respectively. The photoelectron peaks u′ and v', u'' and v'', and u''' and v''' corresponded to the concentration of Ce^{4+}. Meanwhile, the two weak peaks labeled as u' and v' were ascribed as being characteristic of Ce^{3+} [1]. The relative concentration ratio of Ce^{3+} to Ce^{4+} can be calculated from the peak areas of deconvoluted peaks according to Equation (1)

$$r = \frac{A_{U'''} + A_{V'''} + A_{u''} + A_{v''} + A_U + A_V}{A_{U'} + A_{V'}} \quad (1)$$

The chemical valence state of Ce in CeO_2 and Co-Mn/CeO_2 mainly included the oxidation state of Ce^{4+} that coexisted with a relatively small amount of Ce^{3+}. The relative percentage of Ce^{3+}/Ce^{4+} of CeO_2 and Co-Mn/CeO_2 were then calculated to be 13.95% and 29.12%, respectively. As a defect indicator, the higher concentration of Ce^3+ in Co-Mn/CeO_2 indicated the creation of relatively more OVs on the surface of the catalyst [28,40,41].

The O1s results can further confirm the generation of abundant OVs. As shown in Figure 4e, the O1s spectrum of CeO_2 and Co-Mn/CeO_2 could be divided into three major components—529.5, 531.5 and 533.1 eV, which were assigned as lattice oxygen (denoted as O_{latt}), surface oxygen (O_{sur}) and adsorbed oxygen (O_{ads}), respectively. Generally, the surface oxygen species could improve the catalytic process [42,43]. The surface oxygen O_{sur} concentration of CeO_2 and Co-Mn/CeO_2 was 14.83% and 43.21%, respectively, indicating the same order as that of Ce^{3+}. These observations showed that combining Mn and Co with CeO2 not only promoted the formation of more new structure defects, but also improved the concentration of (O_{sur}) species, indicating that the Co-Mn/CeO_2 catalyst can provide more surface-active oxygen species for catalytic ozonation.

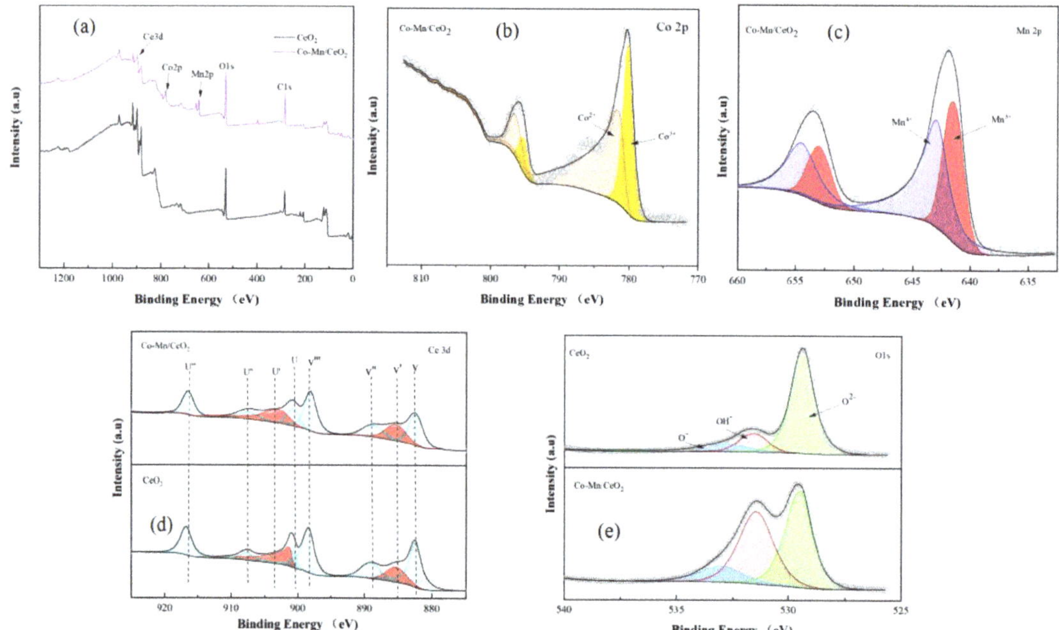

Figure 4. XPS survey spectra of CeO_2 and Co-Mn/CeO_2 (**a**); narrow region scan of Co2p (**b**), Mn2p (**c**), Ce3d (**d**) and O1s (**e**) of XPS spectra.

2.2. Catalytic Activities of Catalysts

To evaluate the performance of Co-Mn/CeO_2 in catalytic ozonation processes, NOR degradation and TOC removal in O_3, CeO_2/O_3 and Co-Mn/CeO_2/O_3 systems were investigated, and the results are shown in Figure 5. As shown in Figure 5a, the degradation efficiency of NOR via single ozonation was only 72.17% after 150 min. The efficiency increased to 76.15% and 87.24% in CeO_2 ozonation and Co-Mn/CeO_2 ozonation processes, respectively.

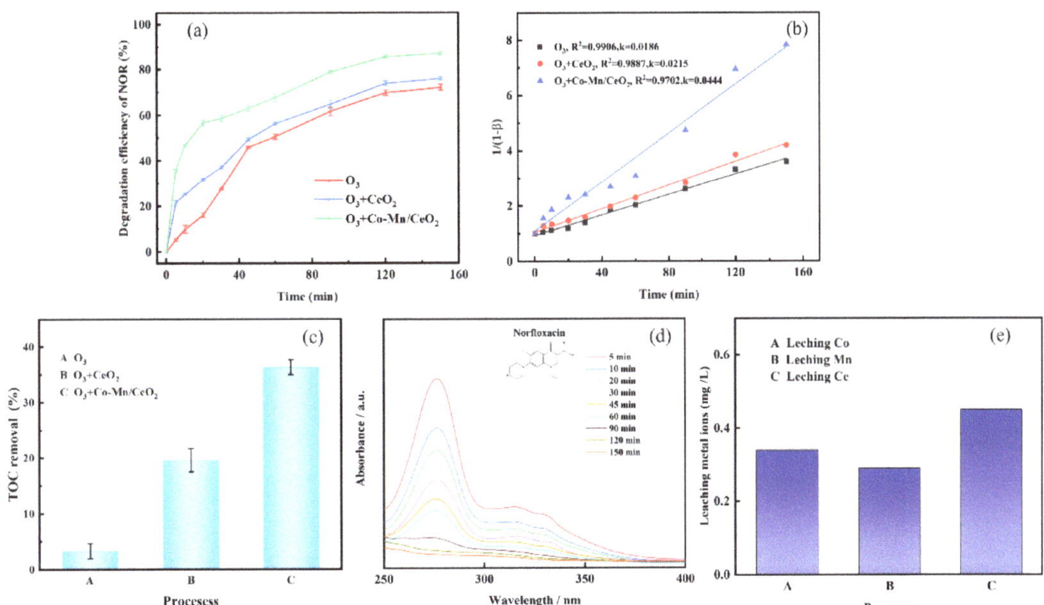

Figure 5. NOR degradation (**a**), quasi-second-order plot of NOR destruction, (**b**) TOC removal (**c**) in different processes, UV–vis spectra of treated water samples at different times (**d**); metal leaching amounts (**e**) (NOR = 60 mg/L; catalyst = 0.60 g/L; O_3 = 200 mL/min; initial pH of 7).

In order to further investigate the ozonation reaction kinetics, the experimental data were fitted with the second-order model (Equation (2)):

$$\frac{1}{1-\beta} = kc_0t + 1 \tag{2}$$

where k is the kinetic rate constant obtained from the fitting results. As shown in Figure 5b, the apparent first-order rate constant k of the NOR degradation was 0.0186 $(mg/L)^{-1}min^{-1}$, 0.0215 $(mg/L)^{-1}min^{-1}$ and 0.0444 $(mg/L)^{-1}min^{-1}$ in O_3, CeO_2/O_3 and Co-Mn/CeO_2/O_3 processes, respectively. It is worth noting that the different removal efficiencies of TOC were achieved with the addition of different catalysts. Moreover, these catalysts could significantly enhance the mineralization of NOR compared to the non-catalytic ozonation processes. Ozone, as a kind of oxidant, reacts easily with NOR, but due to its selective oxidation property, the ozone molecule might not be able to remove some degradation intermediates formed during NOR degradation, resulting in a low mineralization efficiency. As illustrated in Figure 5c, although NOR was effectively removed in 150 min by single ozonation, the removal efficiency of TOC was only about 3.33%. However, in the presence of CeO_2 and Co-Mn/CeO_2 under the same conditions, the removal efficiency of TOC increased to 19.61% and 36.31%, which was 1.9 and 10.9 times higher than that of the single ozonation and CeO_2/O_3 system, respectively. Figure 5d shows that the feature peak of NOR gradually disappeared within 180 min, indicating the complete degradation of NOR during the reaction. In addition, it can be noted in Figure 5e that catalysts had a low dissolution concentration of metal ions in the reaction solution after 150 min of reaction, which is acceptable according to discharge standards.

These results suggest that Co-Mn/CeO_2 had catalytic ozonation activity and can indeed strengthen the degradation of persistent organics. The main reason for this may be attributed to: (1) The doping of Co and Mn lead to the formation of surface defects and OVs, which promote the decomposition of ozone into reactive radicals with stronger oxidation ability, and then achieve a better ozonation effect. (2) The electron transfer between Ce^{3+}/Ce^{4+}, Co^{3+}/Co^{2+} and Mn^{3+}/Mn^{4+} in Co-Mn/CeO_2 made the redox of Ce^{3+}/Ce^{4+} facile during the catalytic oxidation processes, improving the synergistic catalysis of Co, Mn and Ce for the degradation of NOR. This is consistent with the literature that shows Ce^{3+} species as the active sites in the decomposition of ozone into radicals with a more powerful oxidation ability [25].

2.3. *Effect of Operational Conditions*

2.3.1. Effect of Initial Solution pH

Figure 6 presents the effect of the initial solution's pH on NOR removal. The removal rate of NOR gradually increased with the increase in pH from the initial pH of 5.0 to 9.0, and the maximum NOR removal efficiency of 89.61% was achieved when the pH was 9.0. At a higher pH, the abundance of OH− could accelerate the decomposition of ozone into reactive radicals and enhance the generation of active radicals, such as hydroxyl radicals, leading to high NOR removal efficiency (Equations (3)–(7)):

$$O_3 + OH^- \rightarrow HO_4^- \tag{3}$$

$$HO_4^- \leftrightarrow HO_2\cdot + O_2^-\cdot \tag{4}$$

$$O_2\cdot + O_3 \rightarrow O_2 + O_3^-\cdot \tag{5}$$

$$O_3^-\cdot \rightarrow O_2 + O^-\cdot \tag{6}$$

$$O^-\cdot + H_2O \rightarrow \cdot OH + OH^- \tag{7}$$

However, as the initial pH further increased to 11.00, the NOR removal efficiency decreased because the enormous generation of •OH could facilitate the reaction between •OH itself or O^2•, rather than between the intermediate products of NOR degradation. The quasi-second-order kinetics fitting was performed on the removal of NOR molecules within 150 min. The results are shown in Figure 6b. The reaction rates were 0.0233 $(mg/L)^{-1}min^{-1}$, 0.299 $(mg/L)^{-1}min^{-1}$, 0.0439 $(mg/L)^{-1}min^{-1}$, 0.0554 $(mg/L)^{-1}min^{-1}$ and 0.0307 $(mg/L)^{-1}min^{-1}$. The results indicate that a low pH inhibited the reaction and slowed down the oxidation rate of NOR. When the pH value of the initial solution changed and was in the range of 3.00–11.00, NOR was almost removed in all cases. The results indicate that Co-Mn/CeO_2 can work in such a wide pH range.

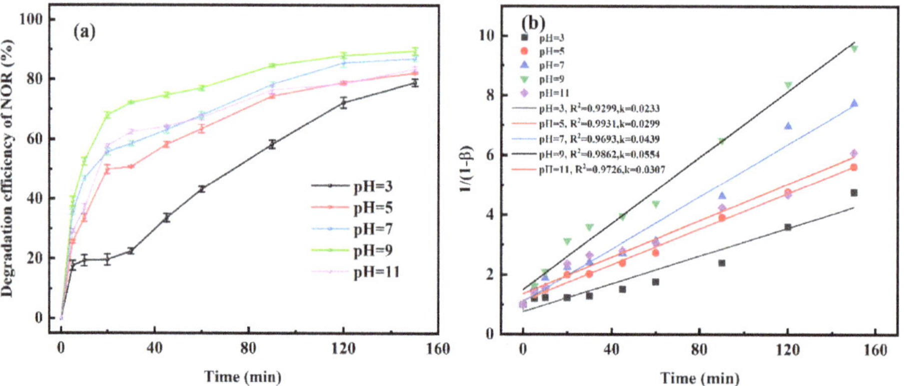

Figure 6. Effect of solution pH on NOR removal (**a**) and the kinetics equations and parameters of quasi-second-order reactions at different pH values (**b**) (if not otherwise specified, NOR = 60 mg/L; O_3 = 200 mL/min; Co-Mn/CeO_2 = 0.60 g/L).

2.3.2. Effect of O_3 Concentration

The increase in O_3 concentration could improve the removal of NOR, as shown in Figure 7. When the applied flow of O_3 was 100, 200 and 300 mL/min, the NOR removal efficiency within 150 min was 27.32%, 87.12% and 88.62%, respectively. The reaction rates were 0.0107 $(mg/L)^{-1}min^{-1}$, 0.0464 $(mg/L)^{-1}min^{-1}$ and 0.1347 $(mg/L)^{-1}min^{-1}$. The increase in NOR removal efficiency was due to the possibility of a higher concentration of O_3 accelerating the transformation of O_3 into the aqueous solution, forming more derived free radicals. However, when O_3 concentration increased from 200 mL/min to 300 mL/min, the reaction rate increased from 0.0464 $(mg/L)^{-1}min^{-1}$ to 0.1347 $(mg/L)^{-1}min^{-1}$. Since excess O_3 could also react with •OH to produce O_2 and H_2O, excess O_3 would compete with pollutants to react with free radicals, resulting in the decrease in oxidants for NOR removal (Equation (8)).

$$O_3 + 2{\cdot}OH \rightarrow 2O_2 + H_2O \tag{8}$$

Therefore, a high O_3 concentration may not always be conducive to the improving NOR removal.

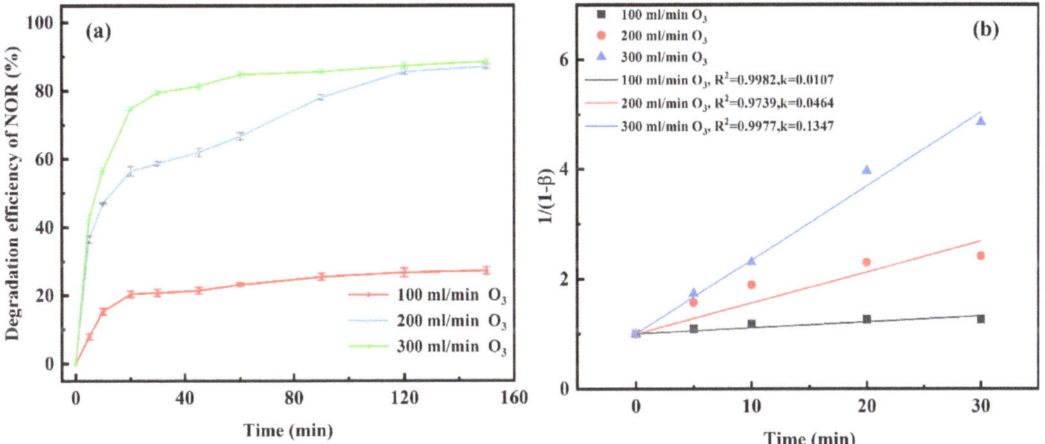

Figure 7. Effect of O_3 concentration on NOR removal (**a**) and the kinetics equations and parameters of quasi-second-order reactions at different O_3 concentrations (**b**) (if not otherwise specified, NOR = 60 mg/L; O_3 = 200 mL/min; Co-Mn/CeO_2 = 0.60 g/L; pH = 7).

2.3.3. Effect of Catalyst Dosage

Figure 8 shows the effect of Co-Mn/CeO_2 dosage on NOR removal. The removal of NOR gradually increased from 77.59% to 87.77% as the catalyst dosage increased from 0.4 to 0.8 g/L within 150 min. The reaction rate increased from 0.0222 $(mg/L)^{-1}min^{-1}$ to 0.0466 $(mg/L)^{-1}min^{-1}$. This might be due to the higher catalyst dose possibly providing more surface areas and available active sites, which could catalyze the disintegration of the ozone to produce more free active radicals in the oxidation process. However, the increase in NOR removal efficiency was only 6%, when Co-Mn/CeO_2 increased from 0.4 g/L to 0.6 g/L. In the presence of an excess catalyst, the concentration of NOR and O_3 per unit area might decrease, which was not conducive to the reaction between NOR and O_3 [28,44]. Hence, the optimized catalyst dosage was chosen as 0.6 g/L in this experiment.

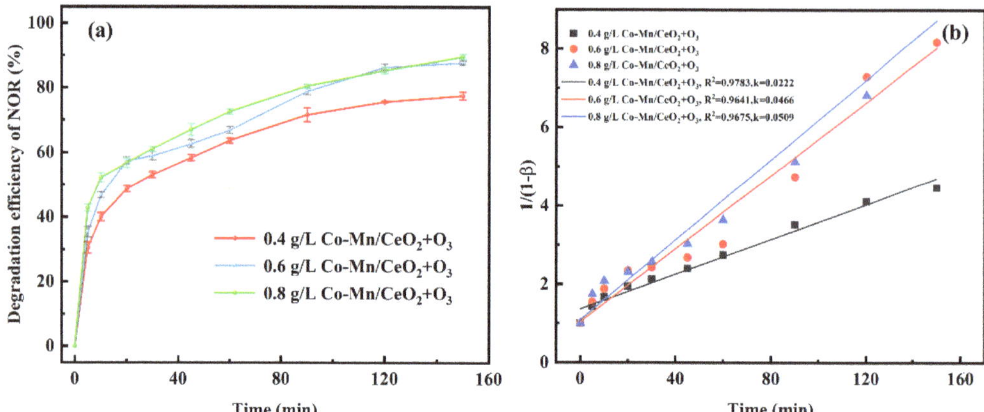

Figure 8. Effect of catalyst dosage on NOR removal (**a**) and the kinetics equations and parameters of quasi-second-order reactions at different catalyst dosages (**b**) (if not otherwise specified, NOR = 60 mg/L; O_3 = 200 mL/min; Co-Mn/CeO_2 = 0.60 g/L; pH = 7).

2.3.4. Effect of Initial NOR Concentration

Figure 9 presents the effect of the NOR initial concentration on NOR removal. When the initial concentration of NOR was 40 mg/L, 60 mg/L and 80 mg/L, the NOR removal efficiency was 86.51%, 87.65% and 79.39%, respectively. The reaction rates were 0.0562 $(mg/L)^{-1}min^{-1}$, 0.0456 $(mg/L)^{-1}min^{-1}$ and 0.0248 $(mg/L)^{-1}min^{-1}$. The results indicate that the degradation rate of norfloxacin was inhibited by the increase in NOR concentration. A higher concentration of pollutants may require more oxidants to be oxidized, and due to incomplete oxidation, intermediates will be produced and accumulate in the catalytic ozonation process, resulting in a low degradation efficiency [45].

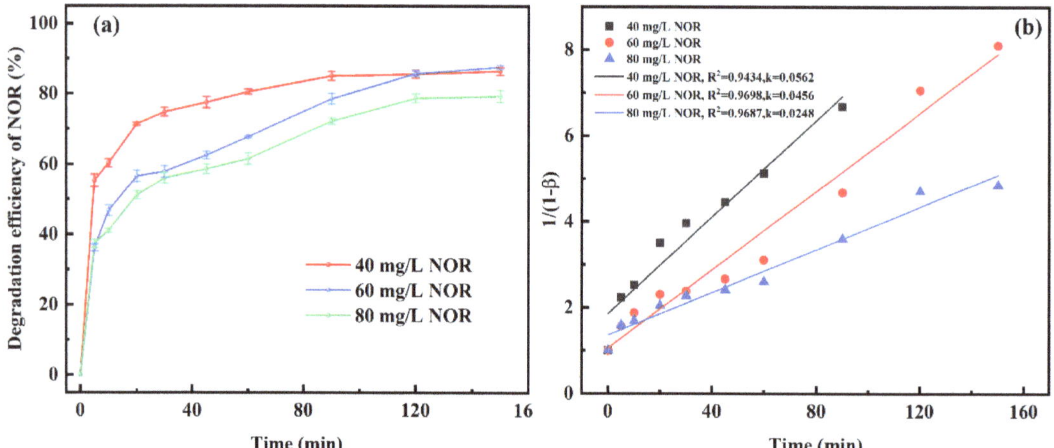

Figure 9. Effect of NOR initial concentration on NOR removal (**a**) and the kinetics equations and parameters of quasi-second-order reactions at different NOR initial concentrations (**b**) (if not otherwise specified, NOR = 60 mg/L; O_3= 200 mL/min; Co-Mn/CeO_2 = 0.60 g/L; pH = 7).

2.4. Catalyst Stability and Reusability

In order to evaluate the stability of Co-Mn/CeO_2 in the catalytic ozonation system, the catalyst was collected after each degradation reaction cycle and reused under the same operating conditions. As presented in Figure 10, the activity of Co-Mn/CeO_2 toward the degradation of NOR does not obviously change after five recycle times. This demonstrates that Co-Mn/CeO_2 had a stable performance in the catalytic ozonation process for the degradation of NOR.

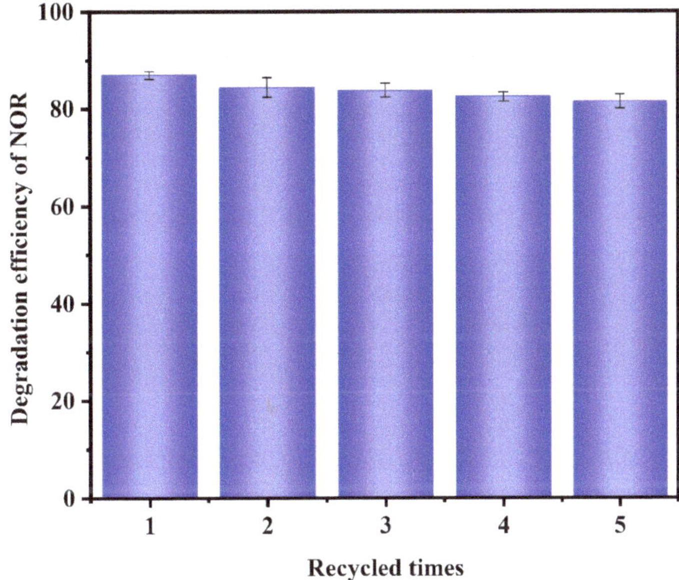

Figure 10. Stability of prepared Co-Mn/CeO_2 for the catalytic ozonation of NOR (if not otherwise specified, NOR = 60 mg/L; O_3 = 200 mL/min; Co-Mn/CeO_2 = 0.60 g/L; pH = 7; reaction time = 150 min).

3. Experimental Procedure

3.1. Materials and Chemicals

Norfloxacin (NOR) was purchased from Meilun Biotechnology Co., Ltd. (Dalian, China). Cobaltous nitrate ($Co(NO_3)_2 \cdot 6H_2O$), cerium nitrate ($Ce(NO_3)_3 \cdot 6H_2O$) and NaOH (Sodium hydroxide) were purchased from Aladdin Chemistry Co., Ltd (Shanghai, China). Manganese nitrate (50% *w/w*) and citric acid were bought from Guangzhou Chemical Reagent Co., Ltd. HCl (hydrochloric acid, 36%) was supplied by Lingfeng Chemical Reagent (Shanghai, China). Ultra-pure water, which was used as the experimental water, was obtained from the Millipore Milli-Q Ultrapure Gradient A10 purification system from Millipore Co., Ltd. (Burlington, MA, USA). All the chemicals and reagents used in the experiment were of analytical purity and could be used directly without further purification.

3.2. Preparation of Catalysts

The Co-Mn/CeO_2 catalyst was prepared by modifying the method described by [25]. Briefly, Co-Mn/CeO_2 was prepared by using the sol-gel method with citric acid as the chelating agent. Nitrate salts of cobalt, manganese and cerium, in addition to citric acid, were dissolved in deionized water with the molar ratio of Co(II): Mn(II): Ce(III): Citric acid = 1:1:1:3. Then, ammonia was added dropwise to adjust the pH to 4.5–5.0. The resulting solution was magnetically stirred at 80 °C until a viscous pale pink gel was formed. The gel was dried in an oven at 80 °C and then calcined in a muffle oven at 350 °C for 2 h.

The obtained black solid was stored in a dryer for further use. By comparison, CeO_2 was also prepared by adding the corresponding nitrate.

3.3. Ozonation Experiments

The ozonation experiment was carried out in a 150 mL glass column batch reactor. A certain amount of catalyst was added to the reactor containing 80 mL of NOR aqueous solution (the NOR test concentration was determined as 60 mg/L according to the relevant literature and experimental conditions [6,46,47], with an initial pH = 7), and then the mixed solution was maintained as a suspension by magnetic stirring. Next, the ozone gas was continuously bubbled to the bottom of the flask through the aeration device. Samples were collected from the ozone reactor within the prescribed time interval and then filtered using membrane filters (0.45 μm) for further analysis. Ozone gas was generated by using a Tonglin 3S-T3 ozone generator as the air source. Within the specified time interval, a certain volume of aliquots was taken from the reactor, and the residual ozone in the tail gas was removed with a sodium thiosulfate solution. Except for the test to investigate the influence of the initial pH value, other tests were conducted without adjusting the initial pH value. All the experiments were repeated at room temperature.

3.4. Characterization of Catalysts

X-ray diffraction (XRD) measurements were performed by using a D8 Discover Bruker diffractometer with Cu Kα radiation (Karlsruhe, Germany). The BET-specific surface areas were determined by using the AUTOSORB-IQ-MP system (Quantachrome, Boynton Beach, FL. USA). X-ray photoelectron spectroscopy (XPS, Waltham, MA, USA) spectra were determined via the Thermo Scientific K-Alpha system. The morphology was characterized by using an FEI Quattro S emission scanning electron microscopy (SEM, Waltham, MA, USA). Energy-dispersive spectroscopic (EDS) data were obtained by using the Bruker Quantax XFlash SDD 6 (Karlsruhe, Germany). An inductively coupled plasma optical emission spectrometry (ICP-OES, optima 8000DV, Waltham, MA, USA) was used to measure the leaching concentration of metal in the solution. The absorbance of NOR was measured with a UV-2700 spectrophotometer (Shimadzu, Kyoto, Japan) at 272 nm.

4. Conclusions

Co-Mn/CeO_2 was first prepared and used as a heterogeneous ozone catalyst. Co-Mn/CeO_2 had a disordered mesostructure and its performance was good in the catalytic ozonation for NOR removal, especially in mineralization. With the addition of Co-Mn/CeO_2, the removal efficiency of TOC significantly increased from 3.33% to 36.31%, compared to single ozonation. The dosage of ozone, the dosage of NOR, the dosage of the catalyst and the solution pH have different effects on the degradation of NOR. In catalytic ozonation, the degradation efficiency of NOR was higher at a basic pH than that at neutral or acid pH values. The removal of NOR was highest at a pH value of 9, Co-Mn/CeO_2 dosage of 0.8 g/L and O_3 concentration of 300 mL/min. Co-Mn/CeO_2 also showed good stability and can be reused five times without significant catalytic activity loss. This research shows an efficient way to modify CeO_2 to remove organic pollution from wastewater through a catalytic ozonation process.

Author Contributions: R.L.: Investigation, Methodology, Validation, Software, Formal analysis, Data curation, Writing—Original draft writing. J.X.: Supervision, Validation, Conceptualization Y.Z.: Conceptualization, Methodology, Writing—Review and editing, Supervision, Formal analysis, Validation, Project administration, Funding acquisition. S.W.: Resources, Supervision. H.Z.: Supervision, Software. L.L.: Supervision, Software. All authors have read and agreed to the published version of the manuscript.

Funding: This research was funded by the Natural Science Foundation of Guangxi Province (2020GX-NSFAA159135), the Foundation (No.2021KF05) of Guangxi Key Laboratory of Clean Pulp & Paper-making and Pollution Control, National Natural Science Foundation of China (NSFC No: 21968005), Innovation Project of Guangxi Graduate Education (YCSW2022055), Guangxi Ba-Gui Scholars Program(2019A33), the foundation of Guangxi Key Laboratory of Clean Pulp & Papermaking and Pollution Control (ZR201702), 2019 Yongjiang Plan Project—Pulp and Paper Industry AOX Ultra-low Emission Technology Development and Application Demonstration (2019003), College of Light Industry and Food Engineering, Guangxi University.

Data Availability Statement: Not applicable.

Conflicts of Interest: The authors declare no conflict of interest.

References

1. Zhang, Y.; Xiao, R.; Wang, S.; Zhu, H.; Song, H.; Chen, G.; Lin, H.; Zhang, J.; Xiong, J. Oxygen vacancy enhancing Fenton-like catalytic oxidation of norfloxacin over prussian blue modified CeO_2: Performance and mechanism. *J. Hazard. Mater.* **2020**, *398*, 122863. [CrossRef]
2. Chen, H.; Wang, J. MOF-derived Co_3O_4-C@FeOOH as an efficient catalyst for catalytic ozonation of norfloxacin. *J. Hazard. Mater.* **2021**, *403*, 123697. [CrossRef]
3. Wang, G.; Zhao, D.; Kou, F.; Ouyang, Q.; Chen, J.; Fang, Z. Removal of norfloxacin by surface Fenton system ($MnFe_2O_4/H_2O_2$): Kinetics, mechanism and degradation pathway. *Chem. Eng. J.* **2018**, *351*, 747–755. [CrossRef]
4. Wang, Y.; Wang, R.; Lin, N.; Xu, J.; Liu, X.; Liu, N.; Zhang, X. Degradation of norfloxacin by MOF-derived lamellar carbon nanocomposites based on microwave-driven Fenton reaction: Improved Fe(III)/Fe(II) cycle. *Chemosphere* **2022**, *293*, 133614. [CrossRef]
5. Larsson, D.G.; de Pedro, C.; Paxeus, N. Effluent from drug manufactures contains extremely high levels of pharmaceuticals. *J. Hazard. Mater.* **2007**, *148*, 751–755. [CrossRef]
6. Li, H.; Chen, J.; Hou, H.; Pan, H.; Ma, X.; Yang, J.; Wang, L.; Crittenden, J.C. Sustained molecular oxygen activation by solid iron doped silicon carbide under microwave irradiation: Mechanism and application to norfloxacin degradation. *Water Res.* **2017**, *126*, 274–284. [CrossRef]
7. Zhang, X.-W.; Wang, F.; Wang, C.-C.; Wang, P.; Fu, H.; Zhao, C. Photocatalysis activation of peroxodisulfate over the supported Fe_3O_4 catalyst derived from MIL-88A(Fe) for efficient tetracycline hydrochloride degradation. *Chem. Eng. J.* **2021**, *426*, 131927. [CrossRef]
8. Yu, D.; Wu, M.; Hu, Q.; Wang, L.; Lv, C.; Zhang, L. Iron-based metal-organic frameworks as novel platforms for catalytic ozonation of organic pollutant: Efficiency and mechanism. *J. Hazard. Mater.* **2019**, *367*, 456–464. [CrossRef]
9. Miao, H.F.; Cao, M.; Xu, D.Y.; Ren, H.Y.; Zhao, M.X.; Huang, Z.X.; Ruan, W.Q. Degradation of phenazone in aqueous solution with ozone: Influencing factors and degradation pathways. *Chemosphere* **2015**, *119*, 326–333. [CrossRef]
10. Dantas, R.F.; Contreras, S.; Sans, C.; Esplugas, S. Sulfamethoxazole abatement by means of ozonation. *J. Hazard. Mater.* **2008**, *150*, 790–794. [CrossRef]
11. Li, P.; Zhan, S.; Yao, L.; Xiong, Y.; Tian, S. Highly porous alpha-MnO_2 nanorods with enhanced defect accessibility for efficient catalytic ozonation of refractory pollutants. *J. Hazard. Mater.* **2022**, *437*, 129235. [CrossRef] [PubMed]
12. Afzal, S.; Quan, X.; Lu, S. Catalytic performance and an insight into the mechanism of CeO_2 nanocrystals with different exposed facets in catalytic ozonation of p-nitrophenol. *Appl. Catal. B Environ.* **2019**, *248*, 526–537. [CrossRef]
13. Nawaz, F.; Cao, H.; Xie, Y.; Xiao, J.; Chen, Y.; Ghazi, Z.A. Selection of active phase of MnO_2 for catalytic ozonation of 4-nitrophenol. *Chemosphere* **2017**, *168*, 1457–1466. [CrossRef] [PubMed]
14. Heuer, J.; Ferguson, C.T.J. Photocatalytic polymer nanomaterials for the production of high value compounds. *Nanoscale* **2022**, *14*, 1646–1652. [CrossRef]
15. Hien, N.T.; Nguyen, L.H.; Van, H.T.; Nguyen, T.D.; Nguyen, T.H.V.; Chu, T.H.H.; Nguyen, T.V.; Trinh, V.T.; Vu, X.H.; Aziz, K.H.H. Heterogeneous catalyst ozonation of Direct Black 22 from aqueous solution in the presence of metal slags originating from industrial solid wastes. *Sep. Purif. Technol.* **2020**, *233*, 115961. [CrossRef]
16. Hama Aziz, K.H. Application of different advanced oxidation processes for the removal of chloroacetic acids using a planar falling film reactor. *Chemosphere* **2019**, *228*, 377–383. [CrossRef]
17. He, Y.; Wang, L.; Chen, Z.; Shen, B.; Wei, J.; Zeng, P.; Wen, X. Catalytic ozonation for metoprolol and ibuprofen removal over different MnO_2 nanocrystals: Efficiency, transformation and mechanism. *Sci. Total Environ.* **2021**, *785*, 147328. [CrossRef]
18. Álvarez, P.M.; Beltrán, F.J.; Pocostales, J.P.; Masa, F.J. Preparation and structural characterization of Co/Al2O3 catalysts for the ozonation of pyruvic acid. *Appl. Catal. B Environ.* **2007**, *72*, 322–330. [CrossRef]
19. Piumetti, M.; Bensaid, S.; Russo, N.; Fino, D. Nanostructured ceria-based catalysts for soot combustion: Investigations on the surface sensitivity. *Appl. Catal. B Environ.* **2015**, *165*, 742–751. [CrossRef]
20. Mann, A.K.P.; Wu, Z.; Calaza, F.C.; Overbury, S.H. Adsorption and Reaction of Acetaldehyde on Shape-Controlled CeO_2 Nanocrystals: Elucidation of Structure–Function Relationships. *ACS Catal.* **2014**, *4*, 2437–2448. [CrossRef]

21. Zheng, M.; Wang, S.; Li, M.; Xia, C. H_2 and CO oxidation process at the three-phase boundary of Cu-ceria cermet anode for solid oxide fuel cell. *J. Power Source* **2017**, *345*, 165–175. [CrossRef]
22. Izu, N.; Itoh, T.; Shin, W.; Matsubara, I.; Murayama, N. The effect of hafnia doping on the resistance of ceria for use in resistive oxygen sensors. *Sens. Actuators B Chem.* **2007**, *123*, 407–412. [CrossRef]
23. Govinda Rao, B.; Jampaiah, D.; Venkataswamy, P.; Reddy, B.M. Enhanced Catalytic Performance of Manganese and Cobalt Co-doped CeO2Catalysts for Diesel Soot Oxidation. *ChemistrySelect* **2016**, *1*, 6681–6691. [CrossRef]
24. Wang, J.; Quan, X.; Chen, S.; Yu, H.; Liu, G. Enhanced catalytic ozonation by highly dispersed CeO_2 on carbon nanotubes for mineralization of organic pollutants. *J. Hazard. Mater.* **2019**, *368*, 621–629. [CrossRef] [PubMed]
25. Chen, X.; Zhan, S.; Chen, D.; He, C.; Tian, S.; Xiong, Y. Grey Fe-CeO2-σ for boosting photocatalytic ozonation of refractory pollutants: Roles of surface and bulk oxygen vacancies. *Appl. Catal. B Environ.* **2021**, *286*, 119928. [CrossRef]
26. Akhtar, J.; Amin, N.S.; Aris, A. Combined adsorption and catalytic ozonation for removal of sulfamethoxazole using Fe_2O_3/CeO_2 loaded activated carbon. *Chem. Eng. J.* **2011**, *170*, 136–144. [CrossRef]
27. Li, S.; Tang, Y.; Chen, W.; Hu, Z.; Li, X.; Li, L. Heterogeneous catalytic ozonation of clofibric acid using Ce/MCM-48: Preparation, reaction mechanism, comparison with Ce/MCM-41. *J. Colloid Interface Sci.* **2017**, *504*, 238–246. [CrossRef]
28. Mo, S.; Li, J.; Liao, R.; Peng, P.; Li, J.; Wu, J.; Fu, M.; Liao, L.; Shen, T.; Xie, Q.; et al. Unraveling the decisive role of surface CeO_2 nanoparticles in the Pt-CeO_2/MnO_2 hetero-catalysts for boosting toluene oxidation: Synergistic effect of surface decorated and intrinsic O-vacancies. *Chem. Eng. J.* **2021**, *418*, 129399. [CrossRef]
29. Zhu, C.; Wei, X.; Li, W.; Pu, Y.; Sun, J.; Tang, K.; Wan, H.; Ge, C.; Zou, W.; Dong, L. Crystal-Plane Effects of $CeO_2\{110\}$ and CeO2{100} on Photocatalytic CO2 Reduction: Synergistic Interactions of Oxygen Defects and Hydroxyl Groups. *ACS Sustain. Chem. Eng.* **2020**, *8*, 14397–14406. [CrossRef]
30. Du, X.; Dai, Q.; Wei, Q.; Huang, Y. Nanosheets-assembled Ni (Co) doped CeO_2 microspheres toward NO + CO reaction. *Appl. Catal. A Gen.* **2020**, *602*, 117728. [CrossRef]
31. Todorova, S.; Kolev, H.; Holgado, J.P.; Kadinov, G.; Bonev, C.; Pereñíguez, R.; Caballero, A. Complete n-hexane oxidation over supported Mn–Co catalysts. *Appl. Catal. B Environ.* **2010**, *94*, 46–54. [CrossRef]
32. Faria, P.C.; Monteiro, D.C.; Orfao, J.J.; Pereira, M.F. Cerium, manganese and cobalt oxides as catalysts for the ozonation of selected organic compounds. *Chemosphere* **2009**, *74*, 818–824. [CrossRef] [PubMed]
33. Song, H.; Hu, F.; Peng, Y.; Li, K.; Bai, S.; Li, J. Non-thermal plasma catalysis for chlorobenzene removal over CoMn/TiO_2 and CeMn/TiO2: Synergistic effect of chemical catalysis and dielectric constant. *Chem. Eng. J.* **2018**, *347*, 447–454. [CrossRef]
34. Zhou, G.; He, X.; Liu, S.; Xie, H.; Fu, M. Phenyl VOCs catalytic combustion on supported CoMn/AC oxide catalyst. *J. Ind. Eng. Chem.* **2015**, *21*, 932–941. [CrossRef]
35. Lu, J.; Sun, J.; Chen, X.; Tian, S.; Chen, D.; He, C.; Xiong, Y. Efficient mineralization of aqueous antibiotics by simultaneous catalytic ozonation and photocatalysis using $MgMnO_3$ as a bifunctional catalyst. *Chem. Eng. J.* **2019**, *358*, 48–57. [CrossRef]
36. Yu, D.; Wang, L.; Yang, T.; Yang, G.; Wang, D.; Ni, H.; Wu, M. Tuning Lewis acidity of iron-based metal-organic frameworks for enhanced catalytic ozonation. *Chem. Eng. J.* **2021**, *404*, 127075. [CrossRef]
37. Yu, H.; Ge, P.; Chen, J.; Xie, H.; Luo, Y. The degradation mechanism of sulfamethoxazole under ozonation: A DFT study. *Environ. Sci. Process Impacts* **2017**, *19*, 379–387. [CrossRef]
38. Zhang, C.; Wang, K.; Xie, K.; Han, X.; Ma, W.; Li, X.; Teng, G. Controllable preparation of hierarchical MnCo bimetallic photocatalyst and the effect of atomic ratio on its photocatalytic activity. *Chem. Eng. J.* **2022**, *446*, 136907. [CrossRef]
39. Rong, S.; Zhang, P.; Liu, F.; Yang, Y. Engineering Crystal Facet of α-MnO_2 Nanowire for Highly Efficient Catalytic Oxidation of Carcinogenic Airborne Formaldehyde. *ACS Catal.* **2018**, *8*, 3435–3446. [CrossRef]
40. Feng, N.; Zhu, Z.; Zhao, P.; Wang, L.; Wan, H.; Guan, G. Facile fabrication of trepang-like CeO_2@MnO_2 nanocomposite with high catalytic activity for soot removal. *Appl. Surf. Sci.* **2020**, *515*, 146013. [CrossRef]
41. Feng, Z.; Ren, Q.; Peng, R.; Mo, S.; Zhang, M.; Fu, M.; Chen, L.; Ye, D. Effect of CeO_2 morphologies on toluene catalytic combustion. *Catal. Today* **2019**, *332*, 177–182. [CrossRef]
42. Mohebali, H.; Moussavi, G.; Karimi, M.; Giannakis, S. Catalytic ozonation of Acetaminophen with a magnetic, Cerium-based Metal-Organic framework as a novel, easily-separable nanocomposite. *Chem. Eng. J.* **2022**, *434*, 134614. [CrossRef]
43. Li, S.; Huang, J.; Ye, Z.; Wang, Y.; Li, X.; Wang, J.; Li, L. The mechanism of Metal-H_2O_2 complex immobilized on MCM-48 and enhanced electron transfer for effective peroxone ozonation of sulfamethazine. *Appl. Catal. B Environ.* **2021**, *280*, 119453. [CrossRef]
44. Qi, F.; Chu, W.; Xu, B. Ozonation of phenacetin in associated with a magnetic catalyst $CuFe_2O_4$: The reaction and transformation. *Chem. Eng. J.* **2015**, *262*, 552–562. [CrossRef]
45. Wang, J.; Bai, Z. Fe-based catalysts for heterogeneous catalytic ozonation of emerging contaminants in water and wastewater. *Chem. Eng. J.* **2017**, *312*, 79–98. [CrossRef]
46. Chen, H.; Zhang, Z.; Hu, D.; Chen, C.; Zhang, Y.; He, S.; Wang, J. Catalytic ozonation of norfloxacin using Co_3O_4/C composite derived from ZIF-67 as catalyst. *Chemosphere* **2021**, *265*, 129047. [CrossRef]
47. Wang, X.; Sun, Y.; Yang, L.; Shang, Q.; Wang, D.; Guo, T.; Guo, Y. Novel photocatalytic system Fe-complex/TiO_2 for efficient degradation of phenol and norfloxacin in water. *Sci. Total Environ.* **2019**, *656*, 1010–1020. [CrossRef]

Article

New Magnetically Assembled Electrode Consisting of Magnetic Activated Carbon Particles and Ti/Sb-SnO_2 for a More Flexible and Cost-Effective Electrochemical Oxidation Wastewater Treatment

Fanxi Zhang [1], Dan Shao [1,*], Changan Yang [1], Hao Xu [2,*], Jin Yang [1], Lei Feng [1], Sizhe Wang [1], Yong Li [1], Xiaohua Jia [1] and Haojie Song [1,*]

1 School of Materials Science and Engineering, Shaanxi Key Laboratory of Green Preparation and Functionalization for Inorganic Materials, Shaanxi University of Science & Technology, Xi'an 710021, China
2 Department of Environmental Engineering, Xi'an Jiaotong University, Xi'an 710049, China
* Correspondence: shaodan@sust.edu.cn (D.S.); xuhao@mail.xjtu.edu.cn (H.X.); songhaojie@sust.edu.cn (H.S.)

Abstract: Magnetic activated carbon particles (Fe_3O_4/active carbon composites) as auxiliary electrodes (AEs) were fixed on the surface of Ti/Sb-SnO_2 foil by a NdFeB magnet to form a new magnetically assembled electrode (MAE). Characterizations including cyclic voltammetry, Tafel analysis, and electrochemical impedance spectroscopy were carried out. The electrochemical oxidation performances of the new MAE towards different simulated wastewaters (azo dye acid red G, phenol, and lignosulfonate) were also studied. Series of the electrochemical properties of MAE were found to be varied with the loading amounts of AEs. The electrochemical area as well as the number of active sites increased significantly with the AEs loading, and the charge transfer was also facilitated by these AEs. Target pollutants' removal of all simulated wastewaters were found to be enhanced when loading appropriate amounts of AEs. The accumulation of intermediate products was also determined by the AEs loading amount. This new MAE may provide a landscape of a more cost-effective and flexible electrochemical oxidation wastewater treatment (EOWT).

Keywords: anode; metal oxide; electrolysis; refractory pollutant; carbon

Citation: Zhang, F.; Shao, D.; Yang, C.; Xu, H.; Yang, J.; Feng, L.; Wang, S.; Li, Y.; Jia, X.; Song, H. New Magnetically Assembled Electrode Consisting of Magnetic Activated Carbon Particles and Ti/Sb-SnO_2 for a More Flexible and Cost-Effective Electrochemical Oxidation Wastewater Treatment. *Catalysts* **2023**, *13*, 7. https://doi.org/10.3390/catal13010007

Academic Editor: Enric Brillas

Received: 19 November 2022
Revised: 14 December 2022
Accepted: 18 December 2022
Published: 22 December 2022

1. Introduction

Industrial organic wastewaters are difficult to be simply treated by conventional biological, physical, and chemical methods due to their complex composition, recalcitrance, and high toxicity [1,2]. Electrochemical oxidation wastewater treatment (EOWT), a cutting-edge advanced oxidation technology, is an effective, adaptable, straightforward, and environmentally friendly approach to treat such refractory wastewaters using clean electrons [3,4].

The anode has a direct impact on the effectiveness and selectivity of EOWT, so choosing an appropriate anode material with as many active sites as possible is crucial in a specific case [5,6]. However, the traditional two-dimensional electrode has insufficient area to provide adequate amounts of active sites, and the advanced three-dimensional electrode has high investment cost and poor recyclability [7,8]. In addition, these anodes are non-variable, which could not provide sustainably high efficiency over a broad range of inlet wastewaters. In view of this, a variable electrode format is needed to avoid electrode updating or system shutting down as possible.

In recent years, we proposed a novel electrode architecture consisting of a two-dimensional titanium-based metal oxide electrode (main electrodes, ME), a number of magnetic catalyst particles with Fe_3O_4 cores and coatings (auxiliary electrodes, AEs), and a permanent magnet (see Scheme 1). The AEs are fixed on the ME surface by magnetic force provided by the permanent magnet. This adjustable and modular electrode format is named the magnetically assembled electrode (MAE) [9,10]. By varying the magnetic force, a flexible and in situ electrode surface modification could be realized. Furthermore, the physically combined electrode has excellent maintainability and recyclability. The synergistic effect of the ME and AEs improves the electrode activity, and the timely replacement of the AEs can achieve flexible tunability and a long lifetime.

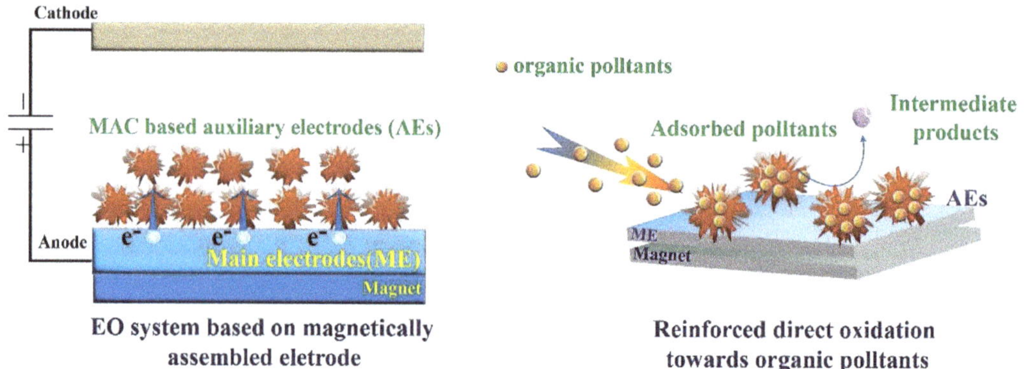

Scheme 1. Structure of the MAE used in this study and the role provided by the MAC in the MAE.

In our earlier research, nearly all kinds of AEs were coated by metal oxides (e.g., Sb-SnO_2 and Pb_3O_4) or conducting polymers (e.g., polyaniline), and the adopted ME coatings were typical dimensional stable anode (DSA) coatings (e.g., PbO_2, Sb-SnO_2, RuO_2-IrO_2-TiO_2, and IrO_2-Ta_2O_5) or others (Pt, graphite) [9–15]. We discovered that increasing the loading amount of AEs can elevate series of electrochemical properties of MAE. In brief, the AEs provide massive active sites, but which are mainly less accessible active sites. Good mass transfer and high current efficiency of pollutant degradation could benefit from these additional active sites [11–15]. Unfortunately, the conductivity of these AEs is insufficient, which may lead to the uneven distribution of anodic potential throughout the AEs layers and the insufficient polarization degree of the outer layers. Therefore, excessive AEs loading is not advised. Fortunately, we also found that ME with good oxygen evolution reaction (OER) activity has an activating effect on AEs and can mitigate the above problems to some extent [13,14].

In this study, we developed a kind of magnetic activated carbon particles (referred to as MAC, i.e., Fe_3O_4/active carbon composites) with stable physicochemical properties, good electrical conductivity, and well-developed pores [7,8]. These MAC particles were expected to further release the potential of MAE for faster electron transfer and more electrochemical area. Ti/Sb-SnO_2 with poor OER activity was used as the ME to reflect the contribution of the new AEs. The effects of the AEs loading amount on the structure–activity of the MAE were thoroughly assessed through material characterizations and electrochemical characterizations. The effects of the AEs loading amount on the EOWT efficiency and selectivity were disclosed by the degradation treatments of azo dye acid red G (ARG), phenol (C_6H_5OH), and sodium lignosulphonate (lignin). This study also aims to enrich the material system of MAE, complete the structure—activity relationship theory of MAE, and provide additional theoretical basis and technical assistance to the application of EOWT towards actual organic wastewaters.

2. Results and Discussion

2.1. Material Characterization

The SEM image (×500) of the ME is shown in Figure 1a. A thick layer of Sb-SnO_2 coating effectively covered most of the titanium substrate and inhibited possible substrate passivation during electrolysis. However, the thermal decomposition process inevitably causes the crack coating morphology. The XRD patterns of the ME and AEs are shown in Figure 1b. Five diffraction peaks of the ME at 2θ degrees of 26.6°, 33.9°, 37.9°, 51.8°, and 54.8° correspond to (110), (101), (200), (211), and (220) facets of tetragonal rutile phase SnO_2, respectively, according to the standard card (JCPDS 41-1445). A diffraction peak of Ti substrate is also identified due to the thin coating and high penetrability of X-ray. In the high depth of field micrograph on the macro level, the surface of ME is relatively flat, and the coating is relatively uniform (Figure 1c).

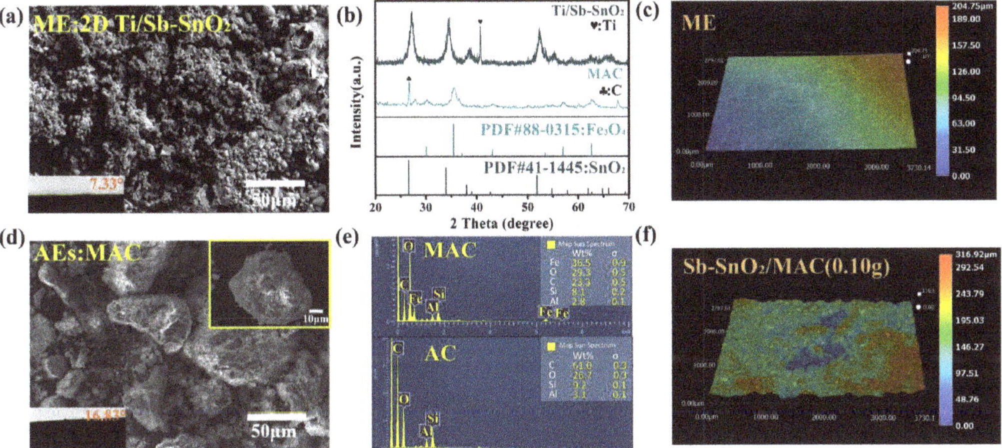

Figure 1. Material characterization results: (**a**) SEM image (×500) of ME (inset: water contact angle); (**b**) XRD of ME and AEs; (**c**) High depth of field micrograph of ME; (**d**) SEM image of (×500) of MAC (insets: SEM image of MAC (×1000) and water contact angle (very low angle, the numbers are just for reference)); (**e**) EDS element mappings of MAC and AC; (**f**) High depth of field micrograph of $Sb\text{-}SnO_2/MAC$(0.10 g).

The SEM image (×500, ×1000) of the MAC AEs are shown in Figure 1d, manifesting their rough surface. XRD result shows that these AEs contains Fe_3O_4 grains, demonstrating Fe_3O_4 is successfully deposited on the AC, making it magnetic (Figure 1b). EDS mappings of the AEs (Figure 1e) also verify that the MAC AEs are successfully prepared, which contain 36.5% of Fe, 29.3% of O, 23.3% of C, 8.1% of Si, and 2.8% of Al comparing with the original AC (Figure 1e, atom %).

The high depth of field micrograph of the $Sb\text{-}SnO_2/MAC$(0.10 g) is shown in Figure 1f. Loading AEs modifies the macroscopic surface roughness of the electrode and also changed the porosity and actual area of the electrode. It is also clear that the AEs are successfully fixed on the ME surface by the magnetic force. The hydrophilicity of the electrode remains well even when after loading the AEs (insets in Figure 1a,d).

2.2. Electrochemical Characterization

The narrow CV curves (0–0.3 V (vs. SCE)) for the ME (2D $Ti/Sb\text{-}SnO_2$), $Sb\text{-}SnO_2/MAC$ (0.01 g), and $Sb\text{-}SnO_2/MAC$(0.10 g) obtained at 0.05 $V{\cdot}s^{-1}$ of potential scan rate are displayed in Figure 2a. The full view of narrow CV curves of all electrodes under different potential scan rates is illustrated in the Supporting Information (Figure S1). It can be seen that both the current density and the curve area significantly increase with the AEs (Figure 2a). A linear fit of the current density values and integral areas obtained from the narrow CV curves (Figure S1) in the non-Faraday region under different potential scan rates were used to determine the double-layer capacitance (C_{dl}) and voltammetric charges (q*), respectively (details are described in the Supporting Information Section and illustrated in Scheme S1). The q* could be categorized as the total voltammetric charge (q_T), outer voltammetric charge (q_o), and inner voltammetric charge (q_i), which correspond to the total number of active sites, easily accessible external active sites, and less accessible internal active sites, respectively. Figure 2b demonstrates that increasing AEs' loading amount alters the electrode's structure, increasing the electrochemical area. For example Sb-SnO_2/MAC(0.10 g) increases the C_{dl} by ~50% compared with the 2D Ti/Sb-SnO_2 electrode. Figure 2c shows adding 0.10 g·cm^{-2} of MAC can increase the number of q_i by 78.5%. The significant increases of C_{dl} and q* could benefit to the mass transfer and adsorption of pollutants, and the AEs can provide more additional catalytic active sites for the electrocatalytic reaction, leading to faster kinetics [11–15]. Therefore, the electrode's direct electron transfer (DET) capability may be improved by the addition of MAC AEs. Figure S2 shows the obvious variations of q* value with the addition of different pollutants, manifesting the effect of adsorbed pollutant on the charge–discharge process of the electrode, especially for MAEs.

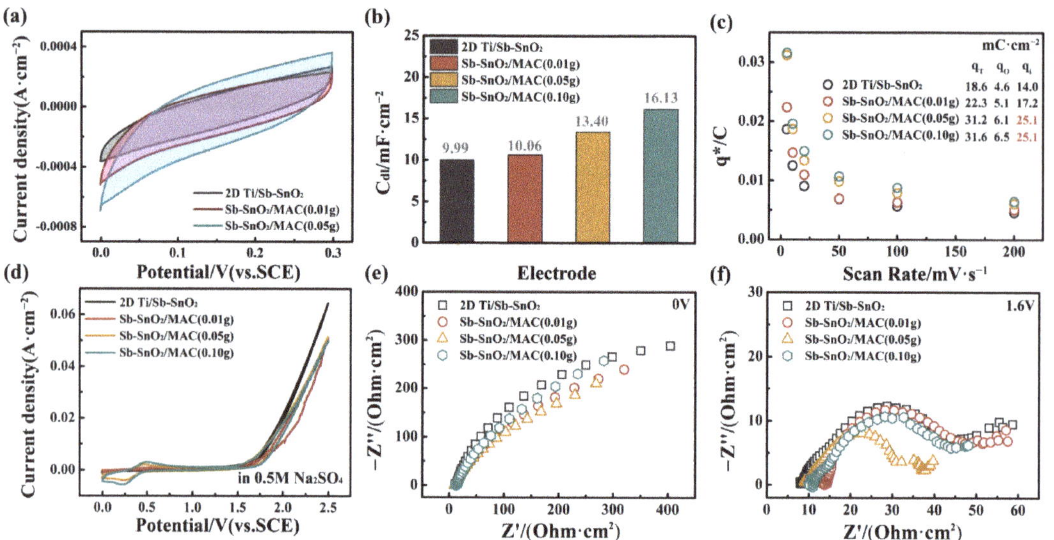

Figure 2. Electrochemical characterization results (in 0.5 M Na_2SO_4 solution): (**a**) Narrow CV curves of 2D Ti/Sb-SnO_2, Sb-SnO_2/MAC(0.01 g), and Sb-SnO_2/MAC(0.10 g) under potential scan rate of 0.005 V·s^{-1}; (**b**) C_{dl} values; (**c**) Voltammetric charges caculated from the narrow CV curves; (**d**) Normal CV curves (0–2.5 V vs. SCE, potential scan rate: 0.01 V·s^{-1}); (**e**) Nyquist plots (equilibrium potential of 0 V (vs. SCE)); (**f**) Nyquist plots (equilibrium potential of 1.6 V (vs. SCE)).

Normal CV curves (0–2.5 V (vs. SCE)) shown in Figure 2d and LSV curves shown in Figure S3 (Supporting Information) could further manifest the effect of MAC AEs on the electrode's electrochemical behaviors. Two-dimensional Ti/Sb-SnO_2 has the onset oxygen evolution potential (OEP) of ~1.80 V (vs. SCE). When AEs are present, the electrode's OEP rises slightly and the response current value in oxygen evolution reaction (OER) zone significantly reduces. Two new redox peaks appeared on MAEs' normal CV curves at ~0.5 V and 0.25 V (vs. SCE), respectively, which may be attributed to the redox of carbon and hydroxyl carbon. This phenomenon indicates that the physically bonded MAC AEs really take part in the electrochemical processes, introducing their own redox reactions before the OER, increasing the electrode's charging/discharging capacity, and inhibiting the OER activity of the electrode.

The enhanced electrode conductivity could also be reflected by the Nyquist plots of MAE in Figure 2e,f. The activated carbon particles could offer excellent conductivity and well-developed pore structure [7,8,16,17], which are beneficial to charge transfer and mass transfer. However, only loading appropriate amount of AEs (i.e., for Sb-SnO_2/MAC(0.05 g)) would achieve the best result.

When three kinds of pollutants are added, the variations of normal CV curves (0–2.5 V (vs. SCE)) shown in Figure 3 could furtherly verify the enhanced DET process as well as the direct oxidation of pollutants introduced by the MAC AEs. The current response in the OER region is significantly decreased, and no current pollutant oxidation peak is found for 2D Ti/Sb-SnO_2 (Figure 3a,c). As a contrast, the current response in the OER region is nearly unchanged, and several pollutant oxidation peaks (or obvious current enhancement) are found between 1.2 V to 1.5 V (vs. SCE) for Sb-SnO_2/MAC(0.10 g) (Figure 3b,d). This result could furtherly verify that prior to the occurrence of OER (low applied potential), the three organic pollutants can undergo direct oxidation or DET process thanks to the introduction of MAC AEs. In addition, the electron transfer or mass transfer rate is faster on MAE than on 2D Ti/Sb-SnO_2 (see results in Figure 2), thus the pollutant oxidation on MAE could compensate the OER current decrease when the applied potential is high.

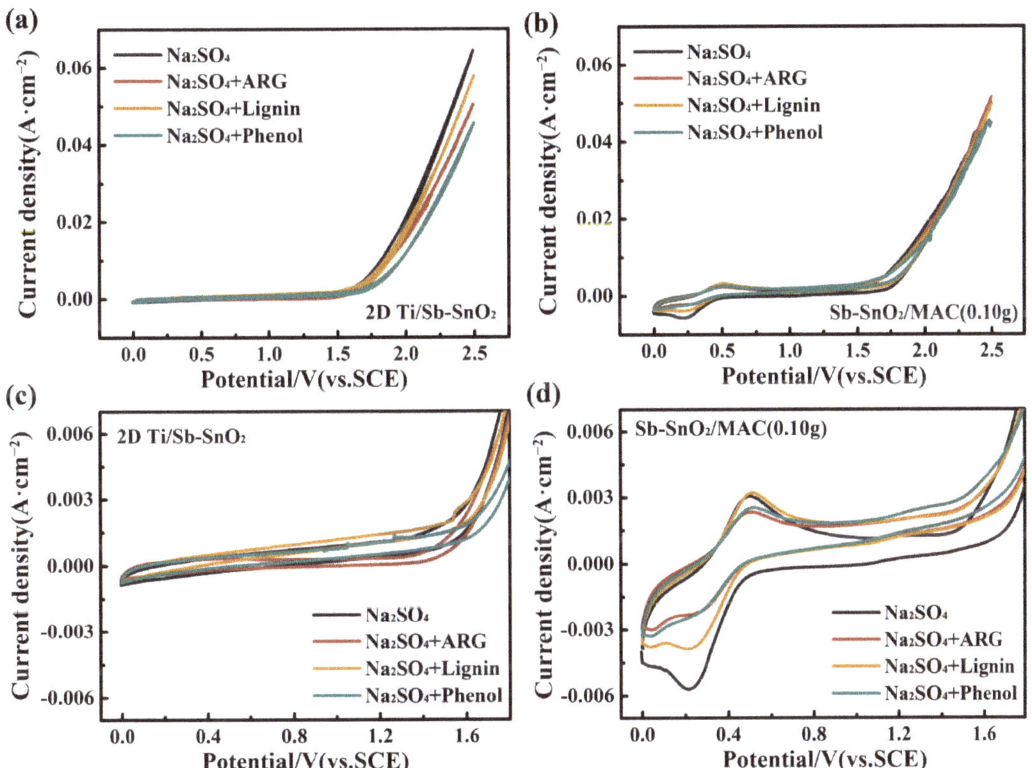

Figure 3. Normal CV curves of the electrodes (0–2.5 V (vs. SCE); scan rate: 0.01 $V{\cdot}s^{-1}$) in 0.5 M Na_2SO_4 solution and 0.5 M Na_2SO_4 solution with 2000 ppm of ARG, lignin or phenol: (**a**) 2D Ti/Sb-SnO_2; (**b**) Sb-SnO_2/MAC(0.10 g); (**c**) Partial enlarged view of 2D Ti/Sb-SnO_2; (**d**) Partial enlarged view of Sb-SnO_2/MAC(0.10 g).

2.3. Pollutant Degradation

2.3.1. Degradation of ARG

The destroying of the azo linkage of ARG is what we focus on due to its toxicity and character of chromophore. Direct oxidation or the DET process is as efficient as the hydroxyl radicals (indirect oxidation) in breaking the azo linkage [18,19]. Therefore, the current densities of both 2 mA·cm^{-2} and 20 mA·cm^{-2} were selected as comparison. It is known from the previous normal CV or LSV curves that only 2 mA·cm^{-2} of current density benefits to direct oxidation (or DET process) while 20 mA·cm^{-2} has reinforced indirect oxidation effect [5,16]. Another reason for choosing these two current densities is that the mass transfer requirements are different under these two conditions.

From Figure 4 it can be found that the 15 min of pre-adsorption process has little effect on ARG removal, indicating MAC has insufficient absorption ability comparing with the traditional AC we anticipated. As seen from Figure 4a, and at low current density (2 mA·cm^{-2}), the EOWT basically follows the zero-order reaction kinetics (kinetics control). Sb-SnO_2/MAC(0.01 g) and Sb-SnO_2/MAC(0.05 g) do not show their superiority under this condition due to low anodic potential. However, Sb-SnO_2/MAC(0.10 g) show 100% higher efficiency than the 2D Ti/Sb-SnO_2 electrode. At high current density (e.g., 20 mA·cm^{-2}, Figure 4b) corresponding to high anodic potential, the kinetics control turns to mass transfer control. All electrodes show improved ARG degradation efficiencies, especially for the three MAEs. For example, Sb-SnO_2/MAC(0.10g) could achieve ARG removal efficiency from ~50% (2 mA·cm^{-2}) to above 90% (20 mA·cm^{-2}) after 3 h treatment, while 2D Ti/Sb-SnO_2 only has minor enhancement.

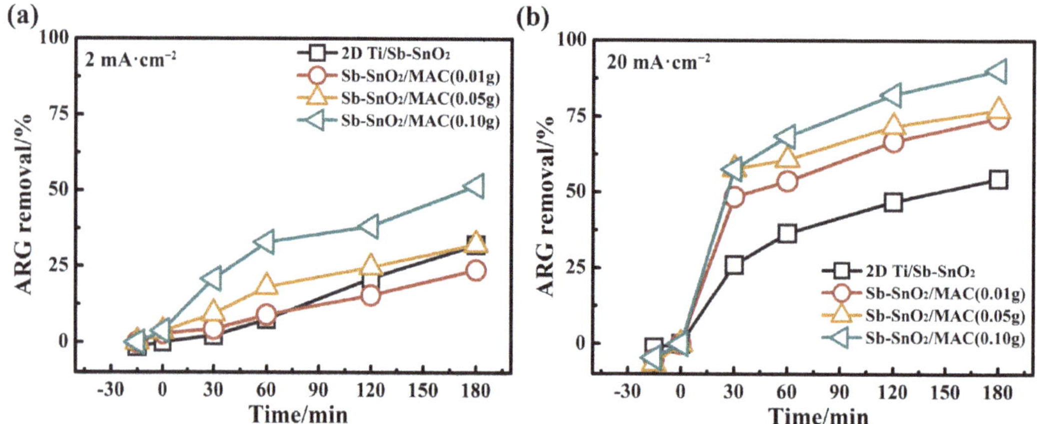

Figure 4. Degradation results of ARG (initial ARG concentration of 200 ppm, anode area of 9 cm^2, solution volume of 250 mL, 125 ppm of Na_2SO_4 as supporting electrolyte, room temperature): (**a**) ARG removal percentage versus time under low current density (2 mA·cm^{-2}); (**b**) ARG removal percentage versus time under high current density (20 mA·cm^{-2}).

The above results reflect the MAC's functions in improving electrode's electrochemical properties (Figures 2 and 3). The above results also demonstrate the superiorities of the MAEs in terms of reinforced direction oxidation (or DET) and improved mass transfer. More importantly, more electrochemical area or active sites brought by MAC AEs also lower the real current density significantly (under galvanostatic mode), which would further inhibit OER side reaction and benefit the DET process. The variations of UV-vis spectra, total dissolved solids (TDS), and pH value during ARG treatment are also placed in the Supporting Information for reference (Figure S5).

2.3.2. Degradation of Phenol

The structure of phenol requires a higher degradation ability of the electrode (ring opening). Phenol has a strong absorption peak at 270 nm in the UV region, which can reflect the phenol content (Figure S6, Supporting Information). From Figure S6 it can be seen that no matter whether under lower or higher current density, the 2D Ti/Sb-SnO_2 electrode could not effectively handle this pollutant. The addition of the MAC AEs could make the solution's color deepen versus electrolysis time (from colorless to tea-brown and finally becoming turbid). A higher AEs loading amount would cause partial UV-vis adsorption peak shift phenomena under higher current density. These optical or spectral phenomena may reveal the effective oxidation of phenol to polymers or other intermediate products by MAEs.

Phenol adsorption is more negligible on these MAEs comparing with ARG (not shown here). After 3 h degradation of 100 ppm phenol, Figure 5 displays the mass-to-charge ratio (m/z) versus GC retention time of the organics present in the samples using bubble plots. In addition, the content of phenol and its intermediate degradation products can be visually reflected by the bubble area (corresponding to the GC peak area; two identified substances (m/z: 218, m/z: 206, which might be bimolecular polymers) and were the main intermediate products. The degradation effects of anodes varied greatly. In summary, loading more MAC AEs leads to more effective phenol degradation and less intermediate products accumulation, which is consistent with the electrochemical characterization results. The most effective electrode for removing phenol is Sb-SnO_2/MAC(0.10 g), with phenol removal efficiency of almost 100% and with the fewest accumulation of intermediate products. Sb-SnO_2/MAC(0.05 g) has phenol removal efficiency of ~80%. However, the performance of Sb-SnO_2/MAC(0.01 g) and 2D Ti/Sb-SnO_2 are similarly poor (only ~50% phenol removal). The above result indicates that the reinforced direct oxidation (or DET) caused by sufficient MAC AEs loading amount is necessary for phenol removal and intermediate products (e.g., polymers) elimination.

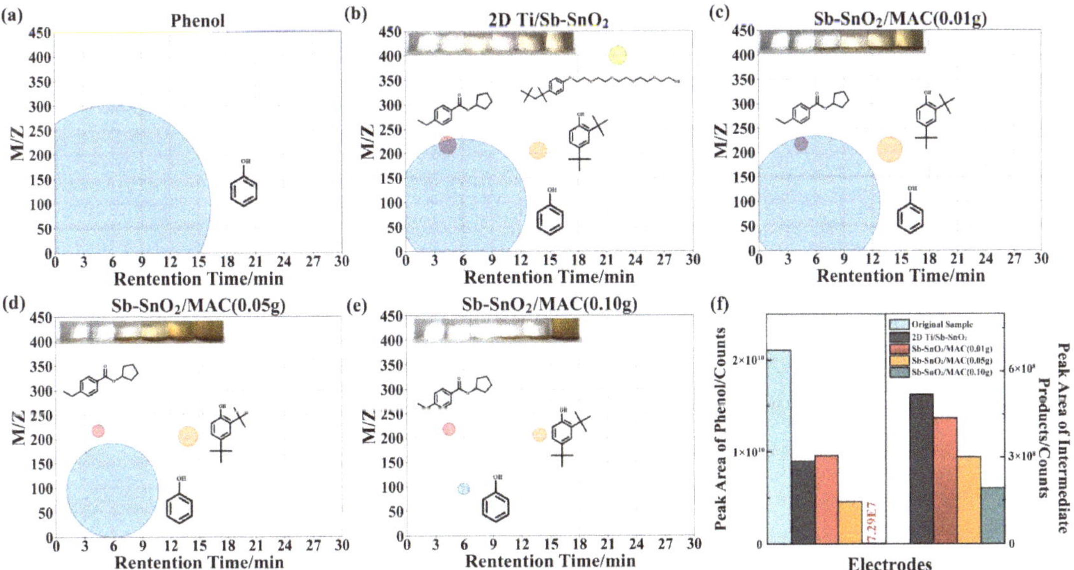

Figure 5. Phenol degradation results under current density of 20 mA·cm^{-2} (initial phenol concentration of 100 ppm, anode area of 9 cm^2, solution volume of 250 mL, supporting electrolyte of 125 ppm of Na_2SO_4, room temperature) before and after 3 h electrolysis: (**a**–**e**) Bubble diagrams of phenol or intermediate products according to their GC peak area ((**a**) Original sample of phenol; (**b**) 2D Ti/Sb-SnO_2 (inset: photos of the solution samples); (**c**) Sb-SnO_2/MAC(0.01 g) (inset: photos of the solution samples); (**d**) Sb-SnO_2/MAC(0.05 g) (inset: photos of the solution samples); (**e**) Sb-SnO_2/MAC(0.10 g) (inset: photos of the solution samples)); (**f**) Accumulative GC peak area columns of phenol and intermediate products.

2.3.3. Degradation of Lignin

Lignin is a series of three-dimensional macromolecules containing various functional groups such as methoxy, carbonyl, aldehyde, etc. It is a more complex organic compound than ARG and phenol, and this organic pollutant is most common in paper wastewater [20,21]. The advisable and cost-effective way to treat lignin wastewater is just to enhance its biodegradability. That is, breaking the linkage bonds (C-C and C-O-C bonds) between the lignin structural units while also avoiding over-oxidation so as to obtain useful small molecular resources such as alcohols, acids, and phenolic derivatives. We found MAE is still applicable to this complex and refractory organic wastewater (Figures 6 and S7). For 2D Ti/Sb-SnO_2, under two different current densities, the UV-vis spectra versus time were nearly unchanged, and the changes in pH and TDS were relatively inconspicuous, suggesting 2D Ti/Sb-SnO_2 could not effectively oxidize lignin in 3 h treatment (Figure S7). In addition, from the comparison between Figure 6a,b, it can be deduced that fewer intermediate products generate on this electrode. When a small amount of MAC AEs was added (Sb-SnO_2/MAC(0.01 g)), the pH and TDS variations were nearly identical to those of the 2D Ti/Sb-SnO_2 electrode. However, the UV-vis spectra change significantly versus time, where the absorbance increases at 290 nm. In addition, this phenomenon is more obvious when loading more MAC AEs (Sb-SnO_2/MAC(0.05 g) and Sb-SnO_2/MAC(0.10 g)), manifesting that MAC AEs could facilitate lignin depolymerization.

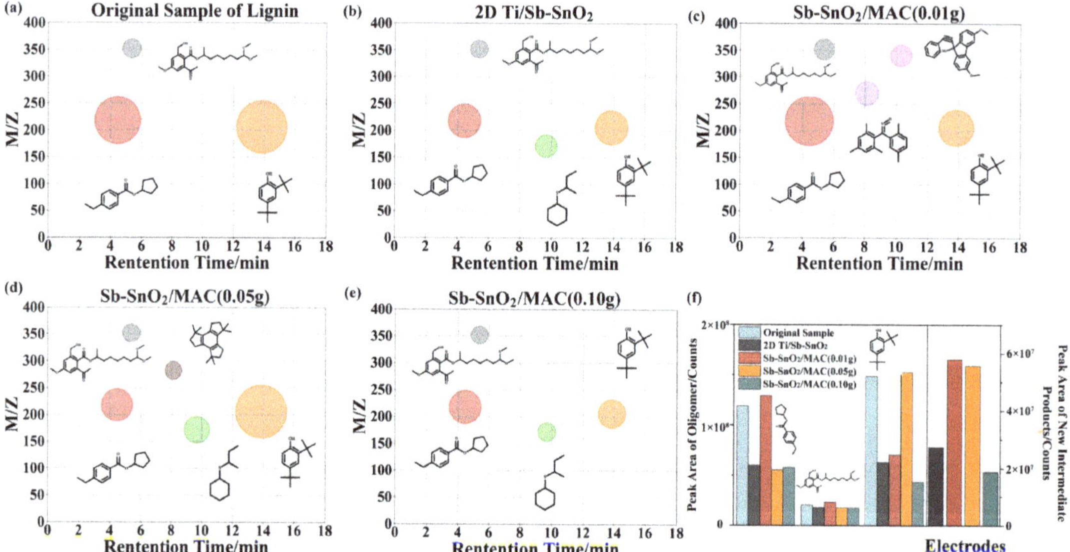

Figure 6. Lignin degradation results under current density of 20 mA·cm^{-2} (initial lignin concentration of 100 ppm, anode area of 9 cm^2, solution volume of 250 mL, supporting electrolyte of 125 ppm of Na_2SO_4, room temperature) before and after 3 h electrolysis (**a–e**) Bubble diagram of original lignin sample or intermediate products according to their GC peak area ((**a**) Original sample of lignin; (**b**) 2D Ti/Sb-SnO_2; (**c**) Sb-SnO_2/MAC(0.01 g); (**d**) Sb-SnO_2/MAC(0.05 g); (**e**) Sb-SnO_2/MAC(0.10 g); (**f**) Accumulative GC peak area columns of oligomers in original sample and degradation samples (left part) and new generated intermediate products after degradation (right part).

The efficiency and selectivity of the electrodes towards lignin and its intermediate products could be furtherly described by the GC-MS result of original lignin solution and treated samples. Lignin macromolecules are over the limit of GC-MS instrument, which could not be detected in the original lignin solution. However, three different oligomers (m/z: 206, m/z: 218, m/z: 352, respectively) are detected successfully (Figure 6a).

The variations of these three compounds are different depending on the electrode. In addition, a different type and amount of new intermediate products are generated by different electrodes (Figure 6b–e). The structures of these intermediate products derived from the reports given by the NIST database software could be used as reference. For example, although 2D Ti/Sb-SnO_2 electrode could not depolymerize lignin effectively (Figure S7), it is good at reducing the amount of the above three oligomers by ~50% in average. Sb-SnO_2/MAC(0.01 g) and Sb-SnO_2/MAC(0.05 g) accumulate some intermediate products (including new compounds as well as the original oligomers) originating from the depolymerization of lignin. However, Sb-SnO_2/MAC(0.10 g) not only depolymerize lignin more effectively (Figure S7) and reduce the original oligomers substantially, but also accumulates fewer new intermediate products. The superiority of Sb-SnO_2/MAC(0.10 g) is more obvious in degrading lignin thanks to the difficulty of treating polymer. As we can imagine, the previously demonstrated better electrochemical properties of this electrode (more active sites, larger surface area, faster charge transfer and mass transfer) may be more valuable in treating more complex and refractory wastewaters.

2.4. Cost-Effectiveness Estimation

The price of powdered AC used this study is only 2.3 ¥·kg^{-1}. The costs of other main agents (e.g., $FeSO_4 \cdot 7H_2O$ or $FeCl_3 \cdot 6H_2O$) used to fabricate MAC could also be negligible. The co-precipitation method to prepare MAC is simple and does not need high-temperature calcination procedures or device. Taking Sb-SnO_2/MAC(0.10 g) with geometric area of 1 m^2 for example, no more than 3 ¥ (loading 1 kg of MAC) of MAC and several magnets (e.g., ~200 ¥) are needed to enlarge electrode's electrochemical area by ~50% (according to Figure 2). The investment of 2D Ti/Sb-SnO_2 in this study is ~1500 ¥·m^{-2} (much lower than noble metal containing electrodes). However, the cost of the new strategy using MAC in this study is only ~27% of the potential additional investment of ~0.5 m^2 of 2D Ti/Sb-SnO_2 (~750 ¥). In addition, the operational cell voltage of galvanostatic electrolysis (or energy consumption) using Sb-SnO_2/MAC is ~10% lower than using 2D Ti/Sb-SnO_2, thanks to the well-conductive carbon-based MAC. Based on the above, loading MAC is cost-effective and advisable.

3. Experiments

3.1. Preparation of MAE

All reagents were analytical pure (Shanghai McLean Biochemical Co., Ltd., Shanghai, China). The ME substrate was a titanium plate with a thickness of 0.5 mm (Baosteel Group, Baoji, China). All solutions were prepared with ultra-pure water (Milli-Q water, Millipore, Milford, MA, USA). The MAC AEs were made using a chemical co-deposition method (details are placed in the Supporting Information Section), where the Fe_3O_4 grains were deposited on or into the active carbon (AC). The preparation of Ti/Sb-SnO_2 ME followed our previous report [22], using electrodeposition and a thermal oxidation method. The AEs were magnetically attracted and fixed on the ME surface by magnetic force. A permanent magnet (NdFeB) was attached to the back side of the ME with tape. The effects of AEs loading amount (0 g·cm^{-2}, 0.01 g·cm^{-2}, 0.05 g·cm^{-2}, and 0.1 g·cm^{-2}, respectively) on MAE performance were studied. Therefore, the assembled four electrodes were named 2D Ti/Sb-SnO_2, Sb-SnO_2/MAC(0.01 g), Sb-SnO_2/MAC(0.05 g), and Sb-SnO_2/MAC(0.10 g), respectively.

3.2. Material Characterizations

Scanning electron microscopy (SEM, JSM-6390A, JEOL, Tokyo, Japan) and X-ray diffraction (XRD, D/MAX-2200PC, Rigaku, Tokyo, Japan; Cu Kα, λ = 0.15406 nm) and a 3D digital microscope (VHX-7000, Keyence, Osaka Japan) were used to characterize the morphology, composition, and structure of ME, AEs, or MAE.

3.3. Electrochemical Characterizations

A typical three-electrode system was used to conduct the electrochemical characterization. The working electrode was the prepared MAE (exposed geometric area of 1 cm^2). The counter electrode was a 9 cm^2 copper plate. A saturated calomel electrode (SCE) served as the reference electrode. Narrow cyclic voltammetry (CV) ranging between 0–0.3 V (vs. SCE) was carried out in a 0.5 M Na_2SO_4 solution (with or without 2000 ppm of various organic pollutants). The scan rates used for narrow cyclic voltammetry (CV) were 0.005 $V \cdot s^{-1}$, 0.01 $V \cdot s^{-1}$, 0.02 $V \cdot s^{-1}$, 0.05 $V \cdot s^{-1}$, 0.1 $V \cdot s^{-1}$, and 0.2 $V \cdot s^{-1}$, respectively. The roughness factors of various electrodes can be calculated through fitting according to previous reports [13–15]. Linear scanning voltammetry (LSV) and the normal CV were performed between 0–2.5 V (vs. SCE) in the same solutions with a scan rate of 1 $mV \cdot s^{-1}$ and 10 $mV \cdot s^{-1}$, respectively. The electrochemical impedance spectroscopy (EIS) tests were conducted (vs. SCE) in the same solutions at equilibrium potentials of 0 V and 1.6 V (vs. SCE) with an amplitude of 5 mV, and a frequency range between 10^5 Hz to 0.1 Hz.

3.4. Pollutant Degradation

As model pollutants, sodium lignosulfonate (lignin, $C_{20}H_{24}Na_2O_{10}S_2$), phenol ($C_6H_5OH$), and the azo dye Red G (ARG, $C_{18}H_{13}N_3Na_2O_8S_2$) were used (structural formulas are shown in Scheme S2 of the Supporting Information). The wastewater (250 mL, 125 ppm Na_2SO_4 electrolyte) contained 200 ppm (for ARG) or 100 ppm (for phenol or lignin) of pollutants. Given that MAC would adsorb pollutants, each set of degradation experiments contained a 15-min resting period (adsorption process) prior to electrolysis to allow the pollutants to be adsorbed as possible. The studied anodic current densities were 2 $mA \cdot cm^{-2}$ and 20 $mA \cdot cm^{-2}$, respectively. The anode (effective electrode area of 9 cm^2) and cathode (copper sheet with the same size) were positioned parallel with a spacing of 15 mm. Room temperature served as the initial reaction temperature. In order to analyze the solution samples, a UV-Vis spectrophotometer (Agilent 8453, Santa Clara, CA, USA) and a multifunctional pH meter (LeiCi Group, Shanghai, China, including function of the total dissolved solids (TDS) determination) were used. The degradation intermediate products were identified by gas chromatography–mass spectrometry (GC–MS, Thermo Fisher, Waltham, MA, USA) using the NIST database.

4. Conclusions

The above results have demonstrated AC as a good alternative of metal oxide as a cost-effective AEs coating material of MAE.

1. MAC AEs improve the MAE's conductivity, increase electrochemical area, increase the number of active sites, facilitate charge transfer and mass transfer, and strengthen the electrode's direct oxidation (or DET) capability. These characteristics are more significant on the Sb-SnO_2/MAC(0.10 g).
2. More importantly, the loading amount of MAC AEs has a significant impact on various pollutant degradation rates and intermediate products accumulations. Sb-SnO_2/MAC(0.10 g) has the best pollutant degradation ability with the lowest amount of accumulated intermediate products.

This study has enriched the material system of MAE, completed the structure–activity relationship theory of MAE, and provided an additional theoretical basis and technical assistance to the application of EOWT towards actual organic wastewaters.

Supplementary Materials: The following supporting information can be downloaded at: https://www.mdpi.com/article/10.3390/catal13010007/s1, Text: Electrode preparation detail and voltammetric charge calculation; Scheme S1: Illustations of the C_{dl} calculation procedures; Scheme S2: Structrual formulas of the three pollutants used in this study; Figure S1: Narrow CV curves at different scan rates in 0.5 M Na_2SO_4 solution; Figure S2: Voltammetric charges obtained from the narrow CV curves (from 0 to 0.3 V (vs. SCE)) at different scan rates in 0.5 M Na_2SO_4 solution and 0.5 M Na_2SO_4 solution with 2000 ppm of ARG, lignin or phenol; Figure S3: LSV curves of the electrodes

(0 to 2.5 V (vs. SCE); scan rate: 0.001 $V \cdot s^{-1}$) in 0.5 M Na_2SO_4 solution and 0.5 M Na_2SO_4 solution with 2000 ppm of ARG, lignin or phenol; Figure S4: Normal CV curves of the electrodes (0–2.5 V (vs. SCE); scan rate: 0.01 $V \cdot s^{-1}$) in 0.5 M Na_2SO_4 solution and 0.5 M Na_2SO_4 solution with 2000 ppm of ARG, lignin or phenol; Figure S5: Other details of ARG degradation; Figure S6: Other details of phenol degradation; Figure S7: Other details of lignin degradation; References [9,23].

Author Contributions: Conceptualization, D.S.; methodology, D.S.; validation, F.Z. and D.S.; formal analysis, F.Z. and D.S.; investigation, F.Z.; resources, D.S., C.Y., H.X., J.Y., L.F., S.W., Y.L., X.J. and H.S.; writing—original draft preparation, F.Z.; writing—review and editing, D.S. and C.Y.; visualization, F.Z.; supervision, D.S., H.X. and H.S.; project administration, D.S.; funding acquisition, D.S. All authors have read and agreed to the published version of the manuscript.

Funding: This research was funded by the National Natural Science Foundation of China (21706153) and Natural Science Basic Research Program of Shaanxi Province (2018JQ2066, 2022JM-065).

Data Availability Statement: Not applicable.

Acknowledgments: The authors acknowledge the financial support from the National Natural Science Foundation of China and Natural Science Basic Research Program of Shaanxi Province.

Conflicts of Interest: The authors declare no conflict of interest.

References

1. Martínez-Huitle, C.A.; Rodrigo, M.A.; Sirés, I.; Scialdone, O. Single and Coupled Electrochemical Processes and Reactors for the Abatement of Organic Water Pollutants: A Critical Review. *Chem. Rev.* **2015**, *115*, 13362–13407. [CrossRef]
2. Radjenovic, J.; Duinslaeger, N.; Avval, S.S.; Chaplin, B.P. Facing the Challenge of Poly- and Perfluoroalkyl Substances in Water: Is Electrochemical Oxidation the Answer? *Environ. Sci. Technol.* **2020**, *54*, 14815–14829. [CrossRef]
3. Martínez-Huitle, C.A.; Panizza, M. Electrochemical oxidation of organic pollutants for wastewater treatment. *Curr. Opin. Electrochem.* **2018**, *11*, 62–71. [CrossRef]
4. Zheng, W.; Liu, Y.; Liu, W.; Ji, H.; Li, F.; Shen, C.; Fang, X.; Li, X.; Duan, X. A novel electrocatalytic filtration system with carbon nanotube supported nanoscale zerovalent copper toward ultrafast oxidation of organic pollutants. *Water Res.* **2021**, *194*, 116961. [CrossRef]
5. Panizza, M.; Cerisola, G. Direct and Mediated Anodic Oxidation of Organic Pollutants. *Chem. Rev.* **2009**, *109*, 6541–6569. [CrossRef]
6. Trellu, C.; Coetsier, C.; Rouch, J.-C.; Esmilaire, R.; Rivallin, M.; Cretin, M.; Causserand, C. Mineralization of organic pollutants by anodic oxidation using reactive electrochemical membrane synthesized from carbothermal reduction of TiO_2. *Water Res.* **2018**, *131*, 310–319. [CrossRef]
7. Ren, W.; Zhang, Q.; Cheng, C.; Miao, F.; Zhang, H.; Luo, X.; Wang, S.; Duan, X. Electro-Induced Carbon Nanotube Discrete Electrodes for Sustainable Persulfate Activation. *Environ. Sci. Technol.* **2022**, *56*, 14019–14029. [CrossRef]
8. Li, X.; Wu, Y.; Zhu, W.; Xue, F.; Qian, Y.; Wang, C. Enhanced electrochemical oxidation of synthetic dyeing wastewater using SnO_2-Sb-doped TiO_2-coated granular activated carbon electrodes with high hydroxyl radical yields. *Electrochim. Acta* **2016**, *220*, 276–284. [CrossRef]
9. Shao, D.; Yan, W.; Li, X.; Xu, H. Fe_3O_4/Sb–SnO_2 Granules Loaded on Ti/Sb–SnO_2 Electrode Shell by Magnetic Force: Good Recyclability and High Electro-oxidation Performance. *ACS Sustain. Chem. Eng.* **2015**, *3*, 1777–1785. [CrossRef]
10. Shao, D.; Zhang, X.; Lyu, W.; Zhang, Y.; Tan, G.; Xu, H.; Yan, W. Magnetic Assembled Anode Combining PbO_2 and Sb-SnO_2 Organically as An Effective and Sustainable Electrocatalyst for Wastewater Treatment with Adjustable Attribution and Construction. *ACS Appl. Mater. Interfaces* **2018**, *10*, 44385–44395. [CrossRef]
11. Zhang, X.; Shao, D.; Lyu, W.; Xu, H.; Yang, L.; Zhang, Y.; Wang, Z.; Liu, P.; Yan, W.; Tan, G. Design of magnetically assembled electrode (MAE) with Ti/PbO_2 and heterogeneous auxiliary electrodes (AEs): The functionality of AEs for efficient electrochemical oxidation. *Chem. Eng. J.* **2020**, *395*, 125145. [CrossRef]
12. Shao, D.; Zhang, X.; Wang, Z.; Zhang, Y.; Tan, G.; Yan, W. New architecture of a variable anode for full-time efficient electrochemical oxidation of organic wastewater with variable Cl^- concentration. *Appl. Surf. Sci.* **2020**, *515*, 146003. [CrossRef]
13. Shao, D.; Zhang, Y.; Lyu, W.; Zhang, X.; Tan, G.; Xu, H.; Yan, W. A modular functionalized anode for efficient electrochemical oxidation of wastewater: Inseparable synergy between OER anode and its magnetic auxiliary electrodes. *J. Hazard. Mater.* **2020**, *390*, 122174. [CrossRef]
14. Zhang, Y.; Zhang, C.; Shao, D.; Xu, H.; Rao, Y.; Tan, G.; Yan, W. Magnetically assembled electrodes based on Pt, RuO_2-IrO_2-TiO_2 and Sb-SnO_2 for electrochemical oxidation of wastewater featured by fluctuant Cl- concentration. *J. Hazard. Mater.* **2022**, *421*, 126803. [CrossRef]
15. Shao, D.; Li, W.; Wang, Z.; Yang, C.; Xu, H.; Yan, W.; Yang, L.; Wang, G.; Yang, J.; Feng, L.; et al. Variable activity and selectivity for electrochemical oxidation wastewater treatment using a magnetically assembled electrode based on Ti/PbO_2 and carbon nanotubes. *Sep. Purif. Technol.* **2022**, *301*, 122008. [CrossRef]

16. Chaplin, B.P. Critical review of electrochemical advanced oxidation processes for water treatment applications. *Environ. Sci. Process. Impacts* **2014**, *16*, 1182–1203. [CrossRef]
17. Zhan, J.; Li, Z.; Yu, G.; Pan, X.; Wang, J.; Zhu, W.; Han, X.; Wang, Y. Enhanced treatment of pharmaceutical wastewater by combining three-dimensional electrochemical process with ozonation to in situ regenerate granular activated carbon particle electrodes. *Sep. Purif. Technol.* **2019**, *208*, 12–18. [CrossRef]
18. Giannakis, S.; Lin, K.-Y.A.; Ghanbari, F. A review of the recent advances on the treatment of industrial wastewaters by Sulfate Radical-based Advanced Oxidation Processes (SR-AOPs). *Chem. Eng. J.* **2021**, *406*, 127083. [CrossRef]
19. Du, X.; Oturan, M.A.; Zhou, M.; Belkessa, N.; Su, P.; Cai, J.; Trellu, C.; Mousset, E. Nanostructured electrodes for electrocatalytic advanced oxidation processes: From materials preparation to mechanisms understanding and wastewater treatment applications. *Appl. Catal. B Environ.* **2021**, *296*, 120332. [CrossRef]
20. Behling, R.; Valange, S.; Chatel, G. Heterogeneous catalytic oxidation for lignin valorization into valuable chemicals: What results? What limitations? What trends? *Green Chem.* **2016**, *18*, 1839–1854. [CrossRef]
21. Rafiee, M.; Alherech, M.; Karlen, S.D.; Stahl, S.S. Electrochemical Aminoxyl-Mediated Oxidation of Primary Alcohols in Lignin to Carboxylic Acids: Polymer Modification and Depolymerization. *J. Am. Chem. Soc.* **2019**, *141*, 15266–15276. [CrossRef] [PubMed]
22. Shao, D.; Yan, W.; Li, X.L.; Yang, H.H.; Xu, H. A Highly Stable Ti/TiHx/Sb-SnO_2 Anode: Preparation, Characterization and Application. *Ind. Eng. Chem. Res.* **2014**, *53*, 3898–3907. [CrossRef]
23. Montilla, F.; Morallón, E.; De Battisti, A.; Vázquez, J.L. Preparation and characterization of antimony-doped tin dioxide electrodes. Part 1. Electrochemical characterization. *J. Phys. Chem. B* **2004**, *108*, 5036–5043. [CrossRef]

catalysts

MDPI

Article

Removal Efficiency and Performance Optimization of Organic Pollutants in Wastewater Using New Biochar Composites

Guodong Wang *, Shirong Zong, Hang Ma, Banglong Wan and Qiang Tian

Yunnan Yuntianhua Co., Ltd., Kunming 650228, China
* Correspondence: wanggd0221@163.com

Abstract: The purpose is to optimize the catalytic performance of biochar (BC), improve the removal effect of BC composites on organic pollutants in wastewater, and promote the recycling and sustainable utilization of water resources. Firstly, the various characteristics and preparation principles of new BC are discussed. Secondly, the types of organic pollutants in wastewater and their removal principles are discussed. Finally, based on the principle of removing organic pollutants, BC/zero valent iron (BC/ZVI) composite is designed, among which BC is mainly used for catalysis. The effect of BC/ZVI in removing tetracycline (TC) is comprehensively evaluated. The research results reveal that the TC removal effect of pure BC is not ideal, and that of ZVI is general. The BC/ZVI composite prepared by combining the two has a better removal effect on TC, with a removal amount of about 275 mg/g. Different TC concentrations, ethylene diamine tetraacetic acid (EDTA), pH environment, tert-butanol, and calcium ions will affect the TC removal effect of BC composites. The overall effect is the improvement of the TC removal amount of BC composites. It reveals that BC has a very suitable catalytic effect on ZVI, and the performance of BC composite material integrating BC catalyst and ZVI has been effectively improved, which can play a very suitable role in wastewater treatment. This exploration provides a technical reference for the effective removal of organic pollutants in wastewater and contributes to the development of water resource recycling.

Keywords: new biochar; compound material; organic pollutants; tetracycline; biochar/zero valent iron

Citation: Wang, G.; Zong, S.; Ma, H.; Wan, B.; Tian, Q. Removal Efficiency and Performance Optimization of Organic Pollutants in Wastewater Using New Biochar Composites. *Catalysts* **2023**, *13*, 184. https://doi.org/10.3390/catal13010184

Academic Editors: Yanbiao Liu and Hao Xu

Received: 11 November 2022
Revised: 13 December 2022
Accepted: 19 December 2022
Published: 13 January 2023

1. Introduction

With the progress of society, the use of various new products has caused a serious impact on the environment. In particular, trace persistent organic pollutants in water bodies pose a serious threat to the normal life activities of human beings and organisms, and the effective removal of these pollutants has become a top priority [1]. General water treatment technology is difficult to work with. As a pyrolysis product of biomass waste, biochar (BC) is gradually applied to treat polluted water bodies. It can effectively improve the removal of organic pollutants in wastewater and improve the comprehensive utilization efficiency of water resources [2]. Although the current treatment technology of organic pollutants in wastewater is not advanced enough, many studies have provided technical support for it.

Pan and Tang (2021) [3] pointed out that water environment pollution and water resource shortage are two major problems of global freshwater resources. Water consumption has increased sharply due to the rapid progress of the social economy, the increase in population, the gradual improvement of people's living standards, and the acceleration of industrialization and urbanization. Sewage discharge has also increased accordingly, aggravating the shortage of freshwater resources and the pollution of the water environment. The current backward sewage treatment facilities and low sewage treatment rate are the main reasons for water environment pollution [3]. Lu et al. (2021) [4] pointed out that water consumption has increased sharply with the development of industrial and agricultural production and improved people's living standards. Moreover, the vast majority of huge amounts of untreated industrial wastewater and urban sewage are discharged into water bodies, causing serious pollution to limited water resources [4]. Xiao et al. (2021) [5], through an in-depth investigation of contemporary urban planning practices and combining with the latest research results of environmental theories, made an empirical analysis of the corresponding problems and proposed that environmental protection countermeasures should be adopted in urban planning was a very important measure [5]. Grosso et al. (2022) [6] pointed out that BC can remain in the soil for hundreds to thousands of years to achieve carbon sequestration and fixation. BC can also improve soil physical and chemical properties and microbial activity, cultivate soil fertility, reduce the loss of fertilizer and soil nutrients, delay the release of fertilizer nutrients, and reduce soil pollution [6]. Hota and Diaz (2021) [7] prepared magnetic hydroxyapatite/BC composites and studied the adsorption kinetics and thermodynamic properties of Pb2+ and solid-liquid separation and recovery properties. Scanning electron microscopy, X-ray diffraction, and Fourier infrared spectrometer were used to characterize and analyze the microstructure of materials before and after composite [7]. Wang et al. (2021) [8] loaded magnetic media such as iron oxide onto the surface of BC to achieve simple solid-liquid separation under the action of the external magnetic field, which is a new hotspot in the development of BC materials in recent years [8]. McKenna et al. (2021) [9] argued that there were potential threats to the human body and ecological environment caused by refractory organic matter in water, and it was of great significance to develop an efficient, environmentally friendly, and low-cost catalytic system for the restoration of such wastewater [9]. Industrial and agricultural discharge, leachate leakage of municipal garbage, and environmental accidents all lead to excessive heavy metals in water bodies. It is difficult to remove heavy metal-polluted wastewater, which has a serious impact on water plants and animals, human production and life, and endangers human health. Therefore, it is necessary to remove pollutants in the environment to a great extent through various means so as to provide a guarantee for sustainable social development [10]. Based on these, firstly, the preparation principle and application concept of novel BC composites are discussed. Secondly, the organic pollutants in wastewater and their treatment ideas are expounded. Finally, the BC/ZVI organic pollutant removal material is designed, and its performance is evaluated comprehensively. On account of the novel BC composites, this exploration studied the removal effect and performance optimization effect of organic pollutants in wastewater, thereby improving the removal effect of organic pollutants in wastewater, comprehensively advancing the protection of water resources, and promoting the sustainable utilization of water resources in society. This exploration provides a reference for enhancing the removal effect of organic pollutants and also makes a contribution to promoting the rational utilization of social resources.

BC has been widely used. In recent years, BC has not only been used to improve soil quality but also for soil pollution control and water pollution control. Therefore, BC has become a kind of functional material [11]. The chemical properties of BC mainly refer to the basic chemical properties of different BC. Among them, the main constituent elements of BC include carbon, hydrogen, oxygen, and nitrogen, while the secondary elements cover potassium, calcium, sodium, and magnesium [12]. The chemical properties that determine the properties of BC also involve the structure of functional groups between diverse elements, such as -OH, -(C=O) OH, -OR, and others. At the same time, with the increase in pyrolysis temperature, the acidic group of BC decreases, and the basic group increases. Thus, BC is mostly alkaline [13]. To prepare the new bio-composite, BC composites, namely, BC/zero valent iron (BC/ZVI), can be made by co-heating by adding hematite and pine powder under the main condition of nitrogen. ZVI, with a particle size of less than 100 nm, is usually prepared by wet or dry methods. ZVI has a high specific surface area and strong reducibility, showing excellent reactivity in the removal of water pollutants [14]. However, due to its small size, it is prone to self-aggregation, and its surface is also prone to oxidation to form surface passivation, reducing the reactivity and availability. The best way to solve this problem is to disperse ZVI on the surface of the carbon material so that more reactive sites are activated, thus improving its adsorption. Meanwhile, BC is the best medium for the fixation and dispersion of ZVI due to its complex pore structure and large surface area [15]. To develop an efficient S-type heterojunction photocatalyst to remove harmful pollutants, Li et al. (2022) designed and developed a novel S-type TaON/Bi_2WO_6 heterojunction nanofibers by in situ growing Bi_2WO_6 nanosheets with oxygen vacancies on TaON nanofibers [16]. In order to treat wastewater containing heavy metals and microorganisms, 1-naphthylamine (-NA, chromophore group) was grafted onto the -NH_2 group in a portion of NH_2-MIL-125(Ti) by photocatalysis, NA/NH_2-MIL-125(Ti) homologous coalescence was prepared, and the optical and structural changes were further studied [17]. S-type heterostructures were prepared by coupling Cd0.5Zn0.5S nanoparticles and Bi2MoO_6 microspheres as effective photocatalysts for antibiotic oxidation [18]. Based on the above theories, the two materials are designed to prepare new BC composites, which can be employed to remove organic pollutants from wastewater and optimize water quality.

2. Results and Discussion

2.1. Specific Surface Area Analysis of BC Composites

The specific surface area of the composites is analyzed and compared with BC composites obtained by other methods, such as precipitation. The results are exhibited in Table 1.

Table 1. The specific surface area of BC composites.

Performance Testing	BC/ZVI	BC	ZVI
The specific surface area (m^2/g)	52.05	-	33.11
The specific surface area of other methods (m^2/g)	39.75	-	-

Table 1 describes that compared with BC and ZVI, BC/ZVI composites have a larger specific surface area. Compared with other methods, the specific surface area of this method is larger, and the effect is more obvious.

2.2. Diffraction Results and Morphological Structure of New Biocarbon Composites

Three BC materials are designed and prepared: pure BC, ZVI, and BC/ZVI. First, the basic properties of the three materials can be analyzed by X-ray diffraction results. The diffraction results of three BC composites are expressed in Figure 1.

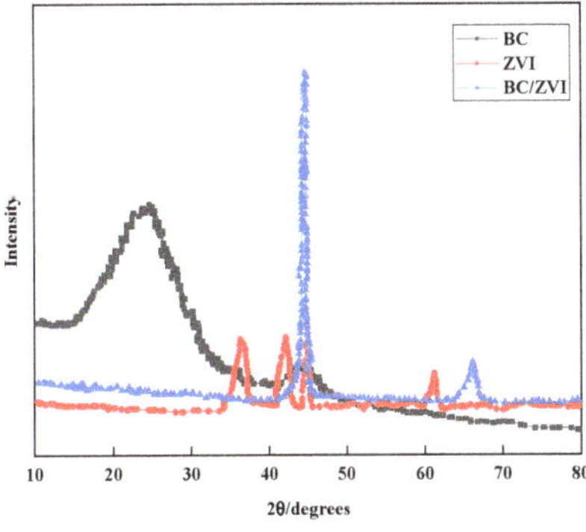

Figure 1. Diffraction results of three BC composites.

Figure 1 shows that when the 2θ value is 24.5° and 44.7°, the characteristic derivative peak of graphite carbon appears, indicating that the graphitization temperature of biomass is higher, and hematite can be calcined by nitrogen at 800 °C. Then, when the 2θ values are 35.8°, 44.3°, and 44.8°, respectively, there are three characteristic peaks, namely FeO, Fe_3O_4, and Fe^0, and there are also very small iron carbide peaks. BC mainly plays a catalytic role in BC composites, so BC mainly acts as a catalyst.

Scanning electron microscope (SEM) and transmission electron microscope (TEM) detection techniques are used to further observe the surface morphology and microstructure of BC composites, as denoted in Figure 2.

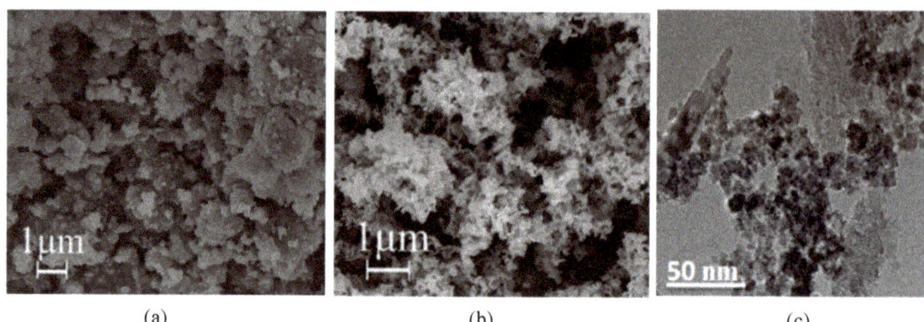

Figure 2. Diffraction results of three BC composites ((**a**): SEM characterization result of ZVI; (**b**): SEM characterization result of BC Composites; (**c**): TEM representation results of BC composites).

In Figure 2, it can be observed that the distribution of black spots in the TEM image represents the presence of iron elements. It can be seen from the SEM image that BC composites show irregular massive morphology, and the formed pore structure is conducive to the adsorption process.

2.3. Evaluation of TC Removal Effect of BC Composite

TC is one of the main organic pollutants that pollute water resources and has an important impact on the protection of water resources. First, BC composites are designed to study the removal effect of TC. At present, ozone and carbon monoxide (CO) are usually used as the main catalysts for the catalytic process in wastewater treatment. Therefore, this exploration designs to compare the performance of the BC catalyst with the above two catalysts. The TC removal effect of three BC composites is displayed in Figure 3.

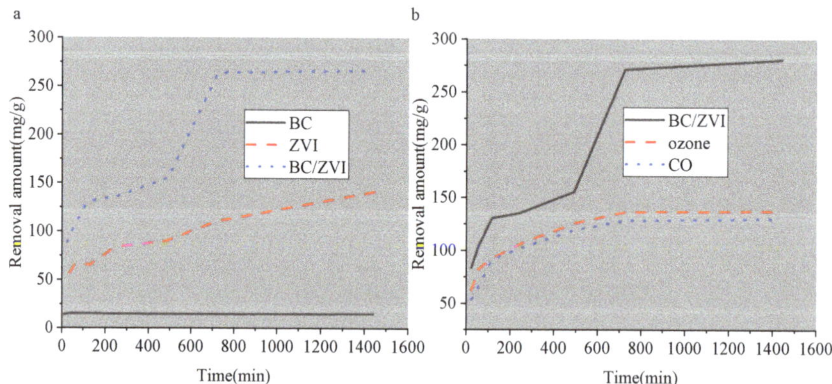

Figure 3. Performance evaluation of BC composite ((**a**) is the TC removal effect of BC composite, and (**b**) is the comparison of TC removal effect under the use of different catalysts).

Figure 3 signifies that the removal effect of pure BC on TC is not ideal, the removal curve is always very stable, and the removal amount is generally maintained at about 20 mg/g. Compared with pure BC, ZVI has improved the removal effect of TC. During the recording process, the removal amount of TC by ZVI always increases steadily, and the final removal amount is about 140 mg/g. The removal effect of BC/ZVI composite is the best, and it is significantly different from the other two materials. Before 800 min, the TC removal amount of composite materials rose rapidly, and after 800 min, it remained stable, with the maximum removal amount of about 275 mg/g. Moreover, when using different catalysts, the composite material using the BC catalyst has a better pollutant removal effect than the other two catalysts.

2.4. Effect of TC Concentration on Properties of BC Composites

In the process of studying the TC removal effect, the TC removal effect of BC composites is evaluated through the use of different concentrations of TC to study the effect of different concentrations on the properties of BC composites. Figure 4 portrays the TC removal effect of pure BC at diverse concentrations.

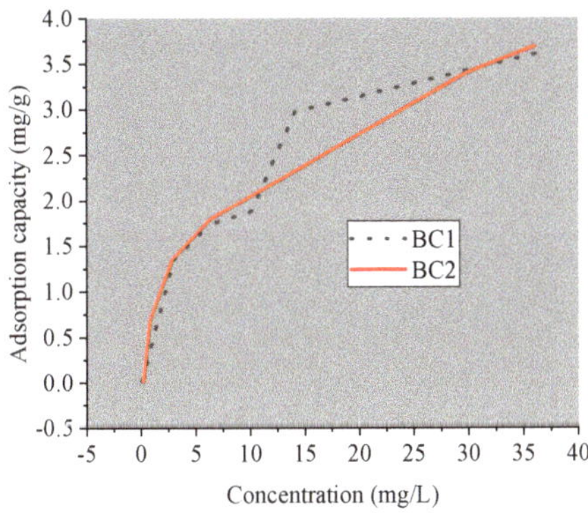

Figure 4. TC removal effect of pure BC at different TC concentrations.

Figure 4 suggests that under different concentrations, the TC removal effect of pure BC is poor, and the removal amount is very low, up to about 3.7 mg/g. It means that the TC removal effect of pure BC is generally poor, and its comprehensive performance can be improved only through optimization. Figure 5 plots the TC removal effect of ZVI and BC/ZVI composites at different concentrations.

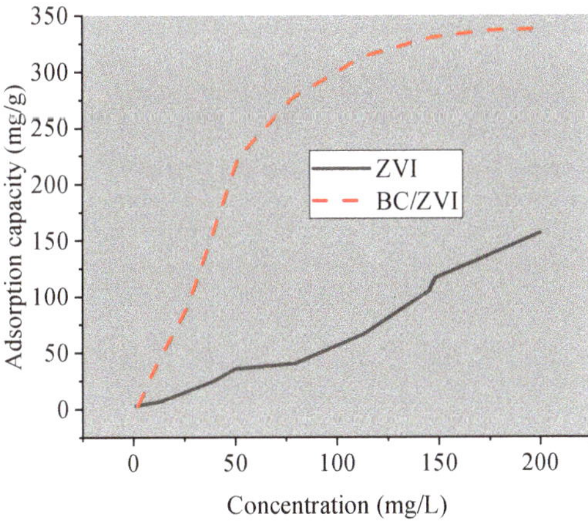

Figure 5. TC removal effect of ZVI and BC/ZVI composites at different concentrations.

Figure 5 details that when the concentration of TC continues to increase, the removal amount of TC by ZVI also continues to increase, showing a rising trend. When the TC concentration is 200 mg/L, the adsorption capacity of ZVI for TC is about 150 mg/g. BC/ZVI composite has a better removal effect on TC. With the increase in TC concentration, the TC adsorption capacity of BC/ZVI composites increases significantly. When the concentration of TC is 200 mg/L, the adsorption capacity of BC/ZVI composite for TC is about 346 mg/g.

2.5. Effects of Different pH Environments on the Properties of BC Composites

pH value has a significant effect on different materials. With the increase in pH value, different BC materials will present diverse adsorption conditions. Thereupon, it is important to study the adsorption effect of BC materials on TC through the change of pH value. Figure 6 demonstrates the adsorption effect of BC material on TC at different pH values.

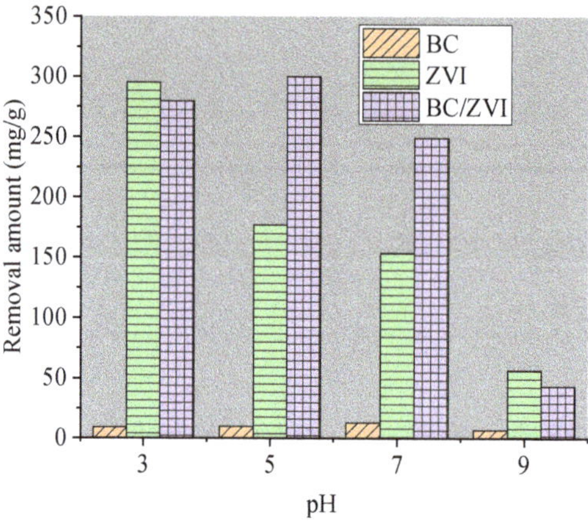

Figure 6. Adsorption effect of BC material on TC at different pH values.

Figure 6 indicates that the TC removal amount of pure BC in different pH environments does not change much, and it is generally stable, with the highest removal amount of about 20 mg/g. The removal amount of TC by ZVI shows a downward trend with the continuous increase in pH value. When pH is 3, its removal of TC is the highest, about 300 mg/g. The removal amount of TC by BC/ZVI composites changes continuously with the increase in pH. At pH 5, the removal amount is the highest, about 300 mg/g, but at pH 9, the removal amount decreases significantly, about 50 mg/g.

2.6. TC Removal Mechanism of BC Composites

The mechanism of TC removal by BC composites designed is mainly that ZVI reacts with oxygen in the solution to generate divalent iron ions and hydroxyl groups, and oxhydryl promotes the degradation of TC. If EDTA is added to the solution, the TC removal effect of the three materials may be affected. The effect of EDTA on the TC removal effect of BC composites is portrayed in Figure 7.

Figure 7 reveals that EDTA has little effect on the TC removal effect of pure BC. Among them, when the EDTA concentration is 1 mM, the TC removal amount of pure BC is the highest, about 10 mg/g. The concentration of EDTA significantly affects the TC removal effect of ZVI and BC/ZVI composites. The maximum removal amount of ZVI is about 210 mg/g when the concentration of EDTA is 5 mM. However, when the TC removal amount of BC/ZVI composite is the highest, the EDTA concentration is 1 mM, and the removal amount is 290 mg/g when the EDTA concentration is 5 mM.

In addition, tert-butanol and calcium ions will greatly impact the TC removal effect of these three materials. Figures 8 and 9 present the effect of tert-butanol and calcium ions on the TC removal effect of the three materials.

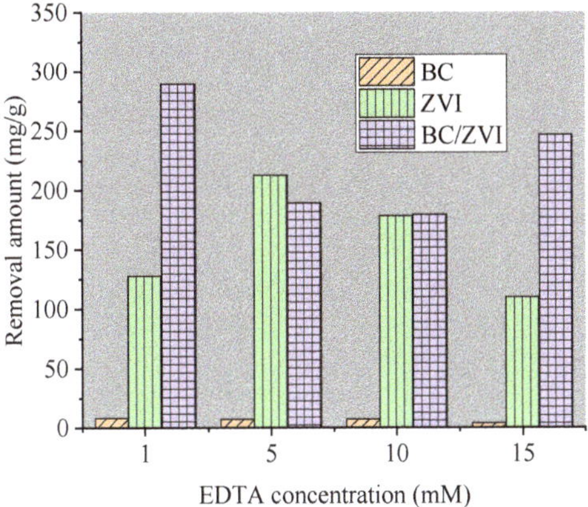

Figure 7. Effect of EDTA on TC removal effect of BC composites.

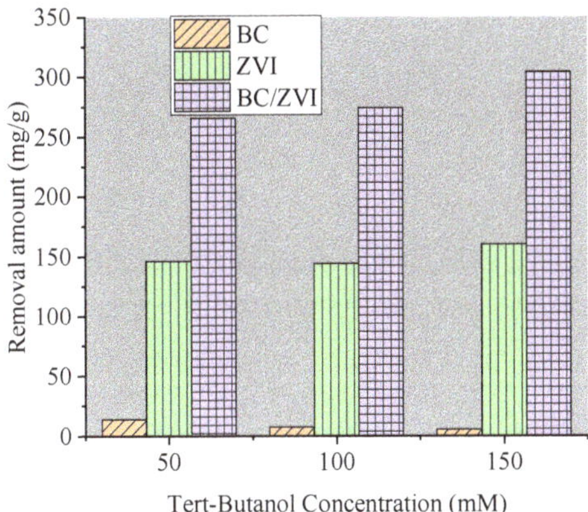

Figure 8. Effect of tert-butanol on TC removal efficiency of three materials.

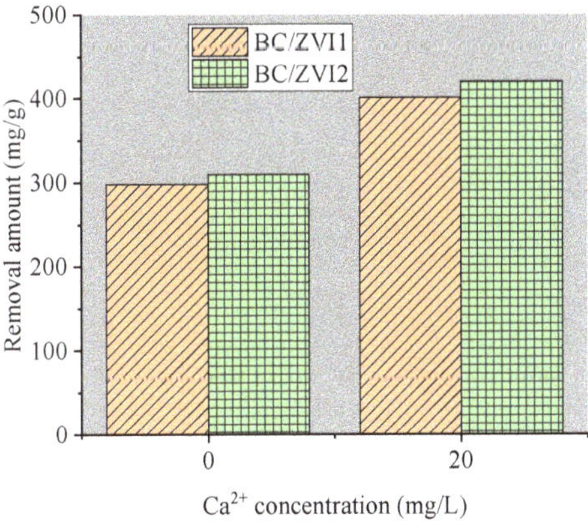

Figure 9. Effect of calcium ion on TC removal efficiency of three materials.

Figure 8 suggests that the tert-butanol has a relatively significant effect on the TC removal effect of the BC/ZVI composites. With the increase in tert-butanol concentration, TC removal of BC/ZVI composites is also increasing.

Figure 9 suggests that adding calcium ions to BC/ZVI composites can significantly improve the TC removal ability of BC/ZVI composites.

2.7. Comparison of Cu(II) Adsorption Capacity of BC Composites

Under the condition of pH = 5 and Cu(II) concentration of 60 mg/L, the adsorption capacity of BC composites on Cu(II) is compared, and the performance is compared with that of BC composites obtained by coprecipitation method, as outlined in Table 2.

Table 2. Comparison of Cu(II) adsorption capacity of BC composites prepared by different methods.

Carbonization Temperature (°C)	Heating Rate (°C/min)	Retention Time (h)	Q_e (mg/g)
A: 500	10	4	16.82
A: 600	10	4	17.71
A: 700	10	4	19.79
A: 700	10	3	19.59
A: 700	10	2	18.82
A: 700	20	4	18.03
A: 700	5	4	18.52
B: 700	10	4	19.02

In Table 2, A represents the experimental method, and B represents the coprecipitation method. It can be seen that under the conditions of 700 °C, 10 °C/min, and 4 h residence time, the adsorption capacity of Cu(II) is the highest, which is 19.79 mg/g. The reason may be that the BC prepared at high temperatures has a larger specific surface area and more aromatic rings to form a π-conjugated bond with Cu(II). The proposed method has a better adsorption capacity of Cu(II) than other methods, and the adsorption capacity is higher.

The isothermal thermodynamic parameters of Cu(II) adsorption on BC composites is revealed in Table 3.

Table 3. The isothermal thermodynamic parameters of Cu(II) adsorption on BC composites.

Sample	Langmuir			Freundlich		
	q_m (mg/g)	P_L (L/mg)	R^2	1/n	P_F (mg/g)	R^2
ZVI	19.28	2.0145	0.9999	0.0159	17.7281	0.9708
BC/ZVI	33.45	0.0997	0.9995	0.1172	17.0025	0.9812

In Table 3, the correlation coefficients of the Langmuir model of BC/ZVI and ZVI are 0.9999 and 0.9995, respectively, which are both higher than those of the Freundlich model of 0.9708 and 0.9812, illustrating that Cu(II) adsorption by materials is monolayer adsorption. The saturated adsorption capacity of BC/ZVI and ZVI of the Langmuir model is 33.45 mg/g and 19.28 mg/g, respectively. BC composites have a higher adsorption capacity.

3. Materials and Methods

3.1. Preparation Process of BC

It is necessary to clarify the raw materials for BC preparation. The commonly used BC raw materials include agricultural waste, and the biomass raw materials mainly cover cellulose, lignin, inorganic ash, hemicellulose, and many other woody fiber substances. In the process of preparing BC, firstly, appropriate biomass raw materials should be selected according to the characteristics of these organisms; secondly, the production process of BC should be controlled [19]. In this process, biomass is mainly used as the control object. By controlling the heating rate of biomass during the process of BC, the biomass raw materials are generated into different substances after pyrolysis. By controlling the pyrolysis process of biomass, the preparation process of BC can be divided into fast decomposition, flash carbonization, pyrolytic gasification, as well as slow pyrolysis. Moreover, slow pyrolysis is the commonly used preparation method of BC. Additionally, preparation conditions are also an important factor affecting the preparation process of BC [20]. Figure 10 displays the main preparation process of BC.

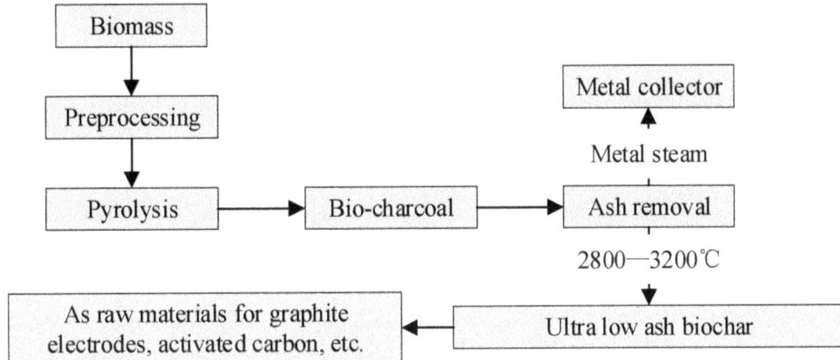

Figure 10. Main preparation process of BC.

Figure 10 shows that the preparation conditions of BC mainly include the holding time at the maximum pyrolysis temperature, the maximum pyrolysis temperature, and the pretreatment of raw materials and carrier gas, among which the effect of the maximum pyrolysis temperature is the most obvious. The pyrolysis process of biomass can be divided into three stages: dehydration, thermal decomposition, and carbonization [21]. BC has a variety of uses, among which the strong adsorption makes BC become the current main material of organic pollutant removal. By controlling the preparation process of BC, the comprehensive performance of BC can be controlled, and the removal effect of BC on organic pollutants can be improved. The preparation process requires little energy and does not require activation. BC has a relatively high yield and suitable pore structure and is widely used due to its easy operation and control [22].

3.2. *Experimental Design*

(1) Experimental Reagent

It includes TC, hydrochloric acid, sodium hydroxide, running water, ferric chloride, tert-butanol, hematite, EDTA, pine biomass, and deionized water with a resistance of 182 Ω.

(2) Preparation of BC Composites

The designed BC preparation process was to weigh 1.0 g hematite and 5.0 g biomass, place them in a beaker in a ratio of 1:5, and add 50 mL of deionized water. After sufficient stirring, the solution was sonicated for 30 min. Then, the solution was heated at 60 °C, and then the mixture was pyrolyzed in a tubular furnace for 1 h. During this period, the temperature was maintained at 800 °C, and 400 mL/min nitrogen was used for protection. Finally, the BC/ZVI composite was taken out and washed three times with ethanol and deionized water. Then, it is put into the oven and dried at 60 °C for 12 h. Next, pure BC was prepared by the same method, and ZVI was prepared by reducing ferric chloride with sodium borohydride as the experimental control material.

(3) Removal Experiment

0.01 g of the test material (BC, BC/ZVI, and ZVI) was weighed and placed in a centrifuge tube. Then, 10 mL TC solution with a concentration of 400 mg/L was added and vibrated for 24 h. Samples were taken at 30, 60, 120, 240, 480, 720, and 1400 min, respectively, and the results were recorded for analysis. Next, different concentrations of TC solution (25, 50, 100, 150, 200, 250, and 300) and different pH (3, 5, 7, and 9) were used for intervention to comprehensively study the removal effect of BC composites on organic pollutants in wastewater.

3.3. Adsorption Kinetic Analysis

(1) Adsorption Kinetics

The dynamic adsorption behavior of Cu(II), Pb(II), and Co(II) on BC composites is studied by fitting dynamic adsorption data, and the quasi-first-order and quasi-second-order kinetic models have been used to fit the data [23], as illustrated in Equations (1) and (2).

$$\lg(q_e - q_t) = \lg q_e - \left(\frac{k_1}{2.303}\right)t \tag{1}$$

$$\frac{t}{q_t} = \frac{1}{k_2 q_e 2} + \frac{t}{q_e} \tag{2}$$

q_t (mg/g) and q_e (mg/g) represent the adsorption capacity at time t and equilibrium, respectively; k_1 (min^{-1}) and k_2 (g $\mathrm{mg}^{-1}\cdot\mathrm{min}^{-1}$) refer to the rate constant of the quasi-first-order kinetic equation and quasi-second-order kinetic equation, respectively, and t (min) stands for the adsorption time.

(2) Adsorption Isotherm

In the isothermal thermodynamic adsorption experiment, Langmuir and Freundlich models are adopted to fit the data [24], whose equations are written as Equations (3) and (4):

$$\frac{c_e}{q_e} = \frac{1}{P_L q_m} + \frac{c_e}{q_m} \tag{3}$$

$$\lg q_e = \lg P_F + \frac{1}{n}\lg C_e \tag{4}$$

q_e (mg/g) expresses the equilibrium adsorption capacity; q_m (mg/g) indicates the saturated adsorption capacity; c_e (mg/L) means the solution concentration after adsorption equilibrium; P_L ($\mathrm{L}\cdot\mathrm{mg}^{-1}$) implies a parameter used by Langmuir to characterize the affinity between adsorbent and adsorbate; P_F (mg/g) stands for the parameter of Freundlich adsorption capacity, and n describes the trend of isotherm change.

4. Conclusions

With society's progress, industrial construction and agricultural development have become society's main tasks, and many problems have occurred in the construction process of these tasks, which seriously impact the environment. In particular, the pollution of water resources has been very serious. Thereupon, this exploration aims to solve the pollution of water resources and promote the recycling and sustainable utilization of water resources. Firstly, the basic concept of BC and the preparation procedure of BC composites are expounded. Secondly, the main organic pollutants that pollute water resources are discussed. Finally, the preparation method of BC composites with BC as the catalyst is designed, and the pollutant removal effect of the prepared BC composites is comprehensively evaluated. Meanwhile, the performance of the BC catalyst is compared with that of other catalysts. The results manifest that the TC removal effect of pure BC is not ideal, while the TC removal effect of ZVI is general. The BC/ZVI composite prepared by combining the two has a better removal effect on TC, and the removal amount is about 275 mg/g. Different TC concentrations, EDTA, pH environment, tert-butanol, and calcium ions will affect the TC removal effect of BC composites. Although a relatively new BC composite material is designed and comprehensively evaluated here, too few indicators are used in the evaluation process, which is not comprehensive enough. Therefore, future research will use more indicators to comprehensively study the removal effect of organic pollutants of new BC composites.

Author Contributions: Conceptualization, H.M. and B.W.; methodology, S.Z.; software, G.W.; validation, G.W., S.Z. and Q.T.; formal analysis, Q.T.; investigation, H.M.; resources, B.W.; data curation, S.Z.; writing—original draft, G.W.; writing—review and editing, S.Z., H.M., B.W. and Q.T.; visualization,

B.W.; supervision, H.M.; project administration, Q.T.; funding acquisition, G.W. All authors have read and agreed to the published version of the manuscript.

Funding: Major Science and Technology Project of Yunnan Province (No. 202102AB080002).

Data Availability Statement: The data used to support the findings of this study are included within the article.

Conflicts of Interest: The authors declare no conflict of interest.

References

1. Ji, M.; Liu, Z.; Sun, K.; Li, Z.; Fan, X.; Li, Q. Bacteriophages in water pollution control: Advantages and limitations. *Front. Environ. Sci. Eng.* **2020**, *15*, 84. [CrossRef]
2. Giorcelli, M.; Bartoli, M.; Sanginario, A.; Padovano, E.; Rosso, C.; Rovere, M.; Tagliaferro, A. High-Temperature Annealed Biochar as a Conductive Filler for the Production of Piezoresistive Materials for Energy Conversion Application. *ACS Appl. Electron. Mater.* **2021**, *3*, 838–844. [CrossRef]
3. Pan, D.; Tang, J. The effects of heterogeneous environmental regulations on water pollution control: Quasi-natural experimental evidence from China. *Sci. Total. Environ.* **2021**, *751*, 141550. [CrossRef]
4. Lu, Q.; Dai, L.; Li, L.; Huang, H.; Zhu, W. Valorization of oxytetracycline fermentation residue through torrefaction into a versatile and recyclable adsorbent for water pollution control. *J. Environ.* **2021**, *9*, 105397. [CrossRef]
5. Xiao, L.; Liu, J.; Ge, J. Dynamic game in agriculture and industry cross-sectoral water pollution governance in developing countries. *Agric. Water Manag.* **2021**, *243*, 106417. [CrossRef]
6. Del Grosso, M.; Cutz, L.; Tiringer, U.; Tsekos, C.; Taheri, P.; de Jong, W. Influence of indirectly heated steam-blown gasification process conditions on biochar physico-chemical properties. *Fuel Process. Technol.* **2022**, *235*, 107347. [CrossRef]
7. Hota, S.K.; Diaz, G. Assessment of Pyrolytic Biochar as a Solar Absorber Material for Cost-Effective Water Evaporation Enhancement. *Environ. Eng. Sci.* **2021**, *38*, 1120–1128. [CrossRef]
8. Wang, M.; Wang, Q.; Li, T.; Kong, J.; Shen, Y.; Chang, L.; Xie, W.; Bao, W. Catalytic Upgrading of Coal Pyrolysis Volatiles by Porous Carbon Materials Derived from the Blend of Biochar and Coal. *ACS Omega* **2021**, *6*, 3800–3808. [CrossRef]
9. McKenna, A.M.; Chacón-Patiño, M.L.; Chen, H.; Blakney, G.T.; Mentink-Vigier, F.; Young, R.B.; Ippolito, J.A.; Borch, T. Expanding the Analytical Window for Biochar Speciation: Molecular Comparison of Solvent Extraction and Water-Soluble Fractions of Biochar by FT-ICR Mass Spectrometry. *Anal. Chem.* **2021**, *93*, 15365–15372. [CrossRef]
10. Pan, X.; Gu, Z.; Chen, W.; Li, Q. Preparation of biochar and biochar composites and their application in a Fenton-like process for wastewater decontamination: A review. *Sci. Total. Environ.* **2020**, *754*, 142104. [CrossRef]
11. Cao, L.; Ding, Q.; Liu, M.; Lin, H.; Yang, D.P. Biochar-supported Cu2+/Cu+ composite as an electrochemical ultrasensitive interface for ractopamine detection. *ACS Appl. Bio Mater.* **2021**, *4*, 1424–1431. [CrossRef] [PubMed]
12. Huang, S.; Ding, Y.; Li, Y.; Han, X.; Xing, B.; Wang, S. Nitrogen and Sulfur Co-doped Hierarchical Porous Biochar Derived from the Pyrolysis of Mantis Shrimp Shell for Supercapacitor Electrodes. *Energy Fuels* **2021**, *35*, 1557–1566. [CrossRef]
13. Lu, Y.; Hussein, A.; Lauzon-Gauthier, J.; Ollevier, T.; Alamdari, H. Biochar as an Additive to Modify Biopitch Binder for Carbon Anodes. *ACS Sustain. Chem. Eng.* **2021**, *9*, 12406–12414. [CrossRef]
14. Zhang, D.; Tong, Z.; Zheng, W. Does designed financial regulation policy work efficiently in pollution control? Evidence from manufacturing sector in China. *J. Clean. Prod.* **2021**, *289*, 125611. [CrossRef]
15. Mandal, S.; Pu, S.; Adhikari, S.; Ma, H.; Kim, D.-H.; Bai, Y.; Hou, D. Progress and future prospects in biochar composites: Application and reflection in the soil environment. *Crit. Rev. Environ. Sci. Technol.* **2020**, *51*, 219–271. [CrossRef]
16. Li, S.; Cai, M.; Liu, Y.; Wang, C.; Lv, K.; Chen, X. S-Scheme photocatalyst TaON/Bi2WO6 nanofibers with oxygen vacancies for efficient abatement of antibiotics and Cr(VI): Intermediate eco-toxicity analysis and mechanistic insights. *Chin. J. Catal.* **2022**, *43*, 2652–2664. [CrossRef]
17. Fu, Y.; Tan, M.; Guo, Z.; Hao, D.; Xu, Y.; Du, H.; Zhang, C.; Guo, J.; Li, Q.; Wang, Q. Fabrication of wide-spectra-responsive NA/NH2-MIL-125(Ti) with boosted activity for Cr(VI) reduction and antibacterial effects. *Chem. Eng. J.* **2023**, *452*, 139417. [CrossRef]
18. Cai, M.; Liu, Y.; Wang, C.; Lin, W.; Li, S. Novel Cd0.5Zn0.5S/Bi2MoO6 S-scheme heterojunction for boosting the photodegradation of antibiotic enrofloxacin: Degradation pathway, mechanism and toxicity assessment. *Sep. Purif. Technol.* **2023**, *304*, 122401. [CrossRef]
19. Wang, Y.; Wei, H.; Wang, Y.; Peng, C.; Dai, J. Chinese industrial water pollution and the prevention trends: An assessment based on environmental complaint reporting system (ECRS). *Alex. Eng. J.* **2021**, *60*, 5803–5812. [CrossRef]
20. Bukhari, Q.U.A.; Silveri, F.; Della Pelle, F.; Scroccarello, A.; Zappi, D.; Cozzoni, E.; Compagnone, D. Water-Phase Exfoliated Biochar Nanofibers from Eucalyptus Scraps for Electrode Modification and Conductive Film Fabrication. *ACS Sustain. Chem. Eng.* **2021**, *9*, 13988–13998. [CrossRef]
21. Feng, Y.; Zhao, D.; Qiu, S.; He, Q.; Luo, Y.; Zhang, K.; Shen, S.; Wang, F. Adsorption of Phosphate in Aqueous Phase by Biochar Prepared from Sheep Manure and Modified by Oyster Shells. *ACS Omega* **2021**, *6*, 33046–33056. [CrossRef]

22. Yang, X.; Zhang, S.; Ju, M.; Liu, L. Preparation and Modification of Biochar Materials and their Application in Soil Remediation. *Appl. Sci.* **2019**, *9*, 1365. [CrossRef]
23. Eltaweil, A.; Mohamed, H.A.; El-Monaem, E.M.A.; El-Subruiti, G. Mesoporous magnetic biochar composite for enhanced adsorption of malachite green dye: Characterization, adsorption kinetics, thermodynamics and isotherms. *Adv. Powder Technol.* **2020**, *31*, 1253–1263. [CrossRef]
24. Khan, Z.H.; Gao, M.; Qiu, W.; Islam, M.S.; Song, Z. Mechanisms for cadmium adsorption by magnetic biochar composites in an aqueous solution. *Chemosphere* **2020**, *246*, 125701. [CrossRef] [PubMed]

Article

Modification of Ti/Sb-SnO_2/PbO_2 Electrode by Active Granules and Its Application in Wastewater Containing Copper Ions

Xuanqi Kang [1,2,†], Jia Wu [2,*,†], Zhen Wei [2], Bo Jia [2], Qing Feng [2], Shangyuan Xu [2] and Yunhai Wang [1,*]

1 State Key Laboratory of Multiphase Flow in Power Engineering, Department of Environmental Science and Engineering, Xi'an Jiaotong University, Xi'an 710049, China

2 Xi'an Taijin New Energy & Materials Sci-Tech Co., Ltd., Xi'an 710016, China

* Correspondence: lhwu123456@126.com (J.W.); wang.yunhai@mail.xjtu.edu.cn (Y.W.); Tel.: +86-15802900957 (J.W.); Fax: +86-29-86968411 (J.W.)

† These authors contributed equally to this work.

Abstract: Active granule (WC/Co_3O_4) doping Ti/Sb-SnO_2/PbO_2 electrodes were successfully synthesized by composite electrodeposition. The as-prepared electrodes were systematically characterized by scanning electron microscopy (SEM), energy-dispersive X-ray spectroscopy (EDS), X-ray diffraction (XRD), X-ray photoelectron spectroscopy (XPS), electrochemical performance, zeta potential, and accelerated lifetime. It was found that the doping of active granules (WC/Co_3O_4) can reduce the average grain size and increase the number of active sites on the electrode surface. Moreover, it can improve the proportion of surface oxygen vacancies and non-stoichiometric PbO_2, resulting in an outstanding conductivity, which can improve the electron transfer and catalytic activity of the electrode. Electrochemical measurements imply that Ti/Sb-SnO_2/Co_3O_4-PbO_2 and Ti/Sb-SnO_2/WC-Co_3O_4-PbO_2 electrodes have superior oxygen evolution reactions (OERs) relative to those of Ti/Sb-SnO_2/PbO_2 and Ti/Sb-SnO_2/WC-PbO_2 electrodes. A Ti/Sb-SnO_2/Co_3O_4-PbO_2 electrode is considered as the optimal modified electrode due to its long lifetime (684 h) and the remarkable stability of plating solutions. The treatment of copper wastewater suggests that composite electrodes exhibit low cell voltage and excellent extraction efficiency. Furthermore, pilot simulation tests verified that a composite electrode consumes less energy than other electrodes. Therefore, it is inferred that composite electrodes may be promising for the treatment of wastewater containing high concentrations of copper ions.

Keywords: Ti/Sb-SnO_2/PbO_2 electrode; copper-containing wastewater; active granule doping; electrocatalytic activity

Citation: Kang, X.; Wu, J.; Wei, Z.; Jia, B.; Feng, Q.; Xu, S.; Wang, Y. Modification of Ti/Sb-SnO_2/PbO_2 Electrode by Active Granules and Its Application in Wastewater Containing Copper Ions. *Catalysts* **2023**, *13*, 515. https://doi.org/10.3390/catal13030515

Academic Editors: Hao Xu and Yanbiao Liu

Received: 3 December 2022
Revised: 29 December 2022
Accepted: 30 December 2022
Published: 3 March 2023

1. Introduction

The development of the nonferrous metal industry has improved social progress and economic development. However, the nonferrous metal industry also produces many pollutants, including wastewater containing copper, zinc, nickel, etc. [1,2]. The discharge of wastewater with high concentrations of metal ions can cause serious environmental problems if it is not properly treated. The technology of metal electrodeposition from solution is environmentally friendly and has attracted considerable attention with respect to resource recovery [3–6]. For example, copper can be extracted from copper-containing wastewater by the electrochemical method, and the product can be widely used in many fields, such as light industry, electrical, national defense, etc. [7–10]. In this procedure, low oxygen evolution overpotential is favorable for energy conservation.

Electrode material is a critical component in the electrochemical process [11–13] and can affect power consumption, production cost, current efficiency, and the quality of the product. Due to the strong oxidizing ability and the corrosivity of sulfuric acid in the electrodeposition process, electrode materials (lead-based alloys, Ti-based metal oxides, Al/PbO_2, and SS/PbO_2) with high conductivity, good stability, mechanical strength, and convenient processing have attracted considerable attention [14–19]. In particular, Ti-based insoluble electrode materials exhibit high corrosion resistance, long service life, excellent electrochemical performance, and electrocatalytic activity, which are highly valued with respect to environmental protection, metallurgy, and resource recovery [20–22]. Usually, Ti-based electrode material contains thinly coated noble (Ti/Ru, Ti/Ir, or Ti/Pt) electrodes and thickly coated Ti/PbO_2 and Ti/MnO_2 electrodes [23–27]. Ti/PbO_2 electrodes possess well-established features such as low cost, ease of synthesis, good chemical stability, and long service life, and have great application potential to copper-containing wastewater. However, the high oxygen evolution overpotential of Ti/PbO_2 electrodes causes energy waste during the copper electrodeposition process [28]. Reducing the oxygen evolution overpotential of the electrode is an important task for many researchers.

On the one hand, Co_3O_4 shows superior electrocatalytic performance towards oxygen evolution reactions (OERs) due to its spinal structure, with a Co^{2+} located in the tetrahedral site and two other Co^{3+} atoms in the octahedral site [29–31]. On the other hand, WC can improve the mechanical properties of coatings and is widely used in the fields of cemented carbide, electrocatalytics, and fuel cells [32,33]. Therefore, in this study, active granule Co_3O_4 and WC were introduced to Ti/Sb-SnO_2/PbO_2 electrodes by electrodeposition with the aim of decreasing the overpotential of oxygen evolution and enhancing the service life of the electrode. An amplified simulated experiment of electrolysis was conducted to investigate the change in voltage and temperature during this process, which is closely related to energy consumption. Furthermore, wastewater containing copper ions and sulfuric acid was employed as a model recyclable resource for the electrochemical extraction of copper. The concentration of copper ions was studied by inductively coupled plasma emission spectroscopy (ICPE), and the extraction efficiency was calculated by the mass of copper deposited on the cathode.

2. Results and Discussion

2.1. Surface Morphology Analysis of Electrodes

Figure 1 shows an SEM of an as-synthesized Ti/Sb-SnO_2/PbO_2 electrode, a Ti/Sb-SnO_2/WC-PbO_2 electrode, Ti/Sb-SnO_2/Co_3O_4-PbO_2 electrode, and Ti/Sb-SnO_2/WC-Co_3O_4-PbO_2 electrode. It can be seen that the Ti/Sb-SnO_2/PbO_2 electrode shows a hill-like surface at the macroscopic level (Figure 1a). However, the hill-like surface becomes smooth after it is modified with active granules (WC/Co_3O_4), which indicates that the addition of active granules may affect the deposition process (Figure 1b–d). The microscopic morphology of the Ti/Sb-SnO_2/PbO_2 electrode displays a typical tetrahedron shape. Some small particles (WC or WC-Co_3O_4) are agglomerated on the surface of the Ti/Sb-SnO_2/WC-PbO_2 and Ti/Sb-SnO_2/WC-Co_3O_4-PbO_2 electrodes, leading to an undulating crest of hillocks. Moreover, it was found that the surface uniformity of Ti/Sb-SnO_2/Co_3O_4-PbO_2 is better than that of Ti/Sb-SnO_2/WC-PbO_2 and Ti/Sb-SnO_2/WC-Co_3O_4-PbO_2, with a denser coating and better coverage. EDS measurements were employed to investigate the composition of particles deposited on the electrode surface (Figure 1e–g). W element was detected on the surface of the Ti/Sb-SnO_2/WC-PbO_2 and Ti/Sb-SnO_2/WC-Co_3O_4-PbO_2 electrodes. Co element was discovered on the surface of the Ti/Sb-SnO_2/Co_3O_4-PbO_2 and Ti/Sb-SnO_2/WC-Co_3O_4-PbO_2 electrodes. These results confirm that spherical Co_3O_4 granules and small WC particles were successfully doped into the composite electrode.

2.2. XRD Structural Characterization

As shown in Figure 2, the XRD pattern of each electrode was obtained to compare the crystal structure and purity of electrodes. Figure 2 shows that bare Ti/Sb-SnO_2/PbO_2 exhibits the reflections of β-PbO_2. The diffraction peaks at 25.4°, 32.0°, 36.2°, 49.0°, 52.1°, 58.9°, 60.7°, and 62.5° are assigned to the (110), (101), (200), (211), (220), (310), (112), and (301) planes of β-PbO_2 (PDF#41-1492), respectively. No peaks corresponding to Sb or SnO_2 were detected, which can be explained by two reasons. One is the low crystallinity of the Sb-SnO_2 layer, and the other is the thick layer of β-PbO_2 coating on the Sb-SnO_2

layer. After doping with WC particles, new peaks appeared at 2θ = 31.5°, 35.6°, and 48.3° on the Ti/Sb-SnO_2/WC-PbO_2 electrode, which are assigned to the (001), (100), and (101) planes of WC (PDF#51-0939), respectively. However, no additional peaks were observed on Ti/Sb-SnO_2/Co_3O_4-PbO_2 and Ti/Sb-SnO_2/Co_3O_4-PbO_2, which can be ascribed to the weak crystallinity, small particle size, and infinitesimal load of active particles. Moreover, the average grain sizes of Ti/Sb-SnO_2/PbO_2, Ti/Sb-SnO_2/WC-PbO_2, Ti/Sb-SnO_2/Co_3O_4-PbO_2, and Ti/Sb-SnO_2/WC-Co_3O_4-PbO_2 calculated by the Debye–Scherrer equation are 56.2 nm, 38.1 nm, 48.7 nm, and 37.5 nm, respectively. The doping of active granules can decrease the grain size, which is consistent with the SEM morphology. A smaller grain size indicates more active sites on the electrode surface, which is favorable for the enhancement of catalytic performance.

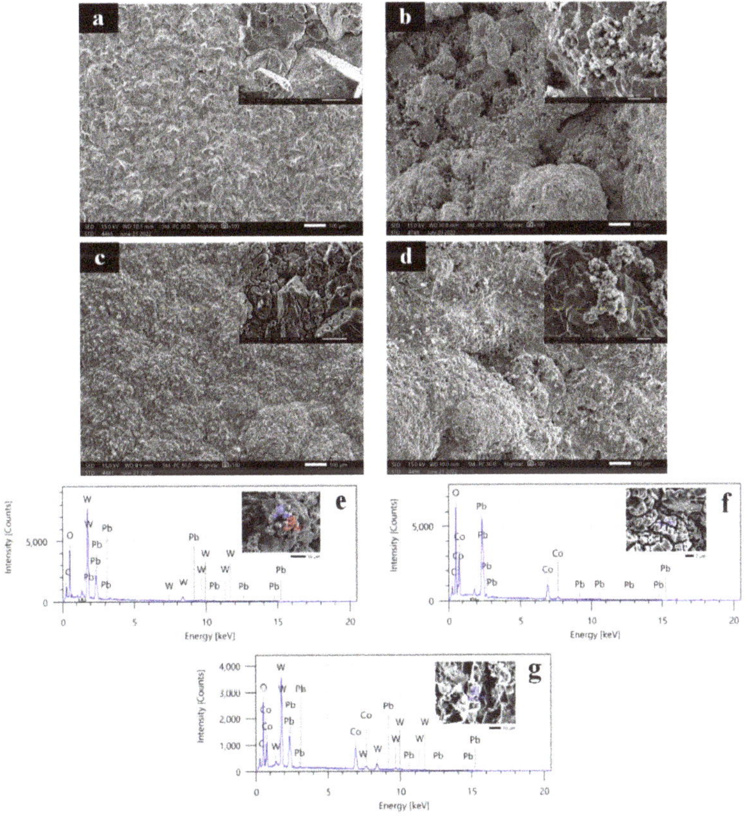

Figure 1. SEM of the electrodes: (**a**) Ti/Sb-SnO_2/PbO_2; (**b**) Ti/Sb-SnO_2/WC-PbO_2; (**c**) Ti/Sb-SnO_2/Co_3O_4-PbO_2; (**d**) Ti/Sb-SnO_2/WC-Co_3O_4-PbO_2. EDS of the electrodes: (**e**) Ti/Sb-SnO_2/WC-PbO_2; (**f**) Ti/Sb-SnO_2/Co_3O_4-PbO_2; (**g**) Ti/Sb-SnO_2/WC-Co_3O_4-PbO_2.

2.3. XPS Analysis

XPS measurements were performed to further analyze the chemical state of the elements on each electrode. In the survey, the XPS spectrum of each electrode and elemental peaks for Pb and O were observed on all electrodes (Figure 3a). A new elemental peak for W and Sn was found on the Ti/Sb-SnO_2/WC-PbO_2 electrode surface. The presence of Sn indicates that part of the coating is too thin; hence, and interlayer Sb-SnO_2 film was detected by XPS. This suggests that the coating on Ti/Sb-SnO_2/WC-PbO_2 is ununiform. An additional peak for Co appeared on Ti/Sb-SnO_2/Co_3O_4-PbO_2, which confirms that

Co_3O_4 was successfully doped in the modified PbO_2 film. Moreover, peaks for W and Co were found on the Ti/Sb-SnO_2/WC-Co_3O_4-PbO_2 electrode, indicating the presence of WC and Co_3O_4. To identify the influence of active granule doping on PbO_2 electrodeposition, XPS analysis of Pb, O, W, and Co was performed; the results are shown in Figure 3b–e.

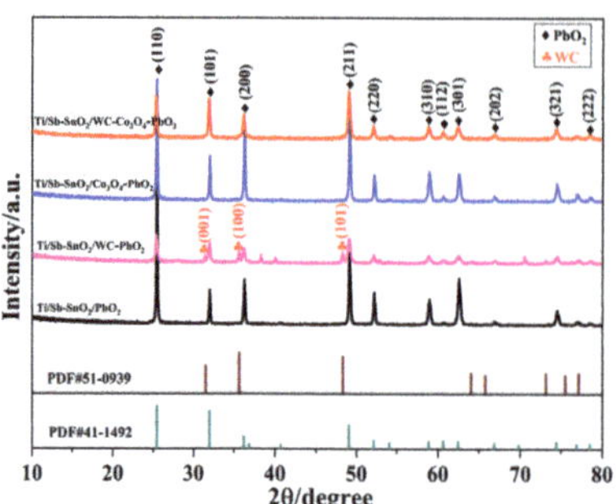

Figure 2. XRD pattern of the different samples.

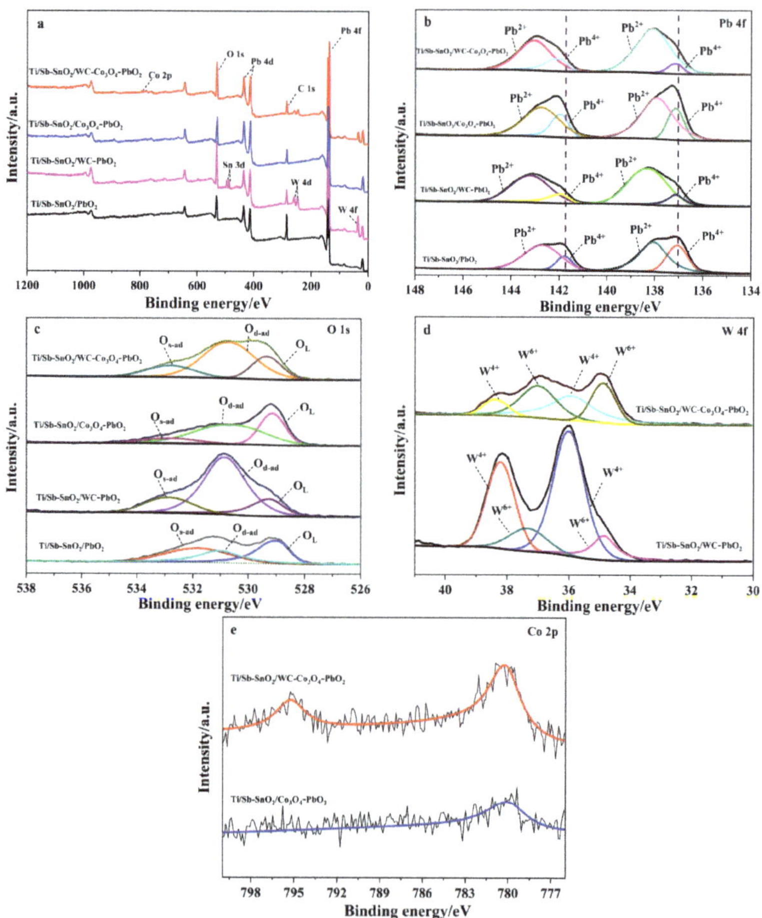

Figure 3. XPS spectra of the as-prepared samples: (**a**) survey, (**b**) Pb 4f; (**c**) O 1s; (**d**) W 4f; (**e**) Co 2p.

In the undoped Ti/Sb-SnO_2/PbO_2 (Figure 3b), the binding energy peaks at 141.79 eV and 137.07 eV correspond to Pb^{4+}, whereas the peaks at 142.70 eV and 138.02 eV are attributed to Pb^{2+} [34], as listed in Table 1. Moreover, the simultaneous presence of Pb^{4+} and Pb^{2+} in the PbO_2 coating implies the formation of non-stoichiometric PbO_2 during the electrodeposition process. After doping with active granules, the peaks of Pb^{4+} on modified electrodes shifted to the higher binding energy, which suggests that active granules have a strong interaction with Pb. The proportion of Pb^{4+} in Ti/Sb-SnO_2/PbO_2, Ti/Sb-SnO_2/WC-PbO_2, Ti/Sb-SnO_2/Co_3O_4-PbO_2, and Ti/Sb-SnO_2/WC-Co_3O_4-PbO_2 is 29.9%, 15.9%, 27.12%, and 15.61%, respectively. The decrease in Pb^{4+} after modification with active granules implies that part of the PbO_2 converts to lower-valence compounds, enhancing the non-stoichiometric PbO_2, which could improve the conductivity of the electrode.

Table 1. The binding energy of Pb 4f on each electrode.

Electrode	Pb^{4+} $4f_{7/2}$/eV	Pb^{4+} $4f_{5/2}$/eV	Pb^{2+} $4f_{7/2}$/eV	Pb^{2+} $4f_{5/2}$/eV
Ti/Sb-SnO_2/PbO_2	137.07	141.79	138.02	142.70
Ti/Sb-SnO_2/WC-PbO_2	137.12	141.94	138.26	143.22
Ti/Sb-SnO_2/Co_3O_4-PbO_2	137.14	141.95	137.89	142.76
Ti/Sb-SnO_2/WC-Co_3O_4-PbO_2	137.13	142.05	138.10	143.05

The O 1s core level shown in Figure 3c can be deconvoluted into four characteristic peaks of lattice oxygen species (528.9–530.4 eV for O_L), surface oxygen vacancies, adsorbed oxygen (530.5–531.7 eV for $O_{d\text{-}ad}$), and surface-adsorbed oxygen species (531.8–532.8 eV for $O_{s\text{-}ad}$), as listed in Table 2. According to the literature [35], the formation of surface oxygen vacancies is closely related to highly oxidative oxygen species and is active for catalysis of OER. The proportions of $O_{d\text{-}ad}$ present on the electrode surface significantly increase after decoration with active granules, which is beneficial to the OER activity.

Table 2. The binding energy of O 1s and its proportions on each electrode.

Electrode	O_L/eV	O_L/%	$O_{d\text{-}ad}$/eV	$O_{d\text{-}ad}$/%	$O_{s\text{-}ad}$/eV	$O_{s\text{-}ad}$/%
Ti/Sb-SnO_2/PbO_2	529.01	27.69%	531.01	20.91%	531.82	29.56%
Ti/Sb-SnO_2/WC-PbO_2	529.25	17.06%	530.84	64.53%	532.83	18.41%
Ti/Sb-SnO_2/Co_3O_4-PbO_2	529.15	34.53%	530.55	54.72%	532.77	10.75%
Ti/Sb-SnO_2/WC-Co_3O_4-PbO_2	529.35	21.16%	530.67	59.91%	532.81	18.93%

Figure 3d shows the W 4f spectrum; the binding energy peaks at 35.96 eV and 38.29 eV are assigned to W^{4+}, whereas the peaks at 34.87 eV and 37.17 eV are ascribed to W^{6+}. This consequence demonstrates the generation of WC. However, W^{6+} also appears on the electrode surface. The formation of W^{6+} can be attributed to the oxidation of WC during composite electrodeposition. Moreover, the XPS spectrum of Co 2p is very weak due to the extremely limited doping of active particles. It was found that two main peaks appeared on the Ti/Sb-SnO_2/WC-Co_3O_4-PbO_2 electrode. The peaks located at 795.26 eV and 780.23 eV are identical to Co $2p_{1/2}$ and Co $2p_{3/2}$, respectively (Figure 3e), which confirms the presence of Co_3O_4.

2.4. Electrochemical Properties

LSV measurements of the as-prepared samples were performed on an electrochemical workstation to investigate the OER performance of the electrodes. The onset potential for oxygen evolution potential (OEP) on Ti/Sb-SnO_2/PbO_2, Ti/Sb-SnO_2/WC-PbO_2, Ti/Sb-SnO_2/Co_3O_4-PbO_2, and Ti/Sb-SnO_2/WC-Co_3O_4-PbO_2 is 2.20 V, 2.12 V, 1.85 V, and 1.80 V (vs. Ag/AgCl), respectively. A low OER means that oxygen is more easily formed during the electrochemical process. Therefore, it is speculated that the modification of active granules (WC/Co_3O_4) can improve the OER properties of electrodes due to the presence of more active sites and increased electrocatalytic activity. Moreover, doping with Co_3O_4 is more efficient than doping with WC. This can be attributed to the special properties of Co_3O_4 and its better OER performance. Furthermore, the non-uniformity of the WC-PbO_2 coating may enhance resistance, hindering the formation of oxygen. As shown in the illustration in Figure 4, when the current density is 200 A/m^2, the potential of Ti/Sb-SnO_2/PbO_2, Ti/Sb-SnO_2/WC-PbO_2, Ti/Sb-SnO_2/Co_3O_4-PbO_2, and Ti/Sb-SnO_2/WC-Co_3O_4-PbO_2 is 2.14 V, 1.96 V, 1.88 V, and 1.75 V (vs. Ag/AgCl), respectively. This implies that Ti/Sb-SnO_2/Co_3O_4-PbO_2 and Ti/Sb-SnO_2/WC-Co_3O_4-PbO_2 need lower potential than Ti/Sb-SnO_2/PbO_2 and Ti/Sb-SnO_2/WC-PbO_2 under the same current density. Low potential is more helpful for saving energy during electrolysis.

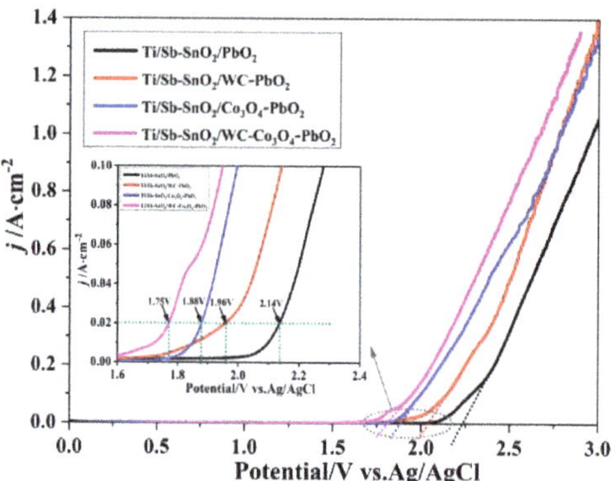

Figure 4. LSV curves of the different electrodes.

Energy conservation has become an important problem due to energy shortages. As is known to all, voltage is closely related to energy consumption. How to reduce the voltage of electrolysis is a crucial issue in the electrolytic industry. Therefore, a simulation experiment of electrolysis under a current density of 200 A/m^2 was conducted to explore the variation in voltage, as displayed in Figure 5. It was found that the Ti/Sb-SnO_2/Co_3O_4-PbO_2 and Ti/Sb-SnO_2/WC-Co_3O_4-PbO_2 electrodes have advantages over the Ti/Sb-SnO_2/PbO_2 and Ti/Sb-SnO_2/WC-PbO_2 electrodes. These results are consistent with the LSV measurements.

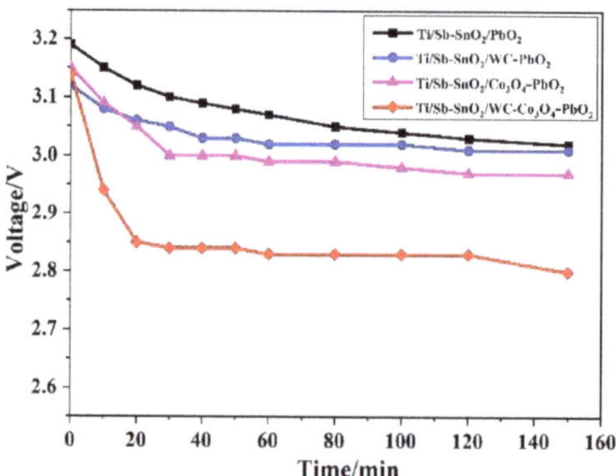

Figure 5. The change in voltage during electrolysis (200 A/m^2; 15% H_2SO_4 aqueous solutions).

Moreover, an accelerated life test was carried out to evaluate the electrochemical stability of the electrodes. Figure 6 shows the time course of cell potential in the accelerated life test. It was observed that the Ti/Sb-SnO_2/PbO_2, Ti/Sb-SnO_2/WC-PbO_2, Ti/Sb-SnO_2/Co_3O_4-PbO_2, and Ti/Sb-SnO_2/WC-Co_3O_4-PbO_2 electrodes exhibited a lifetime of 252 h, 204 h, 684 h, and 592 h, respectively. The lifetime of Ti/Sb-SnO_2/WC-PbO_2 is shorter than that of Ti/Sb-SnO_2/PbO_2, which can be attributed to the non-uniformity of its coating. The lifetimes of Ti/Sb-SnO_2/Co_3O_4-PbO_2 and Ti/Sb-SnO_2/WC-Co_3O_4-PbO_2 are increased by more than two-fold relative to that of Ti/Sb-SnO_2/PbO_2.

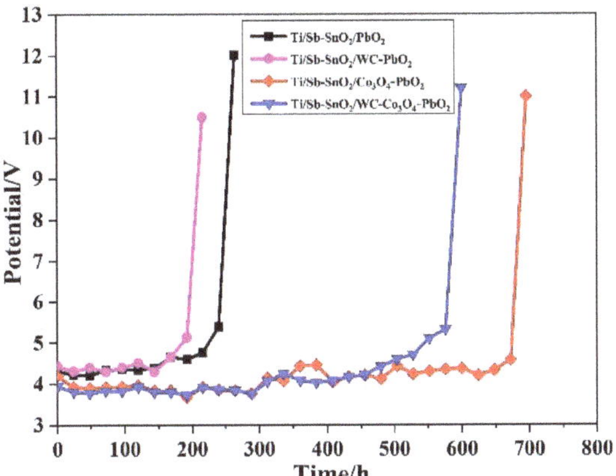

Figure 6. Variation of the cell potential with testing time in an accelerated life test for the as-synthesized samples (10,000 A/m^2; 15% H_2SO_4 aqueous solutions).

2.5. The Application of Modified Electrodes in Wastewater Containing Copper Ions

Deposition solutions containing active granules (WC/Co_3O_4) were tested with a zeta potential analyzer to study their stability [36]. The Zeta potential of deposition solutions with WC, Co_3O_4, and WC-Co_3O_4 are 3.9 mV, 5.5 mV, and 3.4 mV, respectively. It was found that a high absolute value of zeta potential indicates higher stability against coagulation, which implies excellent stability in such systems. Compared to other deposition systems, the plating solution with Co_3O_4 exhibits the best stability. Although Ti/Sb-SnO_2/WC-Co_3O_4-PbO_2 shows the lowest OER property, the stability of the plating solution with WC-Co_3O_4 is the worst. Therefore, considering the lifetime of the electrode and the stability of the solution, we selected the Ti/Sb-SnO_2/Co_3O_4-PbO_2 electrode as the optimal electrode for subsequent experiments.

Electrochemical extraction experiments were carried out to investigate the application of electrodes to wastewater containing copper ions. This system contains two Ti/Sb-SnO_2/Co_3O_4-PbO_2 anodes and one SS cathode with magnetic stirring. As shown in Figure 7a, the color of copper deposited on the electrode becomes dark due to the ion impoverishment of copper ions. Usually, dark copper is generated when the copper ion concentration decreases to a certain value in actual production. Figure 7b displays the mass and concentration of copper over time; the mass of copper increases linearly with time, and the concentration of copper ions decreases with time. The Ti/Sb-SnO_2/Co_3O_4-PbO_2 electrode shows a higher mass and a lower concentration than the Ti/Sb-SnO_2/PbO_2 electrode. The extraction efficiency of Ti/Sb-SnO_2/Co_3O_4-PbO_2 and Ti/Sb-SnO_2/PbO_2 is 99.1% and 83.2%, respectively (Figure 7c). Moreover, the cell voltage of Ti/Sb-SnO_2/Co_3O_4-PbO_2 is lower than that of Ti/Sb-SnO_2/PbO_2, indicating less energy consumption (Figure 7d). These results reveal that the modified Ti/Sb-SnO_2/Co_3O_4-PbO_2 electrode has superior performance in the extraction of copper from wastewater by the electrochemical method.

Laboratory experiments usually slightly deviate from actual working conditions. Therefore, we decided to build a magnifying system that is close to practical working conditions. To that end, an apparatus was manufactured and outsourcing factory consisting of a circular aeration, heat, and electric control (Figure 8a). Amplification tests were performed to research the energy consumption of this device. To simplify the experimental process, we employed 15% H_2SO_4 aqueous solutions to simulate acidic working conditions. The other parameters were similar to those used in the electrochemical extraction experiment. In the pilot simulation tests, the Ti/Sb-SnO_2/Co_3O_4-PbO_2, Ti/Sb-SnO_2/PbO_2, and Pb electrodes were employed as the anode, and SS was used as the cathode, all with dimensions of

200 mm × 290 mm × 4 mm (Figure 8b). It should be noted that Pb electrodes are traditionally used in the metallurgical industry. Ti/Sb-SnO_2/Co_3O_4-PbO_2 had a lower voltage and lower temperature (Figure 8c,d) than Ti/Sb-SnO_2/PbO_2 and Pb. These results suggest that Ti/Sb-SnO_2/Co_3O_4-PbO_2 consumes less energy than Ti/Sb-SnO_2/PbO_2 and Pb electrodes, which is in agreement with the results of the electrochemical extraction experiments. This indicates that Ti/Sb-SnO_2/Co_3O_4-PbO_2 electrodes may replace traditional Pb electrodes in the resource recovery of nonferrous metals.

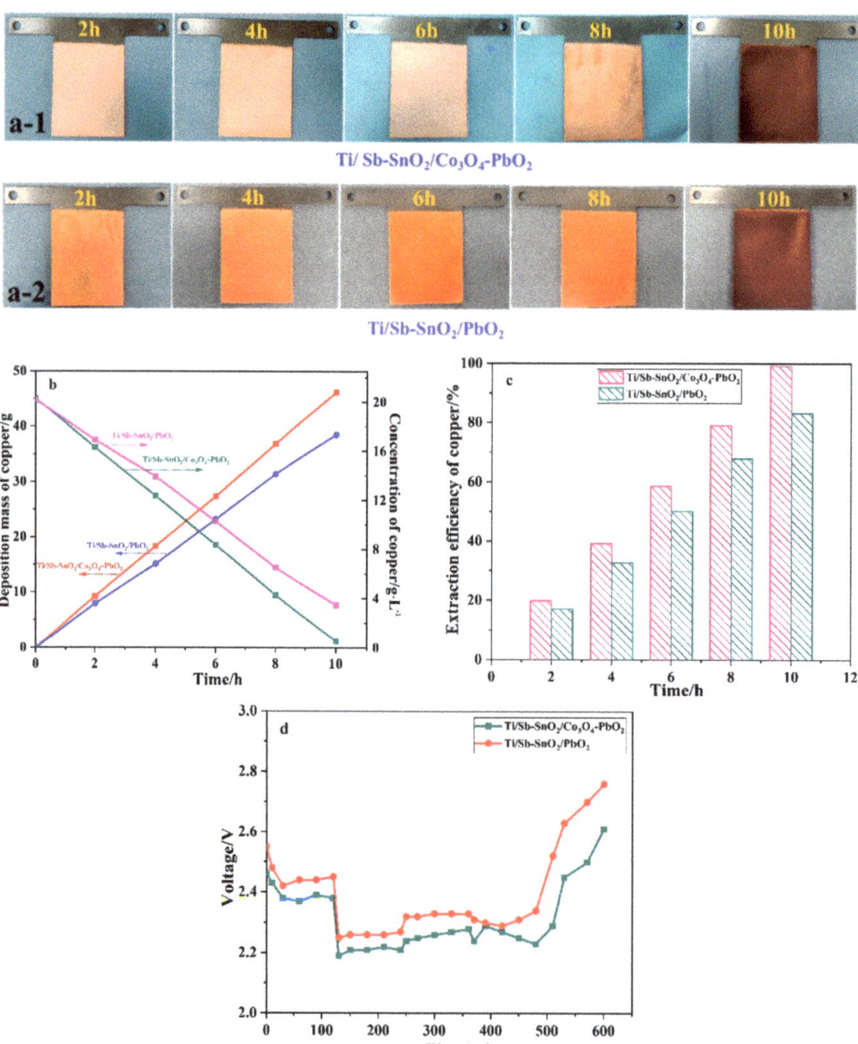

Figure 7. (**a**) Digital images of copper deposited on Ti/Sb-SnO_2/Co_3O_4-PbO_2 (**a-1**) and Ti/Sb-SnO_2/PbO_2 (**a-2**) at different times. (**b**) Variation in the deposition mass and concentration of copper over time. (**c**) The extraction efficiency of copper during the electrochemical process. (**d**) The cell voltage of different electrodes during electrodeposition.

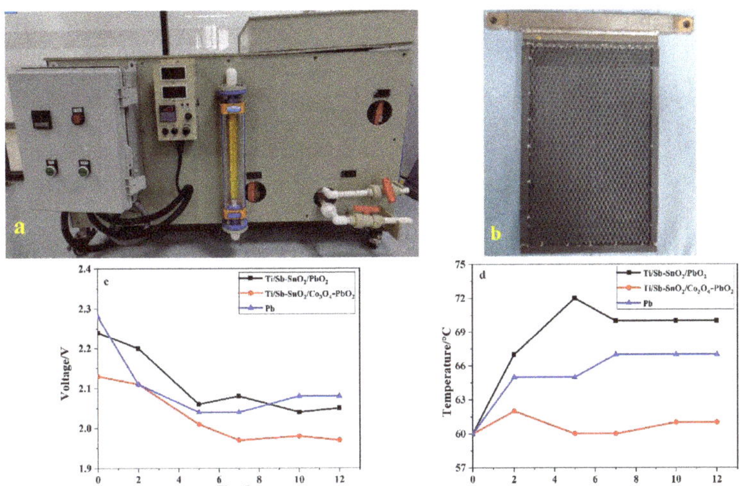

Figure 8. (**a**) Pilot apparatus of the simulation test. (**b**) Part of the sample used in the pilot test. (**c**) The variation in cell voltage during the electrochemical process. (**d**) The change in temperature during the electrochemical process.

3. Experimental Details

3.1. Materials

All chemicals were of analytical grade and were used without any further purification. $SnCl_2{\cdot}2H_2O$ and $SbCl_3$ were provided by Xilong Scientific Co., Ltd. (Chaoshan, China). Ethanol and tert-butyl alcohol were offered by Tianjin Fuyu Fine Chemical Co., Ltd. (Tianjin, China). $Pb(NO_3)_2$, $Cu(NO_3)_2$, NaF, and some additives were obtained from Sinopharm Chemical Reagent Co., Ltd. (Shanghai, China). HNO_3, H_2SO_4, and HCl were produced from Xi'an Sanpu Chemical Reagants Co., Ltd. (Xi'an, China). WC and Co_3O_4 reagents were purchased from MACKLIN (Shanghai, China). All solutions were prepared with deionized water (DI). Titanium foil (Baoji Baite Metal Co., Ltd., Baoji, China, 1.5 mm thickness) was cut into pieces with effective dimensions of 80 mm × 100 mm before experiments.

3.2. Fabrication of Ti/Sb-SnO_2/PbO_2 Electrode by Active Granules

A Ti sheet was subjected to pretreatment before the experiment, including polishing, degreasing, and etching in boiling oxalic acid for 2 h, to obtain a gray surface with uniform roughness. Sb-SnO_2 was introduced as a conductive transition layer between the PbO_2 film layer and the Ti substrate by the thermal decomposition approach. Sb-SnO_2 precursor solutions were prepared by dissolving a $SnCl_4$ and $SbCl_3$ mixture in a mixed solvent at a molar ratio of 10:1. The coating liquids were evenly brushed on the Ti sheet. Then, the Ti sheet was dried at 100 °C and annealed at 450 °C. The brushing, drying, and calcining steps were repeated several times.

PbO_2 coatings with active granules (Co_3O_4 and WC) were synthesized by composite electrodeposition with a current density of 200 A/m^2 for 2 h at 60 °C. The deposition solutions containing $Pb(NO)_3$, $Cu(NO)_3$, HNO_3, active granules (WC/Co_3O_4), and other additives were ultrasonicated for 30 min. The as-prepared electrodes were rinsed thoroughly with DI and are denoted as Ti/Sb-SnO_2/PbO_2, Ti/Sb-SnO_2/WC-PbO_2, Ti/Sb-SnO_2/Co_3O_4-PbO_2, and Ti/Sb-SnO_2/WC-Co_3O_4-PbO_2.

3.3. Characterization

A scanning electron microscope (SEM, JSM-IT200, JEOL, Tokyo, Japan) equipped with an energy-dispersive X-ray spectroscopy (EDS, JEOL) detector was employed to study the surface morphology and composition of the as-prepared samples. X-ray diffraction (XRD,

D8 Advance, Bruker, Karlsruhe, Germany) measurement was conducted on an X'pert PRO MRD diffractometer using a Cu-Ka source (λ = 0.15416 nm) with a scanning angle (2θ) range of 10–80°. Chemical states of Pb, O, W, and Co in the composite electrodes were identified by X-ray photoelectron spectroscopy (XPS, ESCALAB 250Xi, Thermo Fisher Scientific, Waltham, MA, USA) on an Ultra DLD Electron Spectrometer (Al Ka radiation; hν = 1486.71 eV). XPS data were calibrated using the binding energy of C1s (284.8 eV) as the standard and were fitted using commercial software (Thermo Avantage). The zeta potential of the deposition solutions containing active granules (WC/Co_3O_4) was tested on a zeta potential analyzer (Stabino zeta, Microtrac MRB, Dusseldorf, Germany) to study the stability of the solutions.

Electrochemical performance was evaluated on an electrochemical workstation (Corrtest CS2350, Corrtest Instruments Corp., Ltd., Wuhan, China) using the traditional three-electrode system. A Pt sheet is used as counter electrode, a saturated Ag/AgCl electrode is employed as reference electrode, and the as-prepared electrode served as working electrode. Linear sweep voltammetric (LSV) characterization was measured in 15% H_2SO_4 aqueous solutions at a scan rate of 10 mV/s. simulated electrochemical experiment is also tested in 15% H_2SO_4 aqueous solutions to study the change in voltage during electrolysis. Accelerated life tests were carried out to research the stability and lifetime of the as-prepared electrodes in 15% H_2SO_4 aqueous solutions with a current density of 10,000 A/m^2. In this procedure, the experiment was considered finished when the cell voltage exceeded 10 V.

3.4. Electrochemical Treatment of Copper-Containing Wastewater

Wastewater containing copper ions was treated by electrochemical approaches in a large beaker equipped with a water bath and magnetic stirrer, as shown in Figure 9. Two pieces of Ti/Sb-SnO_2/Co_3O_4-PbO_2 electrodes were used as the anode, and stainless steel (SS) was employed as the cathode. The temperature was set to 60 °C, and the current density was set to 200 A/m^2. Simulated wastewater with a high concentration of copper ions (20 g/L Cu^{2+}) was prepared by dissolving $CuSO_4{\cdot}5H_2O$ in 15% H_2SO_4 aqueous solutions. During experiments, samples were taken out of the beaker every 2 h for ICPE measurement, and the SS cathode was weighed every 2 h to calculate the mass of copper deposited on the cathode. The variation in cell voltage during this chemical process was also recorded to compare the energy consumption. The extraction efficiency of copper was calculated as follows:

$$\text{Extraction efficiency}/\% = \frac{m_0 - m_t}{m_0} \times 100\% = \frac{C_0 \times V - m_t}{C_0 \times V} \times 100\% \quad (1)$$

where m_0 and m_t are the mass of copper before and after an electrolysis time of t, respectively; C_0 is the initial concentration of copper ions, which can be obtained by the ICPE technique; and V is the volume of the solutions.

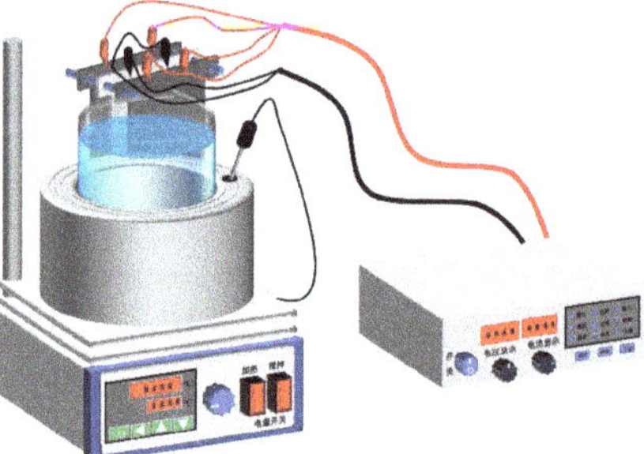

Figure 9. Schematic diagram of treatment of wastewater containing copper ions.

4. Conclusions

In summary, active granule (WC/Co_3O_4) modified Ti/Sb-SnO_2/PbO_2 electrodes were fabricated by composite electrodeposition. EDS and XPS characterization of the as-prepared electrodes suggests that active granules (WC/Co_3O_4) were successfully deposited on the surface of Ti/Sb-SnO_2/WC-PbO_2, Ti/Sb-SnO_2/Co_3O_4-PbO_2, and Ti/Sb-SnO_2/WC-Co_3O_4-PbO_2 composite electrodes. The introduction of active granules (WC/Co_3O_4) can decrease the average grain size and enhance the proportions of $O_{d\text{-}ad}$ present on the electrode surface, leading to more active sites on the electrode surface. Moreover, the excellent conductivity of the electrode caused by the presence of non-stoichiometric PbO_2 was also observed, which is favorable for improving the electron transfer and catalytic activity of the electrode. LSV measurements show that Ti/Sb-SnO_2/Co_3O_4-PbO_2 (1.85 V) and Ti/Sb-SnO_2/WC-Co_3O_4-PbO_2 (1.80 V) have lower OER values than Ti/Sb-SnO_2/PbO_2 (2.20 V) and Ti/Sb-SnO_2/WC-PbO_2 (2.12 V). Ti/Sb-SnO_2/Co_3O_4-PbO_2 is regarded as the optimal modified electrode due to its long service lifetime (684 h) and the good stability of its plating solutions. Wastewater containing copper ions was employed as a model pollutant, and Ti/Sb-SnO_2/Co_3O_4-PbO_2 was used as the anode to study the electrocatalytic activity of the modified electrodes. The experimental results demonstrate that the Ti/Sb-SnO_2/Co_3O_4-PbO_2 composite electrode exhibits remarkable extraction efficiency and low cell voltage. The color of copper deposited on the cathode became dark due to the impoverishment of copper ions. Furthermore, pilot simulation tests were carried out to research the electrodes in practical applications. A low cell voltage further verified that the composite electrodes consume less energy than other electrodes. Therefore, it is deduced that composite electrodes may be promising for the treatment of wastewater containing high concentrations of copper ions. Furthermore, it is expected that Ti/Sb-SnO_2/Co_3O_4-PbO_2 electrodes may substitute traditional Pb electrodes in resource recovery of nonferrous metals. Further studies exploring modified electrodes in specific pilot tests will be conducted in the future.

Author Contributions: Conceptualization, X.K. and S.X.; Methodology, X.K. and B.J.; Software, Q.F.; Validation, X.K., J.W., Z.W. and S.X.; Formal analysis, X.K., J.W. and Z.W.; Investigation, X.K. and Z.W.; Resources, X.K. and Q.F.; Data curation, J.W.; Writing—original draft, J.W.; Writing—review and editing, Y.W.; Supervision, X.K. and Q.F.; Project administration, B.J.; Funding acquisition, Y.W. All authors have read and agreed to the published version of the manuscript.

Funding: The authors gratefully acknowledge the financial support from the National Natural Science Foundation of China (Grant No. 21878242).

Data Availability Statement: All data supporting this study are available from authors upon request.

Acknowledgments: The Northwest Institute for Non-ferrous Metal Research is acknowledged.

Conflicts of Interest: The authors declare no conflict of interest.

References

1. Lin, B.; Zhang, G. Estimates of electricity saving potential in chinese nonferrous metals industry. *Energy Policy* **2013**, *60*, 558–568. [CrossRef]
2. Yang, Y.; Yang, L.; Wang, M.; Yang, Q.; Liu, X.; Shen, J.; Liu, G.; Zheng, M. Concentrations and profiles of persistent organic pollutants unintentionally produced by secondary nonferrous metal smelters: Updated emission factors and diagnostic ratios for identifying sources. *Chemosphere* **2020**, *255*, 126958. [CrossRef]
3. Sorour, N.; Zhang, W.; Gabra, G.; Ghali, E.; Houlachi, G. Electrochemical studies of ionic liquid additives during the zinc electrowinning process. *Hydrometallurgy* **2015**, *157*, 261–269. [CrossRef]
4. Zhang, C.; Duan, N.; Jiang, L.; Xu, F. The impact mechanism of Mn^{2+} ions on oxygen evolution reaction in zinc sulfate electrolyte. *J. Electroanal. Chem.* **2018**, *811*, 53–61. [CrossRef]
5. Karbasi, M.; Eskandar, K.A.; Elaheh, A.D. Electrochemical performance of Pb−Co composite anode during zinc electrowinning. *Hydrometallurgy* **2018**, *183*, 51–59. [CrossRef]
6. Chen, B.; Wang, S.; Liu, J.; Huang, H.; Dong, C.; He, Y.; Yan, W.; Guo, Z.; Xu, R.; Yang, H. Corrosion resistance mechanism of a novel porous Ti/Sn-Sb-RuO_x/β-PbO_2 anode for zinc electrowinning. *Corros. Sci.* **2018**, *144*, 136–144. [CrossRef]
7. Zhang, H.; Chen, J.; Ni, S.; Bie, C.; Zhi, H.; Sun, X. A clean process for selective recovery of copper from industrial wastewater by extraction-precipitation with p-tert-octyl phenoxy acetic acid. *J. Environ. Manag.* **2022**, *304*, 114164. [CrossRef]

8. Zhang, L.; Lu, Z.; Chen, P. An environmentally friendly gradient treatment system of copper-containing wastewater by coupling thermally regenerative battery and electrodeposition cell. *Sep. Purif. Technol.* **2022**, *295*, 121243. [CrossRef]
9. Zhang, X.; Zhang, S. Enhanced copper extraction in the chalcopyrite bioleaching system assisted by microbial fuel cells and catalyzed by silver-bearing ores. *J. Environ. Chem. Eng.* **2022**, *10*, 108827. [CrossRef]
10. Zhou, W.; Liu, X.; Lyu, X.; Gao, W.; Su, H.; Li, C. Extraction and separation of copper and iron from copper smelting slag: A review. *J. Clean. Prod.* **2022**, *368*, 133095. [CrossRef]
11. Kim, H.K.; Jang, H.; Jin, X.; Kim, M.G.; Hwang, S.J. A crucial role of enhanced Volmer-Tafel mechanism in improving the electrocatalytic activity via synergetic optimization of host, interlayer, and surface features of 2D nanosheets. *Appl. Catal. B Environ.* **2022**, *312*, 121391. [CrossRef]
12. Abedini, A.; Valmoozi, A.A.E.; Afghahi, S.S.S. Anodized graphite as an advanced substrate for electrodeposition of PbO_2. *Mater. Today Commun.* **2022**, *31*, 103464. [CrossRef]
13. Kandasamy, K. Electrochemical degradation of sago wastewater using Ti/PbO_2 electrode: Optimisation using response surface methodology. *Int. J. Electrochem. Sci.* **2014**, *10*, 1506–1516.
14. Torres, J.E.; Sierra, A.; Pena, D.Y.; Uribe, I.; Estupinan, H. Corrosion rate in a lead based alloy in a sulfuric acid solution at different temperatures. *Matéria* **2014**, *19*, 183–196.
15. Li, H.; Yuan, T.; Li, R.; Wang, W.; Zheng, D.; Yuan, J. Electrochemical properties of powder-pressed Pb-Ag-PbO_2 anodes. *Trans. Nonferrous Met. Soc. China* **2019**, *29*, 2422–2429. [CrossRef]
16. Ye, W.; Xu, F.; Jiang, L.; Duan, N.; Li, J.; Zhang, F.; Zhang, G.; Chen, L. A novel functional lead-based anode for efficient lead dissolution inhibition and slime generation reduction in zinc electrowinning. *J. Clean. Prod.* **2021**, *284*, 124767. [CrossRef]
17. Wang, X.; Wang, J.; Jiang, W.; Chen, C.; Yu, B.; Xu, R. Facile synthesis $MnCo_2O_4$ modifying PbO_2 composite electrode with enhanced OER electrocatalytic activity for zinc electrowinning. *Sep. Purif. Technol.* **2021**, *272*, 118916. [CrossRef]
18. Chen, S.; Chen, B.; Wang, S.; Yan, W.; He, Y.; Guo, Z.; Xu, R. Ag doping to boost the electrochemical performance and corrosion resistance of Ti/Sn-Sb-RuO_x/α-PbO_2/β-PbO_2 electrode in zinc electrowinning. *J. Alloys Compd.* **2020**, *815*, 152551. [CrossRef]
19. Zhang, F.; Zuo, J.; Jin, W.; Xu, F.; Jiang, L.; Xi, D.; Wen, Y.; Li, J.; Yu, Z.; Li, Z.; et al. Size effect of γ-MnO_2 precoated anode on lead-containing pollutant reduction and its controllable fabrication in industrial-scale for zinc electrowinning. *Chemosphere* **2022**, *287*, 132457. [CrossRef]
20. Gao, G.; Zhang, X.; Wang, P.; Ren, Y.; Meng, X.; Ding, Y.; Zhang, T.; Jiang, W. Electrochemical degradation of doxycycline hydrochloride on Bi/Ce co-doped Ti/PbO_2 anodes: Efficiency and mechanism. *J. Environ. Chem. Eng.* **2022**, *10*, 108430. [CrossRef]
21. Duan, P.; Qian, C.; Wang, X.; Jia, X.; Jiao, L.; Chen, Y. Fabrication and characterization of Ti/polyaniline-Co/PbO_2-Co for efficient electrochemical degradation of cephalexin in secondary effluents. *Environ. Res.* **2022**, *214*, 113842. [CrossRef]
22. Shao, D.; Li, W.; Wang, Z.; Yang, C.; Xu, H.; Yan, W.; Yang, L.; Wang, G.; Yang, J.; Feng, L.; et al. Variable activity and selectivity for electrochemical oxidation wastewater treatment using a magnetically assembled electrode based on Ti/PbO_2 and carbon nanotubes. *Sep. Purif. Technol.* **2022**, *301*, 122008. [CrossRef]
23. Macounová, K.M.; Pittkowski, R.K.; Nebel, R.; Zitolo, A.; Krtil, P. Selectivity of Ru-rich Ru-Ti-O oxide surfaces in parallel oxygen and chlorine evolution reactions. *Electrochimi. Acta.* **2022**, *427*, 140878. [CrossRef]
24. Preez, S.P.; Jones, D.R.; Warwick, M.E.A.; Falch, A.; Sekoai, P.T.; das Neves Quaresma, C.M.; Bessarabov, D.G.; Dunnill, C.W. Thermally stable Pt/Ti mesh catalyst for catalytic hydrogen combustion. *Int. J. Hydrog. Energy* **2020**, *45*, 16851–16864. [CrossRef]
25. Ma, D.; Ngo, V.; Raghavan, S.; Sandoval, S. Degradation of Ir-Ta oxide coated Ti anodes in sulfuric acid solutions containing fluoride. *Corros. Sci.* **2020**, *164*, 108358. [CrossRef]
26. Wang, Y.; Chen, M.; Wang, C.; Meng, X.; Zhang, W.; Chen, Z.; Crittenden, J. Electrochemical degradation of methylisothiazolinone by using Ti/SnO_2-Sb_2O_3/α, β-PbO_2 electrode: Kinetics, energy efficiency, oxidation mechanism and degradation pathway. *Chem. Eng. J.* **2019**, *374*, 626–636. [CrossRef]
27. Chen, X.; Guo, H.; Luo, S.; Wang, Z.; Li, X. Effect of SnO_2 intermediate layer on performance of Ti/SnO_2/MnO_2 electrode during electrolytic-manganese process. *Trans. Nonferrous Met. Soc. China* **2017**, *27*, 1417–1422. [CrossRef]
28. Hakimi, F.; Rashchi, F.; Ghalekhani, A.; Dolati, A.; Astaraei, F.R. Effect of a synthesized pulsed electrodeposited Ti/PbO_2-RuO_2 nanocomposite on zinc electrowinning. *Ind. Eng. Chem. Res.* **2021**, *60*, 11737–11748. [CrossRef]
29. Wang, Y.; Yang, C.M.; Schmidt, W.; Spliethoff, B.; Schüth, E.B.F. Weakly ferromagnetic ordered mesoporous Co_3O_4 synthesized by nanocasting from vinyl-functionalized cubic Ia3d mesoporous silica. *Adv. Mater.* **2004**, *17*, 53–56. [CrossRef]
30. Alhaddad, M.; Ismail, A.A.; Alghamdi, Y.G.; Al-Khathami, N.D.; Mohamed, R.M. Co_3O_4 nanoparticles accommodated mesoporous TiO_2 framework as an excellent photocatalyst with enhanced photocatalytic properties. *Opt. Mater.* **2022**, *134*, 112643. [CrossRef]
31. Dan, Y.; Lu, H.; Liu, X.; Lin, H.; Zhao, J. Ti/PbO_2 + nano-Co_3O_4 composite electrode material for electrocatalysis of O_2 evolution in alkaline solution. *Int. J. Hydrogen Energy* **2011**, *36*, 1949–1954. [CrossRef]
32. He, S.; Xu, R.; Hu, G.; Chen, B. Electrosynthesis and performance of WC and Co_3O_4 co-doped α-PbO_2 electrodes. *RSC Adv.* **2016**, *6*, 3362–3371. [CrossRef]
33. Tang, C.; Lu, Y.; Wang, F.; Niu, H.; Yu, L.; Xue, J. Influence of a MnO_2-WC interlayer on the stability and electrocatalytic activity of titanium-based PbO_2 anodes. *Electrochim. Acta* **2020**, *331*, 135381. [CrossRef]
34. Wan, C.; Zhao, L.; Wu, C.; Lin, L.; Liu, X. Bi^{5+} doping improves the electrochemical properties of Ti/SnO_2-Sb/PbO_2 electrode and its electrocatalytic performance for phenol. *J. Clean. Prod.* **2022**, *380*, 135005. [CrossRef]

35. Zhu, Y.; Zhou, W.; Yu, J.; Chen, Y.; Liu, M.; Shao, Z. Enhancing electrocatalytic activity of perovskite oxides by tuning cation deficiency for oxygen reduction and evolution reactions. *Chem. Mater.* **2016**, *28*, 1691–1697. [CrossRef]
36. Kosmulski, M.; Mączka, E. Zeta potential and particle size in dispersions of alumina in 50-50 *w*/*w* ethylene glycol-water mixture. *Colloid. Surfaces A* **2022**, *654*, 130168. [CrossRef]

 *catalysts*

Article

Degradation of Rhodamine B in Wastewater by Iron-Loaded Attapulgite Particle Heterogeneous Fenton Catalyst

Peiguo Zhou *, Zongbiao Dai, Tianyu Lu, Xin Ru, Meshack Appiah Ofori, Wenjing Yang, Jiaxin Hou and Hui Jin

College of Biology and the Environment, Nanjing Forestry University, Nanjing 210037, China; dzongbiao@163.com (Z.D.); luty.snei@sinopec.com (T.L.); r1227877604@163.com (X.R.); appiahmeshack8@gmail.com (M.A.O.); yyywj7089@163.com (W.Y.); houjiaxin@163.com (J.H.); nmlx1993@163.com (H.J.)

* Correspondence: zhoupeiguo@njfu.edu.cn; Tel.: +86-13913390757

Abstract: The water pollution caused by industry emissions makes effluent treatment a serious matter that needs to be settled. Heterogeneous Fenton oxidation has been recognized as an effective means to degrade pollutants in water. Attapulgite can be used as a catalyst carrier because of its distinctive spatial crystal structure and surface ion exchange. In this study, iron ions were transported on attapulgite particles to generate an iron-supporting attapulgite particles catalyst. BET, EDS, SEM and XRD characterized the catalysts. The particle was used as a heterogeneous catalyst to degrade rhodamine B (RhB) dye in wastewater. The effects of H_2O_2 concentration, initial pH value, catalyst dosage and temperature on the degradation of dyes were studied. The results showed that the decolorization efficiency was consistently maintained after consecutive use of a granular catalyst five times, and the removal rate was more than 98%. The degradation and mineralization effect of cationic dyes by granular catalyst was better than that of anionic dyes. Hydroxyl radicals play a dominant role in RhB catalytic degradation. The dynamic change and mechanism of granular catalysts in catalytic degradation of RhB were analyzed. In this study, the application range of attapulgite was widened. The prepared granular catalyst was cheap, stable and efficient, and could be used to treat refractory organic wastewater.

Keywords: iron-bearing attapulgite; heterogeneous Fenton; granular catalyst; rhodamine B

Citation: Zhou, P.; Dai, Z.; Lu, T.; Ru, X.; Ofori, M.A.; Yang, W.; Hou, J.; Jin, H. Degradation of Rhodamine B in Wastewater by Iron-Loaded Attapulgite Particle Heterogeneous Fenton Catalyst. *Catalysts* **2022**, *12*, 669. https://doi.org/10.3390/catal12060669

Academic Editors: Hao Xu and Yanbiao Liu

Received: 31 May 2022
Accepted: 17 June 2022
Published: 19 June 2022

1. Introduction

With the acceleration of the processes of urbanization and modernization, the economy continues to grow to meet the needs of humanity. Oil extraction, fine chemical industry, production of non-ferrous metals, pesticides, the pharmaceutical industry and cooking, printing and dyeing light industries, paper production, and other areas with high energy consumption and high pollution bring many conveniences to people's lives while causing significant damage to the environment and public health. For example, industrial wastewater often contains a lot of organic pollutants that are difficult to degrade, which not only causes serious environmental pollution but also threatens human health and safety. Therefore, governments worldwide attach great importance to it [1–3].

Among the many industrial water pollution environments, the pollution of printing and dyeing and textile wastewater to the environment is more serious. Wastewater contains many toxic and harmful pollutants that are difficult to be biodegraded [4–6]. Especially the dye wastewater has a great impact on human health and the natural environment. Rhodamine B is a highly water-soluble red dye with high toxicity. Rhodamine B is widely used in textiles, printing and dyeing, food, and other fields because of its low price. This dye is difficult to degrade in water and also causes various diseases, such as cancer [7]. At present, the treatment methods of dye wastewater include adsorption [8,9], biological treatment [10], membrane separation [11], solvent extraction [12], etc. Some treatment methods have drawbacks such as high cost, substandard water quality, and secondary pollution, etc. Considering the advantages and disadvantages of the existing treatment methods of dye wastewater, developing a new and cost-effective approach is a matter of great concern.

Advanced oxidation technology is an innovative treatment method, which contains highly active oxidation free radicals and can effectively degrade different organic pollutants in wastewater [13–15]. To better degrade pollutants in water, advanced oxidation technology has implemented a variety of methods, such as electrochemical oxidation [16–18], ozone oxidation [19], photocatalytic oxidation [20], ultrasonic oxidation [21], catalytic wet oxidation [22], and Fenton oxidation [23]. Fenton oxidation is superior to other advanced oxidation technologies in the simplicity of reactions and operation, rapid and efficient degradation of target pollutants, low cost, and safety for the environment. Hydroxyl radicals produced by Fenton reaction destroy the structure of organic matter and mineralize it into carbon dioxide and water. However, the traditional homogeneous Fenton system has some problems, such as narrow reaction pH range, iron sludge generation, difficult recycling and catalyst reuse, which minimize its application. Researchers fixed Fe^{2+} (Fe^{3+}) or iron oxides (Fe^{II} and Fe^{III}) in the Fenton reagent on the support to form a heterogeneous Fenton system to overcome these drawbacks. Various materials have been studied and developed as heterogeneous Fenton catalyst supports, such as molecular sieve, enriched alumina, silica, mineral clay, transition metal, and iron oxide composites, etc. [24]. Researchers fabricated a heterogeneous catalyst system supported by iron on a modified molecular sieve to treat non-degradable materials in molasses distillation wastewater. Studies showed that the decolorization rate was 90% and the TOC removal rate was 60% after using the sulfuric acid alternative catalyst [25]. Hernandez-Olono et al. studied the degradation of rhodamine B, methyl orange and methylene blue by precipitation on alumina under UV irradiation [26]. It was found that the degradation rates of Rhodamine B, methyl orange and methylene blue on Fe/Al_2O_3 dried under H_2O_2 during UV irradiation were 99.6%, 100%, and 99.1%, respectively. Using roasted red clay as the iron source, a heterogeneous light-Fenton method was constructed to remove C.I. Acid Red 17 (AR17), and the study showed that decolorization efficiency was as high as 94.71% [27]. Hu et al. used four kinds of containing mesoporous silica materials for catalytic treatment of chlorophenol wastewater. They found that Cu-doped SBA-15 with Cu/Si mole ratio of 0.133 had the best catalytic activity for oxidative degradation of 4-chlorophenol. The apparent reaction rate constant was 0.170 min^{-1}, which was 1.3~3.6 times that of other Cu-containing catalysts [28]. Chen et al. prepared Mn-Fe_3O_4/RGO composite as a multiphase photo-Fenton catalyst for degradation of Rhodamine B (RhB) [29]. Under natural pH conditions, the low dosage of catalyst was 0.2 g/L within 80 min, and the degradation rate reached 96.4%. When pH was 2 and 11, the removal rates were 91% and 85%, respectively. After 10 cycles, the catalyst still maintained a high degradation efficiency of about 90%. In addition to the above materials, many scholars have also developed bimetal composite materials as heterogeneous Fenton catalysts to degrade organic pollutants [30–33]. However, these catalysts are complex, expensive, rare and require high cost, which may lead to secondary pollution. Thus, their application in the catalytic advanced oxidation process is limited. Therefore, developing low-cost, environmentally friendly and naturally available heterogeneous Fenton catalysts for water treatment is desirable.

Attapulgite, as a common clay mineral, is abundant in yield and low in price. It also has a large specific surface area, unique spatial crystal structure and surface ion-exchange property. In addition, attapulgite has a good adsorption capacity for organic pollutants in water, which can be used as a cheap natural adsorbent for research in the field of wastewater treatment [34,35]. The rupture of the Si-O bond forms the electronegativity of Si-OH and the loss of coordination water in attapulgite. Therefore, positively coordinated metal ions in promulgating are usually used as catalysts. Cao et al. supported CuO particles on attapulgite and studied the catalytic oxidation of CO by the CuO/ATP catalyst at a low temperature, obtaining a catalytic effect similar to that of metal oxide-supported CuO [36]. Wang et al. used precipitation, impregnation and mechanical blending methods to load

Ni on attapulgite to generate a Ni/ATP catalyst for bio-oil hydrogen production [37]. The results showed that the catalytic hydrogen production rate of Ni/ATP prepared by the precipitation method reached 82%, and the conversion rate of acetic acid reached 85%, which was considerably lower than the Ni/ATP catalyst prepared by the impregnation method. Ma et al. supported platinum nanocatalysts with HCl acidified attapulgite and the results showed remarkable catalytic activity and selectivity for p-chloronitrobenzene, m-chloronitrobenzene, and o-chloronitrobenzene [38].

In this work, iron ions were supported on attapulgite particles and used as heterogeneous Fenton catalysts to degrade RhB dye in wastewater. The basic properties of the granular catalyst were studied. At the same time, the effects of initial concentration, dosage, initial pH value, temperature, and H_2O_2 concentration on the catalytic degradation effect were investigated to find the best catalytic reaction conditions. Under the optimum catalytic reaction conditions, the reusability, stability and degradation effect of different dyes of granular catalysts was studied. Finally, the dynamic change of catalytic degradation of dye RhB by a granular catalyst was studied, the formation and reaction position of a reactive free radical were determined, and the catalytic degradation mechanism of a granular catalyst was analyzed and speculated upon.

2. Results and Discussion

2.1. Catalyst Characterization

2.1.1. BET and EDS Analysis of Catalyst

The chemical elements of the granular catalyst were analyzed. As shown in Table 1, the main elements of the catalyst were Si and O. This was consistent with the XRD pattern. The mass percentage of iron in the catalyst was 12.09%. EDS analysis showed that there was iron in the granular catalyst, which led to the Fenton reaction.

Table 1. EDS spectrograms of the Fe-loaded attapulgite particles and distribution of elements.

Element	O	Mg	Al	Si	K	Ti	Fe	the Total
Weight percentage (wt.%)	54.44	2.3	4.9	24.42	1.21	0.63	12.09	100

Figure 1a shows the catalyst's nitrogen adsorption and desorption isotherms, which were type IV isotherms [39], indicating that the internal pores of the catalyst were mainly mesoporous, with a BET-specific surface area of 119.5409 m^2/g and an average pore size of 11.1603 nm. The catalyst had good adsorption performance.

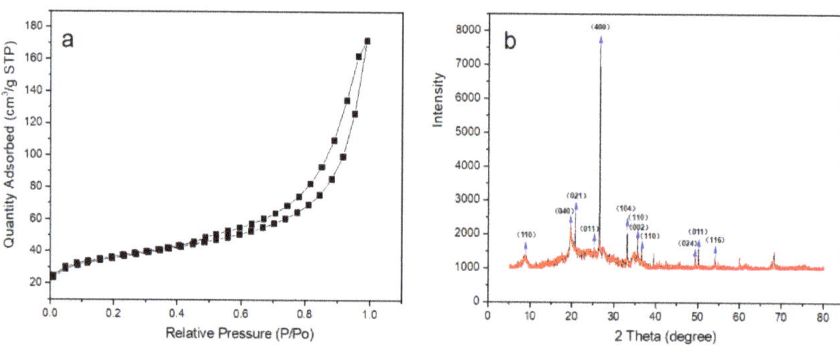

Figure 1. N_2 adsorption/desorption isotherm of granular catalyst at 77 K (**a**) and XRD (**b**).

Figure 1b XRD spectra of iron-loaded catalysts shows that attapulgite characteristic diffraction peaks appeared at $2\theta = 8.5°$, 19.8°, 20.3°, 27.9°, 35.7°, etc. The corresponding crystal plane spacing was D (110) = 10.3974 nm, D (040) = 4.4725 nm, D (021) = 4.3829 nm,

D (400) = 3.1942 nm, D (002) = 2.5139 nm, respectively. The SiO_2 characteristic diffraction peaks appeared at 2θ = 26.7°, 36.6° and 50.2°, and the corresponding crystal plane distances were D (011) = 3.3421 nm, D (110) = 2.4560 nm, D (011) = 1.8171 nm, respectively. In addition, α-Fe_2O_3 cubic spinel characteristic diffraction peaks were found at 2θ = 33.2°, 35.6°, 49.5°, 54.1°, etc. The corresponding crystal plane spacing was D (104) = 2.6989 nm, D (110) = 2.5171 nm, D (024) = 1.8408 nm, D (116) = 1.6943 nm, but the overall peak strength was much weaker than Fe_2O_3 crystal. This could be because only a small part of $Fe(NO_3)_3$ was converted into the α-Fe_2O_3 crystal in the calcination process of the iron-loaded catalyst, while most of the remaining $Fe(NO_3)_3$ could be converted into other crystal iron oxides or only attached to the catalyst surface in the form of some chemical bonds without showing the corresponding crystal structure.

2.1.2. SEM Analysis of Catalyst

Analysis of the particle catalyst before and after the reaction with scanning electron microscopy (SEM) can be seen in Figure 2; Figure 2b shows the particle catalyst after reaction, and Figure 2a shows the reaction front of the catalyst particles. It can be seen that the internal structure has been greatly damaged, the rod between the crystal is an organic whole repeatedly, and the initial needle bar gradually formed a similar layered structure, but most of the channels remain and the catalyst can be recycled many times.

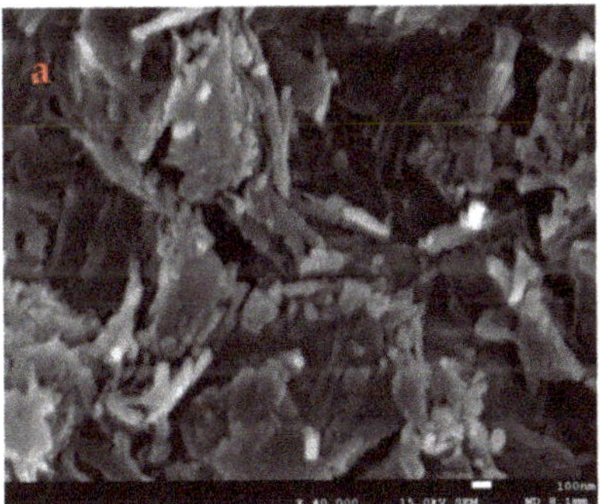

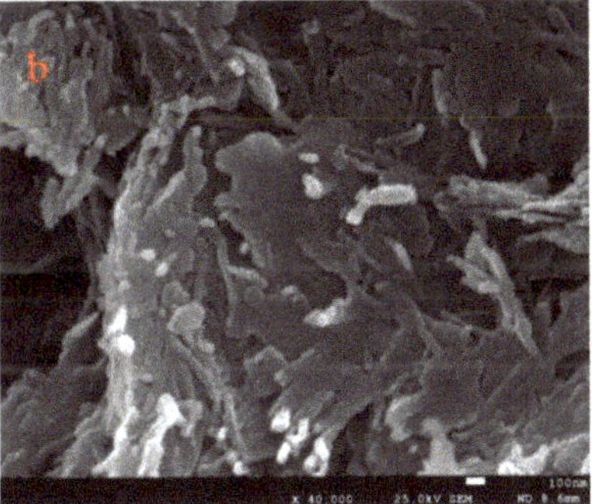

Figure 2. SEM image of catalyst before (**a**) and after (**b**) reaction (×40,000).

2.2. *Influence of Reaction Conditions on Degradation Performance*

2.2.1. Influence of Different Reaction Systems on RhB Degradation

It can be seen from Figure 3a,b that in different reaction systems, the degradation and mineralization effects of RhB were not ideal when a hydrogen peroxide solution was added only at 35 °C and pH 3. After 240 min, the removal rate of RhB and the TOC mineralization rate were only 5.98% and 0.92%, respectively. When only a granular catalyst was added to the reaction solution without H_2O_2, the concentration of RhB and TOC decreased in the same trend, indicating that the reduced RhB concentration was only adsorbed by the granular catalyst, no molecular ring-opening or degradation mineralization occurred, and the adsorption amount reached 39.61% at 240 min. When 0.5 mg/L Fe^{3+} and 196 mmol/L H_2O_2 solutions were added to the reaction solution to form a homogeneous Fenton system, the homogeneous Fenton reaction caused by a small amount of dissolved Fe^{3+} insignificantly contributed to the catalytic degradation of RhB in the system regarding the RhB removal rate and mineralization effect. Therefore, it can be seen that the degradation of dye-RhB by an iron-loaded attapulgite granular catalyst was mainly contributed to by a heterogeneous Fenton reaction, which conformed to the heterogeneous surface catalytic mechanism. When the particle catalyst was added to join the H_2O_2 in the reaction solution, which constituted the similar Fenton system, the reaction temperature and solution pH value also affected their degradation performance. With the increase of temperature, the molecular energy increases and the movement is violent, the amount of active free radical increases, and the oxidative degradation rate is accelerated. At the same time, the influence of pH on the degradation performance was different: in terms of decolorization effect, it was alkaline > acidic > neutral, and in terms of mineralization effect, it was acidic > neutral > alkaline.

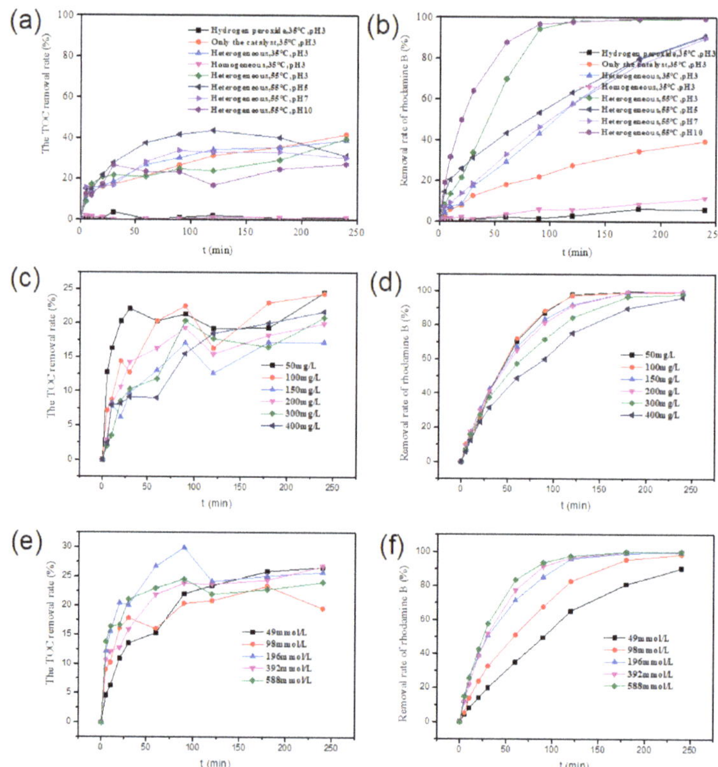

Figure 3. Effects of different reaction systems (**a**,**b**) (c_0 = 100 mg/L, $c_{H_2O_2}$ = 196 mmol/L, catalyst = 10 g/L); (**c**,**d**) RhB initial concentration ($c_{H_2O_2}$ = 196 mmol/L, catalyst = 10 g/L, pH = 10, T = 55 °C); (**e**,**f**) H_2O_2 concentration (c_0 = 200 mg/L, catalyst = 10 g/L, pH = 10, T = 55 °C); on degradation and mineralization rate of Rhodamine B.

2.2.2. Influence of RhB Initial Concentration

Figure 3c,d shows the effect of initial dye concentration on RhB dye degradation. It can be seen that the degradation and decolorization reaction of RhB took place before the mineralization reaction, indicating that the decolorization of the dye was easier than the mineralization, and the lower the initial concentration, the faster the degradation and mineralization rate. When the initial concentration was 50 mg/L, the removal rate of RhB was close to 90% after 90 min, the decolorization was almost complete after 120 min, and the removal rate of TOC reached 24.52% after 240 min. With the increase of RhB's initial concentration, the decolorization and mineralization rates gradually decreased. The removal rate of 400 mg/L RhB solution reached 90% after 180 min, and the TOC removal rate was 21.73% after 240 min. This could be because the higher the concentration of RhB, the more molecules per unit volume. In addition, the generation of reactive free radicals was constant, so the degradation and mineralization rate of RhB decreased with the increase in concentration [40]. Considering the obvious degradation effect, the RhB solution with an initial concentration of 200 mg/L was selected for subsequent experiments.

2.2.3. Influence of Different H_2O_2 Concentrations

In the heterogeneous Fenton system, the concentration of hydrogen peroxide affected the amount of active free radicals [41], and the decolorization and mineralization efficiency of RhB dye in the solution. As can be seen from Figure 3e,f, when the concentration of H_2O_2 in the reaction system increased from 49 mmol/L to 588 mmol/L, the oxidation decolorization and mineralization rate of the RhB solution increased gradually. After 240 min of reaction, the degradation rate of RhB increased from 90.17% to 99.99%. It is verified that the RhB solution can be decolorized to a certain extent by adding a low concentration of H_2O_2 in the heterogeneous Fenton system, while the removal rate of TOC does not increase significantly and is stable at about 25%. Meanwhile, it also indicates that the free radicals generated in the alkaline reaction system only destroy the chromophore group of the RhB molecule. Therefore, dye decolorization took precedence over mineralization because only a small part of it can eventually be mineralized into CO_2 and H_2O. When the concentration of H_2O_2 in the reaction system increased to 196 mmol/L, the decolorization rate reached 95.92% after 120 min, which greatly shortened the reaction time and accelerated the decolorization rate compared with a lower H_2O_2 concentration. The RhB removal rate reached 99.77% after 240 min. This could be because the reaction's amount of reactive free radicals increased with an increase in H_2O_2 concentration, more free radicals attacked the RhB molecule, and the removal rate of RhB and TOC further improved [42]. After that, increasing the concentration of H_2O_2 in the reaction system had little contribution to the oxidative degradation and mineralization effect of RhB, and excessive H_2O_2 in the solution could lead to a self-annihilation reaction or a series of ineffective reactions with the active free radicals generated in the catalytic reaction process [43,44]. As a free radical scavenger, active oxidation substances originally used to attack RhB molecules in the reaction system were consumed, affecting oxidative degradation [45,46]. On the other hand, when the amount of catalyst in the reaction system was fixed and the concentration of H_2O_2 was properly increased, the number of collisions between the H_2O_2 molecule and catalyst increased per unit time and the reaction rate accelerated. Due to the limited amount of catalyst, the collision number of H_2O_2 concentration did not increase any further, which was macroscopically expressed as the RhB removal rate slowing down. Therefore, considering the oxidation decolorization and degradation of RhB, an appropriate H_2O_2 concentration of 196 mmol/L was selected under alkaline conditions.

2.2.4. Influence of Different Catalyst Dosages

As can be seen from Figure 4a,b, under the conditions of the above appropriate initial concentration, pH value and hydrogen peroxide concentration, the degradation and mineralization rate of RhB was significantly accelerated with the increase in catalyst dosage. The removal rate of RhB was only 78.81% after 120 min in the system with a catalyst dosage of 2 g/L. At the same time, the removal rate of RhB in the system with the dosage of 20 g/L reached 98.73%, which was nearly 20% higher. The dosage of the catalyst was positively correlated with the catalytic efficiency of the reaction system to a large extent, indicating that the addition of catalysts provided more adsorption sites on the one hand, which could absorb more RhB molecules. The removal rate of RhB and TOC was improved. On the other hand, the increase in catalyst dosage could provide more reactive sites to a greater extent, and the rate and number of oxidative active free radicals generated by the catalyst and H_2O_2 reaction are greatly improved [47,48], which increased the degradation efficiency of the reaction system. When the amount of catalyst was low, the reaction rate of the heterogeneous Fenton system was slow. With greater dosage, the catalytic reaction is inhibited to a certain extent, and the occurrence of side reactions is promoted, affecting the rate of oxidative degradation. As can be seen from the TOC removal rate, when the catalyst dosage increased, the TOC removal rate in the reaction system increased first and then decreased, because the greater the dosage of the catalysts, the greater the adsorption of RhB molecule in the system at the beginning, and the higher the TOC removal rate. As the reaction progressed, the adsorbed RhB molecule was gradually oxidized and degraded into small molecules by active free radicals. However, the particle catalyst prepared in this experiment had a weak adsorption capacity for intermediate products of these small molecules, so it was desorbed from the pore channel of the particle catalyst to the liquid solution, which showed a temporary decrease in TOC removal rate. Subsequently, these small molecule intermediates were further mineralized and degraded into CO_2 and H_2O, so the TOC removal rate increased again. From an economic perspective, the appropriate catalyst dosage for subsequent experiments was 10 g/L.

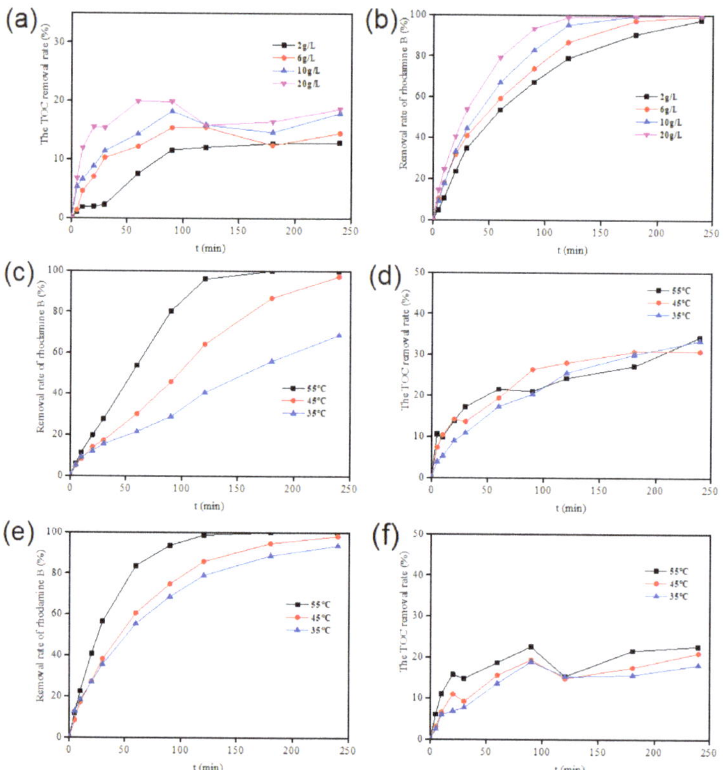

Figure 4. Effects of different reaction temperature (**c**,**d**) (c_0 = 200 mg/L, $c_{H_2O_2}$ = 196 mmol/L, pH = 3, catalyst = 10 g/L), (**e**,**f**) (c_0 = 200 mg/L, $c_{H_2O_2}$ = 196 mmol/L, pH = 10,catalyst = 10 g/L); catalyst dosage (**a**,**b**) (c_0 = 200 mg/L, $c_{H_2O_2}$ = 196 mmol/L, pH = 10, T = 55 °C) on the removal rate of rhodamine B.

2.2.5. Influence of Initial pH Value

In a heterogeneous Fenton reaction system, the change of pH greatly influences the oxidation decolorization and mineralized degradation of dyes [41,49]. As can be seen from Figure 5, in terms of the RhB decoloring effect, it was alkaline condition > acid > neutral conditions, and in terms of the RhB mineralization effect, it was the acid condition > neutral condition > alkaline conditions. According to the literature, the application of the pH range in the mentioned conventional Fenton system has a very big distinction [50].

Under neutral conditions, it can be seen that pH had a negligible effect on the oxidation, decolorization, and degradation of RhB, and the decolorization rate was very slow. The decolorization rate only reached approximately 50% after 120 min of reaction, and the oxidation degradation effect was relatively weak due to the adsorption of the RhB molecule in the solution by a particle catalyst. Under alkaline conditions, with the increase in pH, the decolorization rate of RhB by the granular catalyst increased gradually, and the decolorization rate reached more than 99% in the same reaction time. However, the TOC removal rate was lower than that under neutral and acidic conditions, and the TOC removal rate was about 25%. The results showed that only a small part of RhB molecules adsorbed by the granular catalyst could be eventually mineralized into CO_2 and H_2O. Under the acid condition, the oxidation of RhB was also very quick, and the decolorization rate was improved with the reduction of pH. Finally, with the initial pH value of 3 and a reaction time of 120 min, the decolorization rate reached 96%, which was far higher than that of the neutral condition. Compared to neutral conditions and acidic conditions, the removal rate of TOC experienced a period of growth, and the process of further decrease and increase was consistent with the rule mentioned in the previous sections, which also indicated that the granular catalyst first adsorbed the RhB molecule, then degraded into small molecule intermediates, was desorbed, and finally was mineralized into CO_2 and H_2O, which conformed to the heterogeneous surface catalysis mechanism.

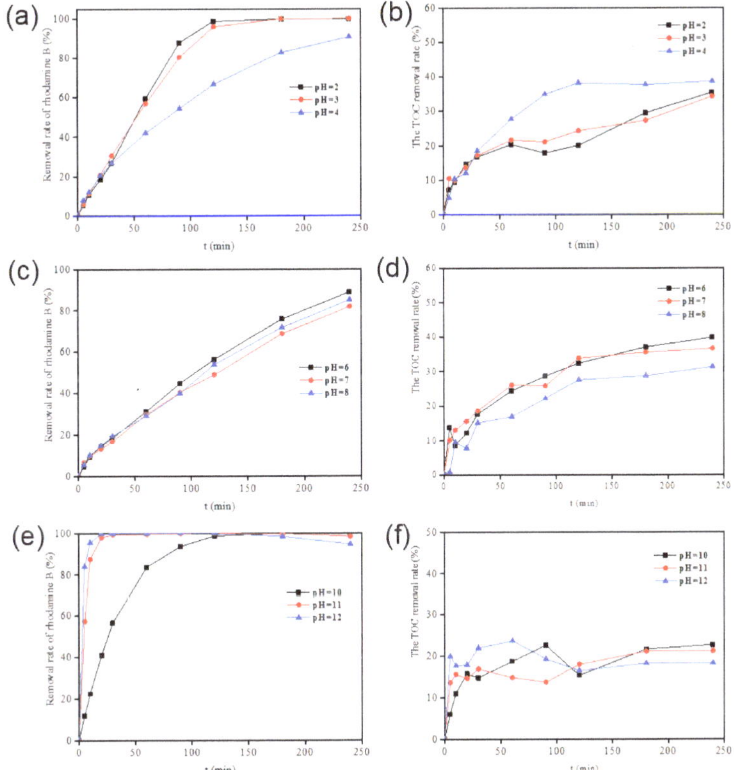

Figure 5. Effects of different pH values on the removal rate and mineralization rate of rhodamine B (c_0 = 200 mg/L, catalyst = 10 g/L, T = 55 °C, $c_{H_2O_2}$ = 196 mmol/L).

On the whole, it can be seen that the decolorization effect of RhB was better in alkaline conditions than in acidic conditions. The mineralization effect of dye was better in acidic conditions than in alkaline conditions because H_2O_2 was relatively stable in acidic conditions. At the same time, H_2O_2 was easy to decompose into peroxide radical (•OOH) and H^+ in alkaline conditions. Under the condition of the constant addition of the granular catalyst and hydrogen peroxide in the system, the rate and number of free radicals generated under acidic conditions were low, while the rate and number of free radicals generated mainly by •OH were high under alkaline conditions, mainly by •OOH. The strong oxidation capacity of •OH can further mineralize the small molecule intermediates into CO_2 and H_2O, while •OOH is difficult to further mineralize these intermediates. Therefore, the mineralization effect under alkaline conditions is inferior to that under acidic conditions. However, the pH of the reaction solution should not be too high or too low. Moreover, excessive acid and alkali environments destroyed the structure of the granular catalyst, resulting in particle dissolution and iron ion loss. On the other hand, the concentration of •OH was inversely proportional to the concentration of $[OH^-]$, and the formation of •OH was inhibited when the pH of the reaction solution was too low [51]. The conversion balance between Fe^{III} and Fe^{II} on the catalyst surface was destroyed, affecting the catalytic reaction [52], and the specific reaction mechanism remains to be explored. Nevertheless, the particle catalyst prepared in this study solved the problem of pH limitation of the traditional homogeneous or heterogeneous system very well [53] and still had a good decolorization effect under alkaline conditions, which was conducive to the extensive application of the heterogeneous Fenton system.

2.2.6. Influence of Temperature on Degradation of Rhodamine B

The effect of Fenton-like oxidation reaction on decolorization efficiency and mineralization of 200 mg/L RhB dye solution at different temperatures was studied. From the acidic (c) and (d) groups with pH 3 and alkaline (e) and (f) groups with pH 10 in Figure 4, the catalyst dosage was 10 g/L and H_2O_2 concentration remained unchanged at 196 mmol/L. As can be seen from Figure 5, regardless of the system, with the increase in temperature, the oxidation degradation rate was accelerated, RhB removal rate and the removal rate of TOC increased. When pH was 3 and reaction temperature was 35 °C, the reaction rate of RhB had an obvious rise after fall, and this trend remained constant. This was because the particle catalyst only had an adsorption effect on RhB at the beginning, or the adsorption rate was much higher than the oxidative degradation rate. With the progress of the reaction, a large amount of H_2O_2 entered into the iron oxide inside the particle catalyst and on the surface of the particle catalyst to form a heterogeneous Fenton system and generate active free radicals. The RhB molecules adsorbed on the pore surface were oxidized and degraded into small molecules that fall off from the surface, and then continued to adsorb RhB molecules in the solution, forming a cycle of adsorption–degradation–desorption–re-adsorption–re-degradation–re-desorption. At that time, the oxidation degradation process was dominant, so the reaction rate in the system increased twice. The energy was required for the formation of active free radicals, so the higher the temperature in the system, the higher the energy [54], the greater the probability of breaking the O-O H_2O_2 bond and the higher the concentration of active free radicals, which significantly accelerated the process of oxidative degradation [55]. Therefore, there was no secondary increase in the reaction rate when the reaction temperature was 55 °C. In actual industrial applications such as printing and dyeing, the wastewater discharge temperature is higher, so the reaction temperature of 55 °C was chosen for the follow-up experiment.

2.3. *Study on Sustainability and Stability of the Granular Catalyst*

It is an ideal research target for heterogeneous Fenton catalysts with certain regeneration, reuse and good stability. The particle catalyst reached a saturation state after 360 min. Therefore, H_2O_2 was added after the particle catalyst was adsorbed for 360 min to investigate the change in the concentration of the RhB dye under the optimal catalytic reaction conditions. As can be seen from Figure 6, only when a granular catalyst was presented in the solution, the RhB residual rate was 50.26% after 360 min. After that, with the increase in adsorption time, the concentration of the RhB dye did not decrease much. At 480 min, the residual rate of RhB was still as high as 47.07%, which indicated that the adsorption of RhB on the granular catalyst reached saturation. Therefore, a certain amount of H_2O_2 was added immediately after the particle catalyst was adsorbed and saturated. It can be seen that under both acidic and alkaline conditions, the concentration of RhB in the solution decreased rapidly, and the residual rate of RhB was less than 1% after 120 min, indicating that after adding H_2O_2, the particle catalyst could effectively degrade and completely decolorize the RhB molecules adsorbed on the surface of particle channels, so the prepared particle catalyst had a certain in situ regeneration performance.

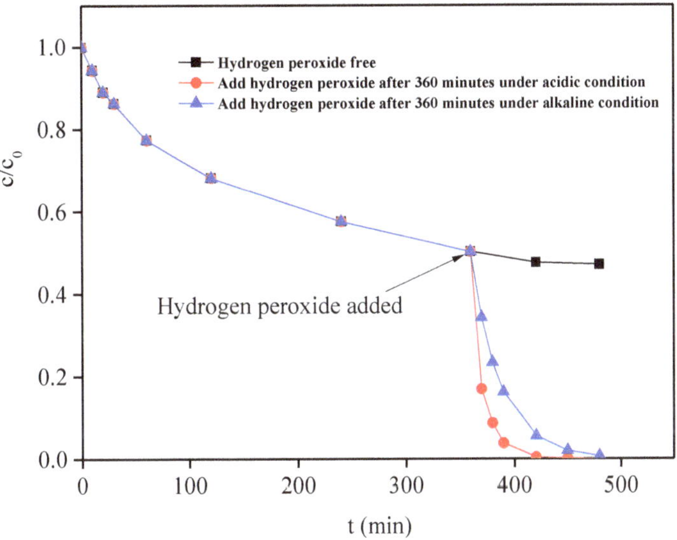

Figure 6. Concentration change of H_2O_2 on Rhodamine B after adsorption equilibrium of granular catalyst.

In practical engineering applications, the catalyst itself must be reusable and have an excellent stable structure [56,57], so the reusability of the annular catalyst is also an important evaluation index. The granular catalyst was put into a 100 mL RhB solution with an initial concentration of 200 mg/L; the catalyst dosage was kept at 10 g/L, the H_2O_2 concentration was 196 mmol/L, the pH value was 3 and 10, and the solution temperature was 55 °C. The reaction was repeated five times, the granular catalyst was removed and washed then dried with deionized water. As can be seen from Figure 7a,b, RhB can be effectively decolorized and degraded by a granular catalyst 180 min after five repeated reactions under acidic and alkaline conditions. Decoloring performance remained the same, and the basic removal rate was stable and constituted over 98%, which showed that the preparation of catalysts had good reusability. However, the mineralization effect of RhB decreased. After the first use of granular catalyst, the mineralization rate of RhB reached 28.11% (pH = 3) and 25.75% (pH = 10). After repeated reuse, the mineralization rate decreased. The mineralization rate of the fifth use was only 22.16% (pH = 3) and 16.28% (pH = 10), as shown in Figure 7c. This was due to the dissolution of a small amount of infirm bound iron ions on the catalyst's surface after each repeated reuse, as shown in Figure 7d. Another reason was that a small amount of organic matter in the previous reaction was adsorbed and remained on the surface of the catalyst channel, resulting in the reduction of active sites and the TOC removal rate.

2.4. Study on the Degradation Effect of Different Dyes by the Granular Catalyst

The above results showed that the granular catalyst had a significant degradation effect on RhB. The feasibility of its application in dye wastewater can be better clarified through the degradation experiment of different dyes. Methylene blue and Congo red were used as the target pollutants to degrade as RhB in the same reaction system. Figure 8 showed that under alkaline conditions, the reaction rate of cationic RhB was faster than that of the acidic conditions, and the TOC removal rate in the acidic conditions was higher than that of alkaline conditions. The other cationic dye (methylene blue) also showed a similar trend; the difference was a good effect on the mineralization of the methylene blue on rhodamine B. For cationic dyes, the alkaline condition of the heterogeneous Fenton system was conducive to the oxidation and decolorization of dyes, while the acidic condition was conducive to the degradation and mineralization of dyes. For anionic dyes like Congo red, the decolorization and mineralization effects were worse than those of cationic dyes, which was because the surface of ferric attapulgite granular catalyst prepared in this experiment was partially replaced by Al^{3+} with Si^{4+} in the 4-valence coordinates, and by Mg^{2+} with Al^{6+} in the 6-times coordinates, resulting in electronegativity on the surface of the granular catalyst. Therefore, it has a certain repulsive effect on anionic dyes.

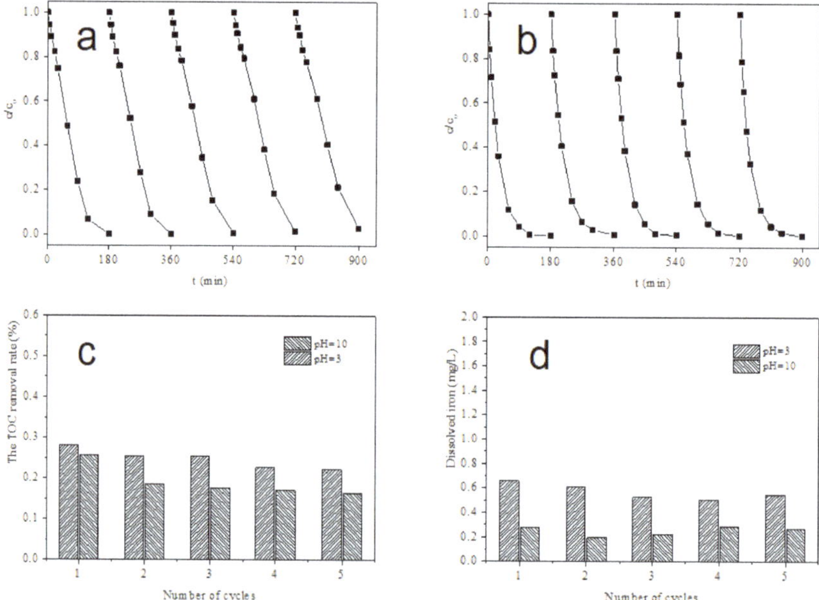

Figure 7. Reusability of the granular catalyst under acidic (**a**) and alkaline (**b**) conditions and mineralization rate of Rhodamine B (**c**) and iron dissolution amount (**d**).

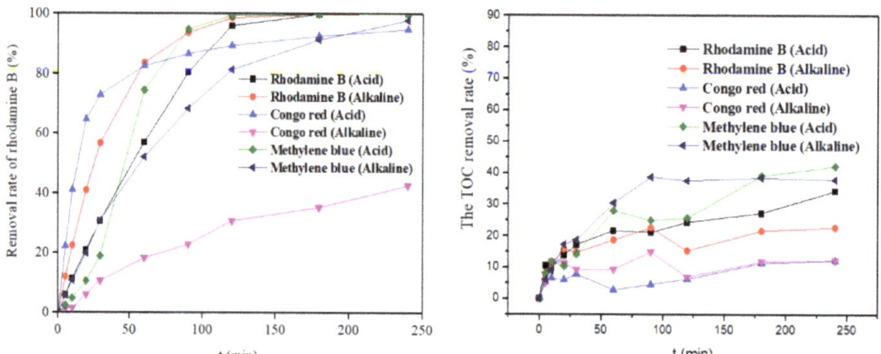

Figure 8. Degradation and mineralization of different dyes (c_0 = 200 mg/L, $c_{H_2O_2}$ = 196 mmol/L, T = 55 °C, catalyst = 10 g/L).

2.5. Dynamic Study on the Catalytic Degradation Process of the Granular Catalyst

2.5.1. Dynamic UV-VIS Spectra of RhB during the Reaction Process

The degradation process of RhB was described qualitatively by UV-VIS spectroscopy. Under the optimal reaction conditions, samples were taken at different times for dilution and placed in a quartz colorimetric dish for full-wavelength scanning within the wavelength range of 250–800 nm, as shown in Figure 9. It can be seen from the figure that the RhB solution had the maximum characteristic absorption peak at 554 nm in the visible region, which was mainly caused by the benzene amino group, carbonyl group, and four ethyl groups in RhB molecular structure [58]. According to relevant literature, with the progress of the reaction process, the degradation of RhB molecule mainly went through n-site diethyl, destruction of the large conjugated structure, ring-opening and mineralization, etc. The macroscopic manifestation was the gradual fading of dye color, and the absorption peak at 554 nm of the UV-VIS spectrum in the figure also gradually weakened. At 180 min, it can be seen that the absorption peak almost disappeared, indicating that the oxidation decolorization of RhB was basically completed and decomposed into small molecule intermediates or completely mineralized into CO_2 and H_2O. This was due to the similar activity of the Fenton process for the production of oxygen free radicals, which directly attacked the molecular structure of the RhB conjugate with a large variety of anthracene (i.e., benzene amine and carbonyl), which led to its cracking and loss of characteristic absorption peak. At the same time, the hydroxylation effect led to the redshift of the absorption peak, and the N-ethyl process caused a blue shift in its absorption peak [59]. Therefore, it can be concluded that the degradation process of the RhB molecule in this experiment is the simultaneous destruction of n-site diethyl and large conjugated oxanthracene structures, and the red-shift and blue-shift effects cancel each other out, thus only showing a decrease in absorbance.

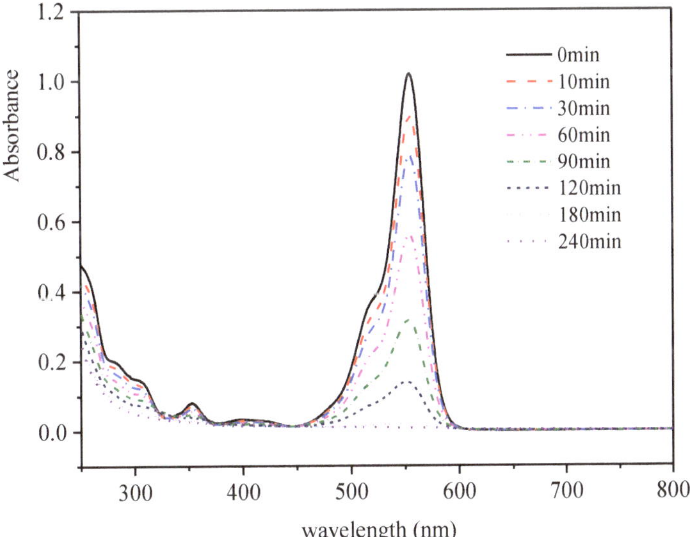

Figure 9. UV-vis spectra of RhB degradation (c_0 = 200 mg/L, $c_{H_2O_2}$ = 196 mmol/L, T = 55 °C, catalyst = 10 g/L, pH = 3).

2.5.2. Determination of Active Substances in the Reaction Process

In order to investigate whether the prepared granular catalyst was a free radical reaction in catalytic oxidation of RhB, the free radical trapping agent ascorbic acid was added in the reaction process to determine the presence of free radicals in the catalytic system according to the degradation of RhB [60]. Active substances that can catalyze reactions in heterogeneous Fenton systems mainly include hydroxyl radical (•OH) [61], peroxide radical (•OOH) [62], and high-valent iron (Fe^{IV}) [63]. To confirm the types of free radicals playing a leading role in the heterogeneous Fenton system, different concentrations of tert-butanol (TBA) and potassium iodide were added as hydroxyl radical trapping agents under the optimal reaction conditions [64–66] to determine the existence of hydroxyl radicals in the reaction process. At the same time, a certain concentration of p-benzoquinone was added to the reaction solution to determine whether there were peroxy radicals in the reaction process.

As can be seen from Figure 10a, the addition of ascorbic acid led to a gradual decline in the degradation rate of RhB. When the amount of ascorbic acid was 2.5 mmol/L, the residual rate of RhB was almost equal to the adsorption rate of the catalyst, which preliminarily proved that the heterogeneous Fenton reaction system was a free radical reaction. As can be seen from Figure 10b,c, the catalytic degradation rate of the reaction solution with tert-butanol was reduced, indicating that a hydroxyl radical existed in the reaction solution and participated in the catalytic degradation process. However, as the concentration of tert-butanol increased, the degradation of RhB was not completely inhibited. The degradation rate was still 74.97% after 240 min, indicating that there were other active substances in the reaction system. After adding 0.2 mmol/L potassium iodide, the degradation rate of RhB decreased, and the degradation rate of RhB was 79.68% after 180 min. Potassium iodide captured hydroxyl radicals on the surface of the granular catalyst, but its inhibition was relatively small compared with tert-butanol, indicating that the granular catalyst catalyzed the degradation of RhB. The RhB molecules were first adsorbed on the surface of the granular catalyst and then degraded, while hydroxyl radicals catalyzed a small part of RhB molecules in the liquid phase. Figure 10d showed that under the acid condition, the degradation of RhB inhibition was limited, indicating that the heterogeneous Fenton system produced a very little amount of oxygen free radicals, from which it can be concluded that the hydroxyl free radical reaction in the process of catalytic degradation of RhB was dominant.

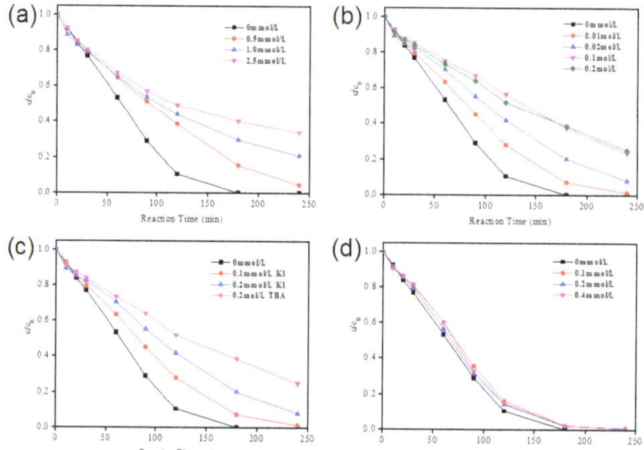

Figure 10. Effects of different types of free radical trapping agents on degradation of rhodamine B: (**a**) Ascorbic acid; (**b**) TBA; (**c**) KI and TBA; (**d**) P-benzoquinone(c_0 = 200 mg/L, $c_{H_2O_2}$ = 196 mmol/L, T = 55 °C, catalyst = 10 g/L, pH = 3).

2.5.3. Formation Rule of Hydroxyl Radical (•OH)

According to the above analysis, hydroxyl radical reaction plays a dominant role in catalytic degradation of RhB. To investigate the rule of hydroxyl free radical, a certain concentration of coumarin solution was used as the trapping agent of the hydroxyl free radical, and the pH of the solution was 3 under the optimal reaction conditions. The influence on the rule of hydroxyl free radical generation was investigated by changing the dosage of the catalyst and the hydrogen peroxide concentration.

Figure 11a,b shows that with an increase in the particle catalyst dosage, the generation of free radicals with similar heterogeneity in the Fenton system gradually increased, and H_2O_2 consumption also increased. This could be caused by the increase in the dosage of the particle catalyst which brought more adsorption sites and its reactivity [47] and increased the adsorption quantity of the catalyst. In addition, it increased the contact area of the catalytic reaction. H_2O_2 was in contact with more iron oxides, and its decomposition rate was accelerated, thus increasing the production of hydroxyl radical.

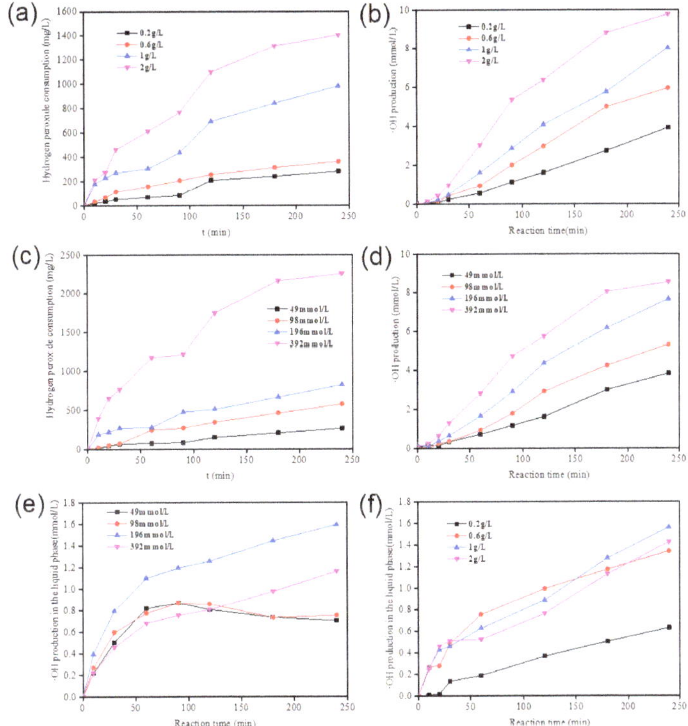

Figure 11. Formation rule of hydroxyl radical under different conditions: (**a**,**c**) The consumption of H_2O_2; (**b**,**d**) •OH production; (**e**) H_2O_2 concentration (T = 55 °C, catalyst = 10 g/L, pH = 3); (**f**) catalyst dosage ($c_{H_2O_2}$ = 196 mmol/L, T = 55 °C, pH = 3).

Figure 11c,d showed that with the increase in H_2O_2 concentration in the solution, the production of hydroxyl radical in the heterogeneous Fenton system also gradually increased, which could be due to the increase in H_2O_2 concentration under the condition of maintaining the same number of adsorption sites and reactive sites. As a result, more H_2O_2 could enter the particle catalyst and contact with the iron oxide on the pore surface, increasing the contact frequency (the number of collisions per unit time) and resulting in a faster decomposition rate of H_2O_2 and production of hydroxyl radicals [67]. However, the generation rule of liquid hydroxyl radicals in the system was slightly different from that of total hydroxyl radicals, as shown in Figure 11e,f. It can be seen that the production amount of liquid hydroxyl radicals was much lower than that of total hydroxyl radicals, indicating that the reaction system was dominated by surface catalytic degradation reaction, supplemented by the liquid radical reaction. With the addition of the catalyst, the amount and rate of formation of liquid hydroxyl radical increased obviously. When the addition of granular catalyst was 1 g/L, the amount of formation of liquid hydroxyl radical reached the maximum. After the addition of the catalyst, the formation rate of liquid hydroxyl radical did not increase. When H_2O_2 concentration was low, the generation rate of hydroxyl radical increased first and then decreased. With the increase in H_2O_2 concentration, hydroxyl radical generation in the liquid phase increased first and then decreased, indicating that the higher the H_2O_2 concentration, the better. A high concentration of H_2O_2 will lead to the greatest quenching of the hydroxyl radical generated in the liquid phase, reducing the reaction efficiency.

2.6. *Speculation on the Mechanism of Catalytic Oxidation of RhB Dye by the Granular Catalyst*

The degradation mechanism of the heterogeneous Fenton system mainly includes the following three kinds: heterogeneous surface catalysis mechanism, homogeneous catalysis mechanism of iron ion dissolution, and catalytic oxidation mechanism of high iron [68]. The heterogeneous surface catalysis mechanism is as follows: when irradiated with visible light, the dye is excited to reduce part of iron oxide Fe^{III} on the catalyst's surface to Fe^{II}, and the catalyst adsorbs the RhB dye and H_2O_2 on the surface of the granular catalyst. The H_2O_2 adsorbed on the surface of the granular catalyst reacts with Fe^{II} and Fe^{III} on the surface to generate active free radicals. The RhB molecules adsorbed on the catalyst's surface are oxidized and degraded into small intermediate products or partially mineralized into CO_2 and H_2O. The products are desorbed from the particle surface, and then continue to adsorb RhB molecules in the solution, forming a cycle of adsorption–degradation–desorption–re-adsorption–re-degradation–re-desorption, until RhB is completely degraded. The oxidative degradation process is dominant. The homogeneous catalytic mechanism of iron ion dissolution is as follows: the heterogeneous Fenton particle catalyst of ferric attapulgite dissolves a small number of iron ions during the reaction process, which forms a homogeneous Fenton system with H_2O_2 in the solution, catalyzes the decomposition of H_2O_2 to generate active free radicals, and degrades RhB. According to the previous experimental results, the Fenton reaction in this part of dissolved iron ions insignificantly contributes to the degradation of RhB in the system, which is a secondary path of degradation reaction. The mechanism of catalytic oxidation with high iron is as follows: during the conversion of Fe^{II} and Fe^{III} on the catalyst surface, some iron oxides are converted into higher Fe^{IV}, and high iron directly oxidizes RhB to small molecular intermediates or partially mineralizes into CO_2 and H_2O [69]. This study took attapulgite particles as the supported and loaded iron ions on the surface of attapulgite to make a heterogeneous granular catalyst, which was used for the degradation of the RhB dye. The mechanism of the degradation process was shown in Figure 12.

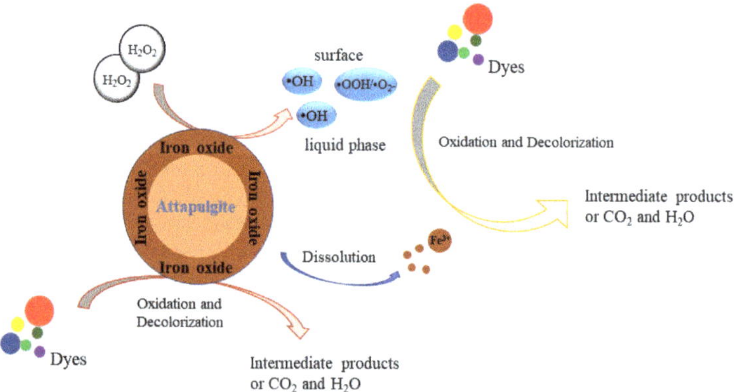

Figure 12. Proposed catalytic mechanism of dye decolorization and oxidation in the system.

3. Experimental Materials and Methods

3.1. Reagents and Materials

Rhodamine B (RhB; $C_{28}H_{31}ClN_2O_3$) was provided by the Tianjin Institute of Chemical Reagents (Tianjin, China). Congo red (CR; $C_{32}H_{22}N_6Na_2O_6S_2$) was purchased from Shanghai Maclin Biochemical Technology Co., Ltd. (Shanghai, China). Methylene blue (MB; $C_{16}H_{20}ClN_3S$) was provided by Sinopharm Chemical Reagents (Shanghai, China). Sodium hydroxide (NaOH), 30% hydrogen peroxide (H_2O_2), nitric acid (HNO_3), hydrochloric acid (HCl), potassium dichromate ($K_2Cr_2O_7$), concentrated sulfuric acid (H_2SO_4), and mercury sulfate (Ag_2SO_4) were purchased from Nanjing chemical reagents (Nanjing, China). Iron trioxide (Fe_2O_3), iron nitrate ($Fe(NO_3)_3{\cdot}9H_2O$), p-benzoquinone ($C_6H_4O_2$), potassium iodide (KI), ascorbic acid ($C_6H_8O_6$), and silver sulfate were provided by Sinopagol Chemical reagents (Nanjing, China). Tert-butyl alcohol ($C_4H_{10}O$) was purchased from Shanghai Lingfeng Chemical reagent (Shanghai, China). Coumarin ($C_9H_6O_2$) was purchased from Aladdin (Shanghai, China). 7-hydroxycoumarin ($C_9H_6O_3$) and potassium titanium oxalate ($K_2TiOC_4O_8{\cdot}2H_2O$) were purchased from McLean (Shanghai, China). All reagents used were analytically pure. Attapulgite (ATP) was found in Xuyi, Jiangsu Province. Attapulgite was prepared into granular supported iron ions as the catalyst.

3.2. Preparation and Characterization of Catalyst

Different additives and attapulgite were mixed evenly in a certain proportion, and a certain quality of deionized water was added into a molded shape for mixing and rubbed into 1–3 mm particles. The mixture was placed in the oven at 110 °C to dry, and then into the tube furnace to calcinate for a certain time. The prepared particles were immersed in a solution of iron ion concentration of 1.6 mol/L, the ratio of solid to liquid was 20:1, the immersion temperature was 80 °C, and the immersion time was 4 h. After filtration and repeated washing several times, the obtained products were dried in a 110 °C oven and then calcined in a tubular furnace at a certain temperature to prepare an iron-bearing attapulgite granular catalyst.

The ASP-2020 specific surface area and pore size analyzer of Micromeritics were used to measure the specific surface area at the degassing station of the analyzer at a rate of 10 °C/min to 300 °C for 10 h at the temperature of liquid nitrogen (77 K). EDS analysis was carried out under the conditions of 40–50,000 times magnification and gold spraying on the surface of dry samples. Ultima IV multifunctional composite X-ray diffractometer $\theta = 5{\sim}80°$ (Cu target Kα, $\lambda = 1.5406$ A) was used to test the samples; the JCPDS file number was 87–2096. Jsm-7600f Thermal field emission scanning electron microscope (JEOL) was used to spray gold (Au) on the surface of the absolute dry samples with a magnification of 40~50,000 times.

3.3. Instruments and Analytical Methods

The absorbance of rhodamine B in the solution was determined by a TU-1810 UV-vis spectrophotometer at 554 nm and its concentration was analyzed. Meanwhile, the solution's Congo red and methylene blue can be determined by ultraviolet spectrophotometer and visible spectrophotometer at 497 nm and 664 nm. The pH of the solution was measured using an EOTECH pH 700 pH meter. A TOC-vcpn (Shimadzu, Japan) total organic carbon analyzer was used to determine the TOC value of the solution. The concentration of Fe^{3+} in the reaction solution was determined by a TAS-990 Super atomic absorption spectrophotometer. The reaction of hydrogen peroxide with potassium titanium oxalate was carried out in an acetic acid-sodium acetate buffer solution to form a stable orange complex. The absorbance of this orange complex was measured at 375 nm and the concentration of hydrogen peroxide (H_2O_2) was analyzed.

4. Conclusions

This study of the use of natural clay minerals content load iron load attapulgite particles of iron catalyst was prepared by using XRD, EDS and BET methods to study the basic properties of the catalyst particles itself, and found that particle catalysts for RhB dye not only have a catalytic degradation effect, but also has a certain adsorption performance and adhere to the surface of the catalyst particles in an iron oxide crystal shape. In the process of reaction, the amount of iron dissolved in the granular catalyst is very low. In the catalytic reaction process, conditions are changed to investigate their effects on the dye RhB catalytic degradation performance and find out the best reaction conditions. The study found that under the condition of alkaline dye, the decolorization rate is greater than the acid condition; the lower the initial concentration and reaction, the greater the concentration of H_2O_2 and catalyst dosing quantity, the higher the reaction temperature, and the greater and faster the dye RhB degradation and mineralization. However, pH value has a great influence on the degradation and mineralization of RhB. In terms of the decolorization effect, alkaline condition > acidic condition > neutral condition, and in terms of the mineralization effect, acidic condition > neutral condition > alkaline condition. Under the optimal catalytic reaction conditions, granular catalyst's reusability, stability and degradation effects on different types of dyes were studied. It was found that the prepared granular catalyst has a charge-negative surface, and the degradation and mineralization effects of cationic dyes are better than that of anionic dyes. It was found that hydroxyl radical reaction dominated the catalytic degradation of RhB dye in a heterogeneous Fenton system, in which surface radical reaction was the main catalyst and liquid radical reaction was the auxiliary one. The study shows a simple and environmentally friendly route that can be scaled up and seems to be a promising approach for industrial wastewater treatment in the future.

Author Contributions: P.Z.; methodology and review, Z.D.; disgining-analysis and writing, T.L. and X.R.; data curation, M.A.O.; writing—original draft preparation, W.Y., J.H. and H.J.; visualization and collection. All authors have read and agreed to the published version of the manuscript.

Funding: This research was funded by the National Key Research and Development Program "Key Technology of Safety Production and Pollution Monitoring of Wood-Based Panel" (2016YFD0600703).

Institutional Review Board Statement: Not applicable, this study did not involve human or animal studies.

Informed Consent Statement: Not applicable, this study did not involve human or animal studies.

Data Availability Statement: This study did not report any data.

Conflicts of Interest: The authors declare that they have no conflict of interest.

References

1. Rosales, E.; Pazos, M.; Sanroman, M. Advances in the Electro-Fenton Process for Remediation of Recalcitrant Organic Compounds. *Chem. Eng. Technol.* **2012**, *35*, 609–617. [CrossRef]
2. Zhang, H.; Ran, X.; Wu, X.; Zhang, D. Evaluation of electro-oxidation of biologically treated landfill leachate using response surface inversely proportional to the concentration of methodology. *J. Hazard. Mater.* **2011**, *188*, 261–268. [CrossRef] [PubMed]
3. Ramirez-Sosa, D.R.; Castillo-Borges, E.R.; Mendez-Novelo, R.I.; Sauri-Riancho, M.R.; Barcelo-Quintal, M.; Marrufo-Gomez, J.M. Determination of organic compounds in landfill leachates treated by Fenton-Adsorption. *Waste Manag.* **2013**, *33*, 390–395. [CrossRef] [PubMed]
4. Qin, X.; Bai, L.; Tan, Y.; Li, L.; Song, F.; Wang, Y. β-cyclodextrin-Crosslinked polymeric adsorbent for simultaneous removal and stepwise recovery of organic dyes and heavy metal ions: Fabrication, performance and mechanisms. *Chem. Eng. J.* **2019**, *372*, 1007–1018. [CrossRef]
5. He, K.; Chen, G.; Zeng, G.; Chen, A.; Huang, Z.; Shi, J.; Huang, T.; Peng, M.; Hu, L. Three-dimensional graphene supported catalysts for organic dyes degradation. *Appl. Catal. B-Environ.* **2018**, *228*, 19–28. [CrossRef]
6. Karim, M.A.H.; Aziz, K.H.H.; Omer, K.M.; Salih, Y.M.; Mustafa, F.; Rahman, K.O.; Mohammad, Y. Degradation of aqueous organic dye pollutants by heterogeneous photo-assisted Fenton-like process using natural mineral activator: Parameter optimization and degradation kinetics. *IOP Conf. Ser. Earth Environ. Sci.* **2022**, *958*, 012011. [CrossRef]
7. Natarajan, T.S.; Thomas, M.; Natarajan, K.; Bajaj, H.C.; Tayade, R.J. Study on UV-LED/TiO_2 process for degradation of Rhodamine B dye. *Chem. Eng. J.* **2011**, *169*, 126–134. [CrossRef]
8. Zhai, S.; Li, M.; Wang, D.; Ju, X.; Fu, S. Cyano and acylamino group modification for tannery sludge bio-char: Enhancement of adsorption universality for dye pollutants. *J. Environ. Chem. Eng.* **2021**, *9*, 104939. [CrossRef]

9. Liu, X.; Tian, J.; Li, Y.; Sun, N.; Mi, S.; Xie, Y.; Chen, Z. Enhanced dyes adsorption from wastewater via Fe_3O_4 nanoparticles functionalized activated carbon. *J. Hazard. Mater.* **2019**, *373*, 397–407. [CrossRef]
10. Hasana, M.M.; Hasan, M.N.; Awual, M.R.; Islam, M.M.; Shenashen, M.A.; Iqbal, J. Biodegradable natural carbohydrate polymeric sustainable adsorbents for efficient toxic dye removal from wastewater. *J. Mol. Liq.* **2020**, *319*, 114356. [CrossRef]
11. Yin, H.; Zhao, J.; Li, Y.; Huang, L.; Zhang, H.; Chen, L. A novel Pd decorated polydopamine-SiO_2/PVA electrospun nanofiber membrane for highly efficient degradation of organic dyes and removal of organic chemicals and oils. *J. Clean. Prod.* **2020**, *275*, 122937. [CrossRef]
12. Awasthi, A.; Datta, D. Application of Amberlite XAD-7HP resin impregnated with Aliquat 336 for the removal of Reactive Blue-13 dye: Batch and fixed-bed column studies. *J. Environ. Chem. Eng.* **2019**, *7*, 103502. [CrossRef]
13. Nagel-Hassemer, M.E.; Carvalho-pinto, C.R.S.; Matias, W.G.; Lapolli, F.R. Removal of coloured compounds from textile industry effluents by UV/H_2O_2 advanced oxidation and toxicity evaluation. *Environ. Technol.* **2011**, *32*, 1867–1874. [CrossRef] [PubMed]
14. Aziz, K.H.H.; Mahyar, A.; Miessner, H.; Mueller, S.; Kalass, D.; Moeller, D.; Khorshid, I.M.; Rashid, M.A. Application of a planar falling fifilm reactor for decomposition and mineralization of methylene blue in the aqueous media via ozonation, Fenton, photocatalysis and non-thermal plasma: A comparative study. *Process Saf. Environ. Prot.* **2018**, *113*, 319–329. [CrossRef]
15. Javaid, R.; Qazi, U.Y.; Kawasaki, S.I. Highly effificient decomposition of Remazol Brilliant Blue R using tubular reactor coated with thin layer of PdO. *J. Environ. Manag.* **2016**, *180*, 551–556. [CrossRef]
16. Chanikya, P.; Nidheesh, P.V.; Babu, D.S.; Gopinath, A.; Kumar, M.S. Treatment of dyeing wastewater by combined sulfate radical based electrochemical advanced oxidation and electrocoagulation processes. *Sep. Purif. Technol.* **2021**, *254*, 117570. [CrossRef]
17. Guo, H.; Yang, H.; Huang, J.W.; Tong, J.; Liu, X.Y.; Wang, Y.W.; Qiao, W.C.; Han, J.G. Theoretical and experimental insight into plasma-catalytic degradation of aqueous p-nitrophenol with graphene-ZnO nanoparticles. *Sep. Purif. Technol.* **2022**, *295*, 121362. [CrossRef]
18. Wang, Y.W.; Huang, W.J.; Guo, H.; Puyang, C.D.; Han, J.G.; Li, Y.; Ruan, Y.X. Mechanism and process of sulfamethoxazole decomposition with persulfate activated by pulse dielectric barrier discharge plasma. *Sep. Purif. Technol.* **2022**, *287*, 120540. [CrossRef]
19. Wen, C.; Xu, X.; Fan, Y.; Xiao, C.; Ma, C. Pretreatment of water-based seed coating wastewater by combined coagulation and sponge-iron-catalyzed ozonation technology. *Chemosphere* **2018**, *206*, 238–247. [CrossRef]
20. Pan, G.; Jing, X.; Ding, X.; Shen, Y.; Xu, S.; Miao, W. Synergistic effects of photocatalytic and electrocatalytic oxidation based on a three-dimensional electrode reactor toward degradation of dyes in wastewater. *J. Alloy Compd.* **2019**, *809*, 151749. [CrossRef]
21. Shen, Y.; Xu, Q.; Wei, R.; Ma, J.; Wang, Y. Mechanism and dynamic study of reactive red X-3B dye degradation by ultrasonic-assisted ozone oxidation process. *Ultrason. Sonochem.* **2017**, *38*, 681–692. [CrossRef] [PubMed]
22. Bulca, O.; Palas, B.; Atalay, S.; Ersoz, G. Performance investigation of the hybrid methods of adsorption or catalytic wet air oxidation subsequent to electrocoagulation in treatment of real textile wastewater and kinetic modelling. *J. Water Process. Eng.* **2021**, *40*, 101821. [CrossRef]
23. Sreeja, P.H.; Sosamony, K.J. A Comparative Study of Homogeneous and Heterogeneous Photo-Fenton Process for Textile Wastewater Treatment. *Procedia Technol.* **2016**, *24*, 217–223. [CrossRef]
24. Wang, J.; Tang, J. Fe-based Fenton-like catalysts for water treatment: Preparation, characterization and modification. *Chemosphere* **2021**, *276*, 130177. [CrossRef]
25. Arimi, M.M. Modified natural zeolite as heterogeneous Fenton catalyst in treatment of recalcitrants in industrial effluent. *Prog. Nat. Sci.* **2017**, *27*, 275–282. [CrossRef]
26. Hernandez-Olono, J.T.; Infantes-Molina, A.; Vargas-Hernandez, D.; Dominguez-Talamantes, D.G.; Rodriguez-Castellon, E.; Herrera-Urbina, J.R.; Tanori-Cordova, J.C. A novel heterogeneous photo Fenton Fe/Al_2O_3 catalyst for dye degradation. *J. Photochem. Photobiol. A Chem.* **2021**, *421*, 113529. [CrossRef]
27. Khataee, A.; Salahpour, F.; Fathinia, M.; Seyyedi, B.; Vahid, B. Iron rich laterite soil with mesoporous structure for heterogeneous Fenton-like degradation of an azo dye under visible light. *J. Ind. Eng. Chem.* **2015**, *26*, 129–135. [CrossRef]
28. Hu, H.; Miao, K.; Luo, X.; Guo, S.; Yuan, X.; Pei, F.; Qian, H.M.; Feng, G. Efficient Fenton-like treatment of high-concentration chlorophenol wastewater catalysed by Cu-Doped SBA-15 mesoporous silica. *J. Clean. Prod.* **2021**, *318*, 128632. [CrossRef]
29. Chen, Z.; Zheng, Y.; Liu, Y.; Zhang, W.; Wang, Y.; Guo, X.; Tang, X.; Zhang, Y.; Wang, Z.; Zhang, T. Magnetic Mn-Doped Fe_3O_4 hollow Microsphere/RGO heterogeneous Photo-Fenton Catalyst for high efficiency degradation of organic pollutant at neutral pH. *Mater. Chem. Phys.* **2019**, *238*, 121893. [CrossRef]
30. Han, X.; Gou, L.; Tang, S.; Cheng, F.; Zhang, M.; Guo, M. Enhanced heterogeneous Fenton-like degradation of refractory organic contaminants over Cu doped (Mg,Ni)$(Fe,Al)_2O_4$ synthesized from laterite nickel ore. *J. Environ. Manag.* **2021**, *283*, 111941. [CrossRef]
31. Suligoj, A.; Ristic, A.; Drazic, G.; Pintar, A.; Logar, N.Z.; Tusar, N.N. Bimetal Cu-Mn porous silica-supported catalyst for Fenton-like degradation of organic dyes in wastewater at neutral pH. Catal. *Today* **2020**, *358*, 270–277. [CrossRef]
32. Yamaguchi, R.; Kurosu, S.; Suzuki, M.; Kawase, Y. Hydroxyl radical generation by zero-valent iron/Cu (ZVI/Cu) bimetallic catalyst in wastewater treatment: Heterogeneous Fenton/Fenton-like reactions by Fenton reagents formed in-situ under oxic conditions. *Chem. Eng. J.* **2018**, *334*, 1537–1549. [CrossRef]

33. Goncalves, R.G.L.; Mendes, H.M.; Bastos, S.L.; Dagostino, L.C.; Tronto, J.; Pulcinelli, S.H.; Santilli, C.V. Fenton-like degradation of methylene blue using Mg/Fe and MnMg/Fe layered double hydroxides as reusable catalysts. *Appl. Clay Sci.* **2020**, *187*, 105477. [CrossRef]
34. Yin, H.; Kong, M. Simultaneous removal of ammonium and phosphate from eutrophic waters using natural calcium-rich attapulgite-based versatile adsorbent. *Desalination* **2014**, *351*, 128–137. [CrossRef]
35. Zhang, Z.; Wang, W.; Wang, A. Highly effective removal of Methylene Blue using functionalized attapulgite via hydrothermal process. *J. Environ. Sci.* **2015**, *33*, 106–115. [CrossRef]
36. Cao, J.; Shao, G.; Wang, Y.; Liu, Y.; Yuan, Z. CuO catalysts supported on attapulgite clay for low-temperature CO oxidation. *Catal. Commun.* **2008**, *9*, 2555–2559. [CrossRef]
37. Wang, Y.; Chen, M.; Liang, T.; Yang, Z.; Yang, J.; Liu, S. Hydrogen Generation from Catalytic Steam Reforming of Acetic Acid by Ni/Attapulgite Catalysts. *Catalysts* **2016**, *6*, 172. [CrossRef]
38. Ma, H.; Sun, K.; Li, Y.; Xu, X. Ultra-chemoselective hydrogenation of chloronitrobenzenes to chloroanilines over HCl-acidified attapulgite-supported platinum catalyst with high activity. *Catal. Commun.* **2009**, *10*, 1363–1366. [CrossRef]
39. Kosmulski, M. pH-dependent surface charging and points of zero charge III Update. *J. Colloid Interface Sci.* **2006**, *298*, 730–741. [CrossRef]
40. Bai, C.; Gong, W.; Feng, D.; Xian, M.; Zhou, Q.; Chen, S.; Ge, Z.; Zhou, Y. Natural graphite tailings as heterogeneous Fenton catalysts for the decolorization of rhodamine B. *Chem. Eng. J.* **2012**, *197*, 306–313. [CrossRef]
41. Li, L.; Lai, C.; Huang, F.; Cheng, M.; Zeng, G.; Huang, D.; Li, B.; Liu, S.; Zhang, M.; Qin, L.; et al. Degradation of naphthalene with magnetic bio-char activate hydrogen peroxide: Synergism of bio-char and Fe–Mn binary oxides. *Water Res.* **2019**, *160*, 238–248. [CrossRef] [PubMed]
42. Li, X.; Zhang, Y.; Xie, Y.; Zeng, Y.; Li, P.; Xie, T.; Wang, Y. Ultrasonic-enhanced Fenton-like degradation of bisphenol A using a bio-synthesized schwertmannite catalyst. *J. Hazard. Mater.* **2018**, *344*, 689–697. [CrossRef] [PubMed]
43. Cleveland, V.; Bingham, J.P.; Kan, E. Heterogeneous Fenton degradation of bisphenol A by carbon nanotube-supported Fe_3O_4. *Sep. Purif. Technol.* **2014**, *133*, 388–395. [CrossRef]
44. Yin, D.; Zhang, L.; Zhao, X.; Chen, H.; Zhai, Q. Iron-glutamate-silicotungstate ternary complex as highly active heterogeneous Fenton-like catalyst for 4-chlorophenol degradation. *Chin. J. Catal.* **2015**, *36*, 2203–2210. [CrossRef]
45. Lopez-Lopez, C.; Martin-Pascual, J.; Martinez-Toledo, M.V.; Munio, M.M.; Hontoria, E.; Poyatos, J.M. Kinetic modelling of TOC removal by H_2O_2/UV, photo-Fenton and heterogeneous photocatalysis processes to treat dye-containing wastewater. *Int. J. Environ. Sci. Te.* **2015**, *12*, 3255–3262. [CrossRef]
46. Rusevova, K.; Kopinke, F.D.; Georgi, A. Nano-sized magnetic iron oxides as catalysts for heterogeneous Fenton-like reactions-Influence of Fe (II)/Fe (III) ratio on catalytic performance. *J. Hazard. Mater.* **2012**, *241*, 433–440. [CrossRef]
47. Cheng, S.; Liu, F.; Shen, C.; Zhu, C.; Li, A. A green and energy-saving microwave-based method to prepare magnetic carbon beads for catalytic wet peroxide oxidation. *J. Clean. Prod.* **2019**, *215*, 232–244. [CrossRef]
48. Acisli, O.; Khataee, A.; Soltani, R.D.C.; Karaca, S. Ultrasound-assisted Fenton process using siderite nanoparticles prepared via planetary ball milling for removal of reactive yellow 81 in aqueous phase. *Ultrason. Sonochem.* **2017**, *35*, 210–218. [CrossRef] [PubMed]
49. Hassan, A.; Al-Kindi, G.; Ghanim, D. Green synthesis of bentonite-supported iron nanoparticles as a heterogeneous Fenton-like catalyst: Kinetics of decolorization of reactive blue 238 dye. *Water Sci. Eng.* **2020**, *13*, 286–298. [CrossRef]
50. Wang, C.; Cao, Y.; Wang, H. Copper-based catalyst from waste printed circuit boards for effective Fenton-like discoloration of Rhodamine B at neutral pH. *Chemosphere* **2019**, *230*, 278–285. [CrossRef]
51. Krysa, J.; Mantzavinos, D.; Pichat, P.; Poulios, I. Advanced oxidation processes for water and wastewater treatment. *Environ. Sci. Pollut. Res.* **2018**, *25*, 34799–34800. [CrossRef] [PubMed]
52. Zhang, P.; Huang, W.; Ji, Z.; Zhou, C.; Yuan, S. Mechanisms of hydroxyl radicals production from pyrite oxidation by hydrogen peroxide: Surface versus aqueous reactions. *Geochim. Cosmochim. Acta* **2018**, *238*, 394–410. [CrossRef]
53. Feng, J.; Hu, X.; Yue, P.; Qiao, S. Photo Fenton degradation of high concentration Orange II (2mM) using catalysts containing Fe: A comparative study. *Sep. Purif. Technol.* **2009**, *67*, 213–217. [CrossRef]
54. Hashemian, S. Fenton-like oxidation of malachite green solutions: Kinetic and thermodynamic study. *J. Chem.* **2013**, *2013*, 809318. [CrossRef]
55. Demirezen, D.A.; Yıldız, Y.S.; Yılmaz, D.D. Amoxicillin degradation using green synthesized iron oxide nanoparticles: Kinetics and mechanism analysis. Environ. Nanotechnol. *Monit. Manag.* **2019**, *11*, 100219. [CrossRef]
56. Park, J.; Wang, J.; Xiao, R.; Tafti, N.; Delaune, R.D.; Seo, D.C. Degradation of Orange G by Fenton-like reaction with Fe-impregnated biochar catalyst. *Bioresour. Technol.* **2018**, *249*, 368–376. [CrossRef]
57. Zhang, X.; Guo, Y.; Shi, S.; Liu, E.; Li, T.; Wei, S.; Li, Y.; Li, Y.; Sun, G.; Zhao, Z. Efficient and stable iron-copper montmorillonite heterogeneous Fenton catalyst for removing Rhodamine B. *Chem. Phys. Lett.* **2021**, *776*, 138673. [CrossRef]
58. Cruz, M.D.L.; Pérez, U.M.G. Photocatalytic properties of BiVO 4 prepared by the co-precipitation method: Degradation of rhodamine B and possible reaction mechanisms under visible irradiation. *Mater. Res. Bull.* **2010**, *45*, 135–141. [CrossRef]
59. Wang, N.; Hu, Q.; Du, X.; Xu, H.; Hao, L. Study on decolorization of Rhodamine B by raw coal fly ash catalyzed Fenton-like process under microwave irradiation. *Adv. Powder Technol.* **2019**, *30*, 2369–2378. [CrossRef]

60. Boulebd, H. Comparative study of the radical scavenging behavior of ascorbic acid, BHT, BHA and Trolox: Experimental and theoretical study. *J. Mol. Struct.* **2019**, *1201*, 127210. [CrossRef]
61. Sun, S.; Zeng, X.; Lemley, A.T. Kinetics and mechanism of carbamazepine degradation by a modified Fenton-like reaction with ferric-nitrilotriacetate complexes. *J. Hazard. Mater.* **2013**, *252*, 155–165. [CrossRef] [PubMed]
62. Panda, N.; Sahoo, H.; Mohapatra, S. Decolourization of Methyl Orange using Fenton-like mesoporous Fe_2O_3-SiO_2 composite. *J. Hazard. Mater.* **2011**, *185*, 359–365. [CrossRef] [PubMed]
63. Jian, H.; Yang, F.; Gao, Y.; Zhen, K.; Tang, X.; Zhang, P.; Wang, Y.; Wang, C.; Sun, H. Efficient removal of pyrene by biochar supported iron oxide in heterogeneous Fenton-like reaction via radicals and high-valent iron-oxo species. *Sep. Purif. Technol.* **2021**, *265*, 118518. [CrossRef]
64. Chen, L.; Li, X.; Zhang, J.; Fang, J.; Huang, Y.; Wang, P.; Ma, J. Production of Hydroxyl Radical via the Activation of Hydrogen Peroxide by Hydroxylamine. *Environ. Sci. Technol.* **2015**, *49*, 10373–10379. [CrossRef] [PubMed]
65. Chen, L.; Ma, J.; Li, X.; Zhang, J.; Fang, J.; Guan, Y.; Xie, P. Strong Enhancement on Fenton Oxidation by Addition of Hydroxylamine to Accelerate the Ferric and Ferrous Iron Cycles. *Environ. Sci. Technol.* **2011**, *45*, 3925–3930. [CrossRef]
66. He, Z.; Song, S.; Ying, H.; Xu, L.; Chen, J. p-aminophenol degradation by ozonation combined with sonolysis: Operating conditions influence and mechanism. *Ultrason. Sonochem.* **2007**, *14*, 568–574. [CrossRef]
67. Javaid, R.; Qazi, U.Y.; Kawasaki, S.I. Efficient and Continuous Decomposition of Hydrogen Peroxide Using a Silica Capillary Coated with a Thin Palladium or Platinum Layer. *Bull. Chem. Soc. Jpn.* **2015**, *88*, 976–980. [CrossRef]
68. Khan, I.; Saeed, K.; Zekker, I.; Zhang, B.L.; Hendi, A.H.; Ahmad, A.; Ahmad, S.; Zada, N. Review on Methylene Blue: Its Properties, Uses, Toxicity and Photodegradation. *Water* **2022**, *2*, 242. [CrossRef]
69. Khan, I.; Saeed, K.; Ali, N.; Khan, I.; Zhang, B.L.; Sadiq, M. Heterogeneous photodegradation of industrial dyes: An insight to different mechanisms and rate affecting parameters. *J. Environ. Chem. Eng.* **2020**, *8*, 104364. [CrossRef]

Article

Evaluation of Fe^{2+}/Peracetic Acid to Degrade Three Typical Refractory Pollutants of Textile Wastewater

Jiali Yu [1,†], Shihu Shu [1,†], Qiongfang Wang [2], Naiyun Gao [3] and Yanping Zhu [1,*]

1 College of Environmental Science and Engineering, Donghua University, Shanghai 201620, China; gloria11071219@163.com (J.Y.); shushihu@dhu.edu.cn (S.S.)
2 College of Chemistry and Chemical Engineering, Shanghai University of Engineering Science, Shanghai 201600, China; wqfang2010@126.com
3 State Key Laboratory of Pollution Control Reuse, Tongji University, Shanghai 200092, China; gaonaiyun@126.com
* Correspondence: yanpingzhu@dhu.edu.cn
† These authors contributed equally to this work.

Abstract: In this work, the degradation performance of $Fe^{2+}/PAA/H_2O_2$ on three typical pollutants (reactive black 5, ANL, and PVA) in textile wastewater was investigated in comparison with Fe^{2+}/H_2O_2. Therein, $Fe^{2+}/PAA/H_2O_2$ had a high removal on RB5 (99%) mainly owing to the contribution of peroxyl radicals and/or Fe(IV). Fe^{2+}/H_2O_2 showed a relatively high removal on PVA (28%) mainly resulting from ·OH. $Fe^{2+}/PAA/H_2O_2$ and Fe^{2+}/H_2O_2 showed comparative removals on ANL. Additionally, $Fe^{2+}/PAA/H_2O_2$ was more sensitive to pH than Fe^{2+}/H_2O_2. The coexisting anions (20–2000 mg/L) showed inhibition on their removals and followed an order of $HCO_3^- > SO_4^{2-} > Cl^-$. Humic acid (5 and 10 mg C/L) posed notable inhibition on their removals following an order of reactive black 5 (RB5) > ANL > PVA. In practical wastewater effluent, PVA removal was dramatically inhibited by 88%. Bioluminescent bacteria test results suggested that the toxicity of $Fe^{2+}/PAA/H_2O_2$ treated systems was lower than that of Fe^{2+}/H_2O_2. RB5 degradation had three possible pathways with the proposed mechanisms of hydroxylation, dehydrogenation, and demethylation. The results may favor the performance evaluation of $Fe^{2+}/PAA/H_2O_2$ in the advanced treatment of textile wastewater.

Keywords: peracetic acid; advanced oxidation; reactive dyes; aniline; polyvinyl alcohol

Citation: Yu, J.; Shu, S.; Wang, Q.; Gao, N.; Zhu, Y. Evaluation of Fe^{2+}/Peracetic Acid to Degrade Three Typical Refractory Pollutants of Textile Wastewater. *Catalysts* **2022**, *12*, 684. https://doi.org/10.3390/catal12070684

Academic Editors: Hao Xu and Yanbiao Liu

Received: 28 May 2022
Accepted: 14 June 2022
Published: 22 June 2022

1. Introduction

The textile industry was one of the most water-consuming and key industrial branches, especially in developing countries. In China, the amount of textile wastewater, around 80% of which is from the printing and dyeing process [1], ranked third among all the 41 industries and accounted for 10.1% [2]. Generally, the discharged textile wastewater was treated either by on-site treatment plants in the factory or a combination of factories and urban wastewater treatment plants (WWTPs) to meet the wastewater discharge standard. However, the increasingly stringent discharge standard forces the advanced tertiary treatment urgent for the enhanced removal of refractory pollutants.

Reactive dyes, aniline (ANL), and polyvinyl alcohol (PVA) are three types of typical refractory pollutants in printing and dyeing wastewater and are of increasing environmental concern [3]. Reactive dye is the most important dyeing class for cellulosic fibers [4], up to 10–50% of which would flow into wastewater in the dyeing process [5]. ANL is an important intermediate in syntheses of benzidine azo dyes [2] as well as a product derived from the biotransformation of azo dyes [6]. ANL has been listed as a priority pollutant by the Environmental Protection Agency of the United States due to its carcinogenic and mutagenic effects [7]. In the latest amended Discharge Standard of Water Pollutants for Dyeing and Finishing of Textile Industry (GB 4287-2012) in China [8], the discharge limit of ANLs was regulated as undetected, which was actually difficult for factories to meet at affordable expenses. PVA, a water-soluble refractory polymer, is widely used in the sizing process of cotton blended fabrics and was lost to the effluent during the desizing process [9]. The discharge of PVA may deteriorate the receiving body of water via causing the lack of dissolved oxygen in the aquatic environment and the release of harmful metals from the sediment [10]. The removal of PVA via the conventional biological process was challenging due to its poor biodegradability [11].

The advanced oxidation process (AOP) is widely adopted as the tertiary treatment for the removal of low-level refractory organic pollutants from the secondary effluent of industrial textile wastewater [12–19]. Recently, Fe^{2+}/peracetic acid (PAA) has emerged as a potential alternative to the conventional Fe^{2+}/hydrogen peroxide (H_2O_2) Fenton AOP [20,21]. PAA has a high disinfection efficiency and less formation of harmful disinfection byproducts (DBPs) compared to those chlorine-based disinfectants [22]. Thus, it is recommended as an attractive disinfectant for secondary and tertiary wastewater effluents in many countries, (e.g., Canada and parts of Europe) [23–25]. In fact, the PAA solution is an equilibrium mixture of PAA, H_2O_2, and acetic acid [26], and the PAA-based Fenton system was defined as $Fe^{2+}/PAA/H_2O_2$. During the $Fe^{2+}/PAA/H_2O_2$ process, PAA played a key role within the initial 5 s and H_2O_2 became the dominant oxidant afterward due to the much higher reaction rate of PAA with Fe^{2+} (>650 times) compared to that of H_2O_2 [27]. Activation of PAA by Fe^{2+} may primarily generate ·OH, $CH_3C(O)O\cdot$, and Fe(IV) according to reactions (1–5) [27,28]. Moreover, ·OH would also react with PAA and H_2O_2 to generate secondary radicals, (e.g., $CH_3C(O)\cdot$, $CH_3C(O)OO\cdot$, and $HO_2\cdot$) via reactions (6–9), and the reaction rate of ·OH with PAA was much higher compared with H_2O_2 [27,29]. That said, the proportion of PAA and H_2O_2 would affect the distribution of organic radicals and ·OH in the $Fe^{2+}/PAA/H_2O_2$ process. In previous research [30–33], PAA-based AOP has exhibited structural selectivity in the removal performance of target pollutants and shown comparable or even superior performance compared to H_2O_2-based AOP.

$$CH_3C(O)OOH + Fe^{2+} \rightarrow CH_3C(O)O\cdot + Fe^{3+} + OH^- \tag{1}$$

$$CH_3C(O)OOH + Fe^{2+} \rightarrow CH_3C(O)O^- + Fe^{3+} + \cdot OH \tag{2}$$

$$CH_3C(O)OOH + Fe^{2+} \rightarrow CH_3C(O)OH + Fe^{IV}O^{2+} \tag{3}$$

$$H_2O_2 + Fe^{2+} \rightarrow \cdot OH + Fe^{3+} + OH^- \tag{4}$$

$$H_2O_2 + Fe^{2+} \rightarrow H_2O + Fe^{IV}O^{2+} \tag{5}$$

$$CH_3C(O)OOH + \cdot OH \rightarrow CH_3C(O)OO\cdot + H_2O \tag{6}$$

$$CH_3C(O)OOH + \cdot OH \rightarrow CH_3C(O)\cdot + H_2O + O_2 \tag{7}$$

$$CH_3C(O)OOH + \cdot OH \rightarrow CH_3C(O)OH + HO_2\cdot \tag{8}$$

$$H_2O_2 + \cdot OH \rightarrow HO_2\cdot + H_2O \tag{9}$$

Among the studies about the active dye degradation by PAA-based AOPs, H_2O_2 and PAA-related organic radicals played dominant roles in the removal of methyl blue and Brilliant Red X-3B, respectively [21,34]. Co(II)-mediated PAA oxidation in previous work has shown a minor contribution of $CH_3C(O)OO\cdot$ to the degradation of ANL [30]. In addition, the degradation of PVA has been evaluated mainly in H_2O_2-based AOPs with the reaction rate constant of PVA with ·OH in the order of 10^6-10^7 $M^{-1}s^{-1}$ [35], while scarcely in PAA-based AOP. Considering the different structural characteristics of the three typical pollutants, there is a need to comparatively evaluate their degradation efficacies by $Fe^{2+}/PAA/H_2O_2$ and assess the contributions of radicals produced, respectively, by PAA and H_2O_2.

The objectives of this study are to investigate: (1) the effectiveness of $Fe^{2+}/PAA/H_2O_2$ to degrade reactive black 5 (RB5), ANL, and PVA in comparison to Fe^{2+}/H_2O_2; (2) the impact of operating conditions, (i.e., Fe^{2+} and PAA dosages, coexisting ions and natural organic matter); (3) contributions of ·OH and PAA-related radicals in their degradation under different pH conditions; (4) the acute toxicity alteration during the process; and (5) degradation intermediates and possible pathways. The main novelty of this work is to investigate the relationship between $Fe^{2+}/PAA/H_2O_2$ and pollutant structure, obtaining the applicability of the process in textile water treatment.

2. Materials and Methods

2.1. Chemicals and Materials

RB5, ANL, sodium thiosulfate, potassium iodide, acetic acid, ferrous sulfate ($FeSO_4{\cdot}7H_2O$), phosphoric acid, sodium chloride, and sodium sulfate were of analytical grade and obtained from Shanghai Titan Scientific Co., Ltd. (Shanghai, China). Ammonium molybdate ($(NH_4)_2MoO_4$), sodium hydroxide (NaOH), sulfuric acid, and *p*-chlorobenzoic acid (*p*CBA) were purchased from Sinopharm Chemical Reagent Co. (Shanghai, China). Tert-butanol (TBA), methanol (MeOH), and 5,5-dimethyl-1-pyrrolinr N-oxide (DMPO) were of chromatographic grade and purchased from Fisher Scientific (Fair Lawn, NJ, USA). H_2O_2 (30%, w.t.) solution, PVA (Type 1788) with an average molecular weight of 46,000 g/mol (hydrolysis degree of 88%), humic acid (HA), N, N-Diethyl-p-phenylenediamine (DPD), and 5,5-dimethyl-1-pyrrolinr N-oxide (DMPO) were purchased from Sigma-Aldrich (Shanghai, China). All solutions were prepared with ultrapure water from a Millipore Milli-Q water system (Direct-Q3 UV). The secondary effluent taken from a municipal wastewater treatment plant (Songjiang District, Shanghai, China) was used as the real wastewater with dissolved organic carbon (DOC) of around 4.5 mg C/L after filtered through a 0.45 μm membrane. PAA solution, containing PAA:H_2O_2 at a molar ratio of 1.34:1, was freshly prepared according to the reaction (Equation (10)) and stored at 4 °C [36].

$$CH_3COOH + H_2O_2 \overset{H_2SO_4}{\leftrightarrow} CH_3C(O)OOH + H_2O \tag{10}$$

2.2. Experimental Procedures

All experiments were conducted in a 200 mL glass reactor with constant magnetic stirring at room temperature (20 ± 1 °C). The reaction solution contained designated concentrations of target pollutants, the initial pH of which was adjusted to 3.0, 4.0, 5.0, 6.0, and 7.0 by sodium hydroxide (1 M) or sulfuric acid (1 M). Reactions were initiated by adding different dosages of PAA and $FeSO_4{\cdot}7H_2O$ simultaneously. Samples (1–2.5 mL) were withdrawn within 10 min at predetermined intervals and immediately quenched by excessive sodium thiosulfate ($[Na_2S_2O_3]/[PAA]_0$ molar ratio >10) for the analysis of target compounds. Meanwhile, PAA decay was also monitored by taking samples periodically without adding any quenching agent. To explore the contribution of direct PAA oxidation and the radicals produced from H_2O_2 contained in PAA solution, additional trials were also conducted by adding PAA only or Fe^{2+}/H_2O_2 (H_2O_2 dosage equal to the concentration of H_2O_2 in PAA solution). Quenching tests were performed by spiking 100 mM TBA or MeOH to the reaction solution before the addition of PAA and Fe^{2+}. The concentrations of TBA or MeOH were high enough to quench reactive radicals. To quantify the steady-state concentration of ·OH in $Fe^{2+}/PAA/H_2O_2$ system under different pH conditions, the ·OH probe (*p*CBA) was spiked to the reaction solution and its time-dependent degradation was also analyzed.

The effect of water matrices on the degradation of the three target pollutants in $Fe^{2+}/PAA/H_2O_2$ system was assessed by adding Cl^- (0–2000 mg/L), SO_4^{2-} (0–2000 mg/L), HCO_3^- (0–2000 mg/L), and HA (0–10 mg C/L), respectively, to the reaction solution. The degradation tests were also conducted in practical wastewater effluent. Samples were also taken for oxidized products or DOC analysis at the beginning and end of each test. All experiments were conducted at least in duplicate, and the error bars represented the standard deviation.

2.3. Analytical Methods

The ANL and *p*CBA concentrations were measured by high-performance liquid chromatography (HPLC, Thermo Scientific UltiMate DioNEX 300, Waltham, MA, USA) coupled with a Symmetry-C18 column (5 μm, 4.6 mm × 250 mm) and a UV detector. The mobile phase for ANL was a 65:35 (v/v) mixture of methanol and ultrapure water at a flow rate of 1 mL/min. The mobile phase for pCBA was a mixture of methanol and phosphoric acid (70:30, v/v %) with a flow rate of 1 mL/min. The injection volumes of ANL and *p*CBA samples were 10 and 100 μL, respectively. Both ANL and pCBA were analyzed at the wavelength of 230 nm. The RB5 concentration was measured with an ultraviolet spectrophotometer (UV1800) at a wavelength of 598 nm. The PVA concentration was measured using the modified colorimetric method [37]. Briefly, eight 25 mL volumetric flasks were prepared with each containing either 0, 0.05, 0.1, 0.2, 0.3, 0.4, 0.5, and 0.6 mL of 0.5 g/L standard PVA solution and diluted to 10 mL. Then, 5 mL of 4% boric acid and 2 mL of I_2-KI (1.27 g/L I_2 and 25 g/L of KI) were added. After equilibration for 5 min, the solutions were diluted to 25 mL and measured at a wavelength of 690 nm.

Electron paramagnetic resonance (EPR, EMXnano231, Bruker, Rheinstetten, Germany) was used to determine the reactive species with DMPO as the spin trapping agent, with further details presented in SI. The oxidized products of RB5 by $Fe^{2+}/PAA/H_2O_2$ system were analyzed by HPLC-MS (Q Exactive Focus, Thermo Fisher Scientific, Waltham, MA, USA), and ANL was detected by GC-MS (Thermo Fisher Scientific, Waltham, MA, USA), the details of which were provided in SI. The PAA stock solution was regularly calibrated using titration method [7]. Concentrations of PAA and H_2O_2 in PAA solution were determined according to the Hach DPD method [38]. H_2O_2 concentration in the absence of PAA was measured using a triiodide absorbance method [39]. Acute ecotoxicity was assessed by the change of bioluminescence intensity bioluminescent with Vibrio fischeri bacteria in toxicity analyzer (HACH, Ames, IA, USA) [40], with the details of the method given in Supplementary Materials.

3. Results and Discussions

3.1. Process Degradation Efficiency Assessment

The degradation behaviors of $Fe^{2+}/PAA/H_2O_2$ towards the three pollutants were comparatively evaluated at initial pHs of 3.0, 4.0, 5.0, 6.0, and 7.0 compared with PAA only and Fe^{2+}/H_2O_2. As shown in Figure 1a, 90% of RB5 was removed through an initial fast degradation (94%) within 5 s, followed by a slow degradation in the $Fe^{2+}/PAA/H_2O_2$ system, while a minor RB5 degradation (<5%) occurred in PAA only system due to the slow reaction rate. In the Fe^{2+}/H_2O_2 system, RB5 removal decreased by 6% in the first 5 s and 75% in the whole process compared with $Fe^{2+}/PAA/H_2O_2$, attributed to reactive oxidative species (ROS) generated from PAA in addition to H_2O_2. By contrast, ANL removal in the $Fe^{2+}/PAA/H_2O_2$ system (47%) was comparable to that (39%) in the Fe^{2+}/H_2O_2 system, which suggested that the degradation efficiency of ANL in the former was mainly attributed to the presence of H_2O_2 other than PAA. Interestingly, the PVA degradation efficiency was lower in the $Fe^{2+}/PAA/H_2O_2$ system compared with the Fe^{2+}/H_2O_2 system. This was likely because PAA reacted much faster with Fe^{2+} than H_2O_2 and thus contributed to its preferential consumption of Fe^{2+}. On the other hand, the presence of PAA may convert $\cdot OH$ to C− via the reaction of $\cdot OH$ with PAA according to Equation (10). In comparison with previously reported results (Table 1), this process showed a relatively rapid removal of these three pollutants, especially RB5 [41–46].

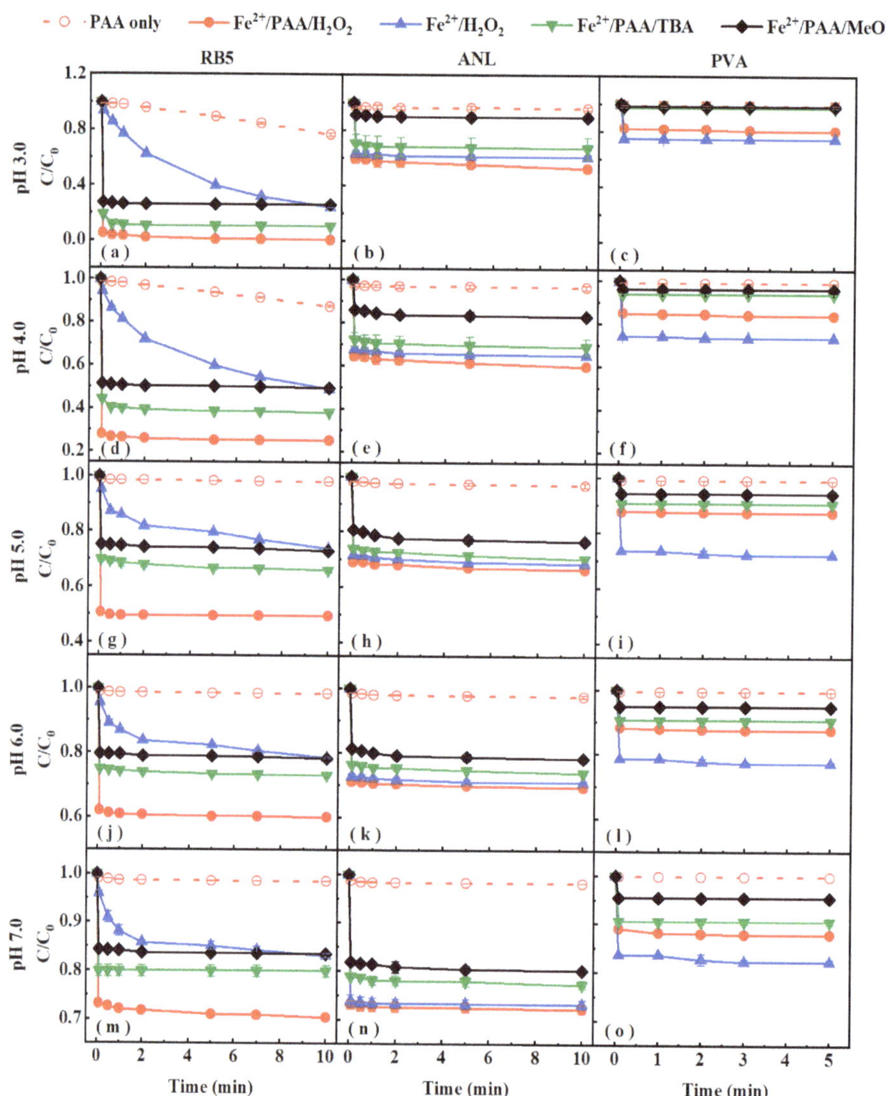

Figure 1. Time-dependent degradation of (**a**,**d**,**g**,**j**,**m**) RB5, (**b**,**e**,**h**,**k**,**n**) PVA and (**c**,**f**,**i**,**l**,**o**) ANL by different processes in a pH range of 3.0–7.0. Conditions: $[PAA]_0$ = 15 mg/L, $[H_2O_2]_0$ = 5 mg/L, $[RB5]_0$ = 20 mg/L, $[PVA]_0$ or $[ANL]_0$ = 10 mg/L, initial pH = 3.0, 5.0, and 7.0, T = 21 ± 1 °C.

Table 1. Comparison of different AOPs in removing RB5, ANL, and PVA with $Fe^{2+}/PAA/H_2O_2$ system.

AOPs	Pollutants	Pollutant Concentration (mg/L)	Reaction Time (min)	Initial pH	Catalyst Dose (mg/L)	Oxidant Dose	Removal Efficiency (%)	References
Ag_3PO_4/Visible light	RB5	50	120	11.0	500	150 W	91	[41]
O_3/Co-Ce-O	RB5	100	80	7.0	1000	60 LPH	96	[42]

Table 1. *Cont.*

AOPs	Pollutants	Pollutant Concentration (mg/L)	Reaction Time (min)	Initial pH	Catalyst Dose (mg/L)	Oxidant Dose	Removal Efficiency (%)	References
Fe_3O_4/PMS	RB5	50	60	7.0	250	614.76 mg/L	94.86	[43]
AmGO/UV-A	RB5	100	120	8.0	5000	40 W	75	[44]
	RB5	20	10	3.0	1.1		94	
Fe^{2+}/PAA/H_2O_2	ANL	10	10	3.0	2.2	15/5 mg/L	47	This work
	PVA	10	5	3.0	2.2		20	
UV/SPC	ANL	93.13	120	6.8	314	17.85 mw/cm^2	54.25	[45]
UV/$NiFe_2O_4$	PVA	25	140	6.0	300	15 W	94.3	[46]

Note: PS, persulfate; PMS, peroxymonosulfate; AmGO, amino-Fe_3O_4-functionalized graphene oxide; SPC, sodium percarbonate; LPH, Litres per hour.

The ·OH was identified in Fe^{2+}/PAA/H_2O_2 system via EPR. Figure S1 showed that the characteristic peak of the DMPO-HO· spin adduct signal appeared in the spectrum, suggesting the existence of ·OH in the system. To differentiate between the contributions of ·OH and other ROSs, (i.e., peroxyl radicals and Fe(IV)), TBA was used to quench ·OH ($k_{\cdot OH/TBA}$ = (3.8–7.6) $\times 10^8$ $M^{-1}s^{-1}$) [47], and MeOH was used as a quencher for both ·OH ($k_{\cdot OH/MeOH}$ = 9.16×10^9 $M^{-1}s^{-1}$) [48] and acetyl(per)oxyl radicals, (i.e., $CH_3COO\cdot$ and $CH_3C(O)OO\cdot$) [30]. As shown in Figure 1a–c, TBA significantly inhibited the removal of ANL and PVA by 15% and 17%, respectively, but showed negligible influence on RB5 removal. By contrast, MeOH inhibited the removal of RB5, ANL, and PVA by 25%, 37%, and 17%, respectively. These results further indicated that PVA degradation mainly depended on ·OH, ANL mainly on both ·OH and other ROSs, while RB5 was mainly on other ROSs compared to ·OH.

The influence of pH on the degradation of these three pollutants was investigated in a range of 3.0–7.0. Figure 1 showed that their degradation efficiencies declined with the increase in pH, the extent of which followed an order of RB5 > ANL > PVA. The pH effects for RB5 and PVA were more significant in Fe^{2+}/PAA/H_2O_2 compared with Fe^{2+}/H_2O_2, while similar for ANL. These results indicated that other ROSs were more susceptible to pH than ·OH, likely due to the higher reaction rate constants in Equations (1)–(3) (16,000–110,000 M^{-1} S^{-1}) compared with Equations (4) and (5) (63–76 M^{-1} S^{-1}) [27]. Therein, the former could be inhibited more significantly at the elevated pH mainly because of the higher OH^- concentration. Meanwhile, the higher pH could result in a decline in Fe^{2+} according to reaction 11.

$$Fe^{2+} + OH^- \rightarrow Fe(OH)_2 \quad (11)$$

3.2. Effects of PAA and Fe^{2+} Dosages on Pollutants' Removal

The effects of PAA and Fe dosages on the degradation efficiencies of the three pollutants were evaluated. As shown in Figure 2a–c, their removals increased with the PAA dosage rising from 5 to 15 mg/L likely attributed to the increased number of radicals. As the PAA dosage further rose to 30 mg/L, their removals either remained stable (RB5) or decreased, which may be explained by the quenching effect of PAA and/or H_2O_2 on radicals [30]. These results indicated that the optimal PAA dosage was 15 mg/L.

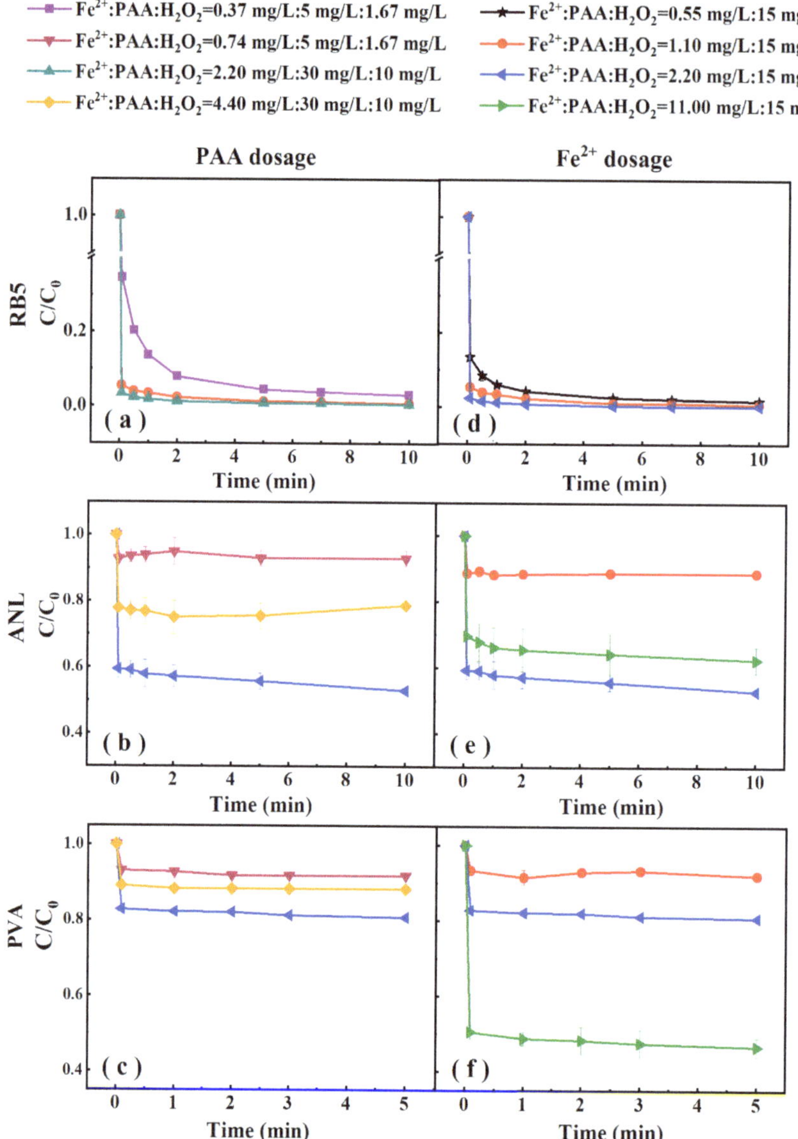

Figure 2. Effects of PAA and Fe^{2+} dosages on the degradation of (**a**,**d**) RB5, (**b**,**e**) ANL and (**c**,**f**) PVA. Conditions: $[RB5]_0$ = 20 mg/L, $[PVA]_0$ and $[ANL]_0$ = 10 mg/L, initial pH = 3.0, T = 21 ± 1 °C.

Moreover, the influence of Fe^{2+} dosage on their removals was studied with the PAA dosage of 15 mg/L. Figure 2d–f showed that RB5 removal was relatively stable against Fe^{2+} dosage from 0.55 to 2.20 mg/L, likely owing to the sufficient Fe^{2+} even at a low dosage (0.55 mg/L) for the activation of PAA/H_2O_2. ANL removal increased from 11% to 38% with the elevated Fe^{2+} dosage, implying the enhanced activation of Fe^{2+} on PAA/H_2O_2. While PVA removal increased initially (Fe dosage < 2.20 mg/L) and then remained unchanged. These results demonstrated that the dosages of both Fe^{2+} and PAA/H_2O_2 had a significant influence on the removal of these pollutants.

3.3. Effects of Coexisting Inorganic Anions and Humic Acid on Pollutants' Removal

3.3.1. Effects of Coexisting SO_4^{2-}, Cl^-, and HCO_3^-

Considering that textile wastewater generally had high contents of SO_4^{2-}, Cl^-, and HCO_3^-, their effects on the removal of the three pollutants were investigated with the anions' concentrations of 0–2000 mg/L. Figure 3 demonstrated that all these anions inhibited their removals, the extent of which followed an order of $HCO_3^- > SO_4^{2-} > Cl^-$. The inhibition of HCO_3^- was probably because of the quenching effect of HCO_3^- on ·OH to generate less reactive $HCO_3^{\cdot}$ according to reaction (12) with a high reaction rate constant (>10^8 $M^{-1}s^{-1}$) [49]. In addition, HCO_3^- may consume Fe^{2+} to form nonreactive Fe^{2+}-HCO_3^- complexes [20], and probably cause a pH increase simultaneously. Among these three pollutants, PVA was the most sensitive against these anions, indicating the possibly easier quenching of ·OH by these anions than other ROSs. As for RB5, SO_4^{2-} at 2000 mg/L or HCO_3^- at 200 and 2000 mg/L significantly decreased the removals from 98 to 82.5% or from 98.6 to approximately 81.2%, respectively. The inhibition of SO_4^{2-} and Cl^- was likely because they could convert ·OH and peroxyl radicals to $SO_4^{\cdot-}$ and chlorine-containing radicals which may show relatively weak oxidative capacity towards these pollutants [36].

$$HCO_3^- + \cdot OH \rightarrow HCO_3^{\cdot} + OH^- \quad (12)$$

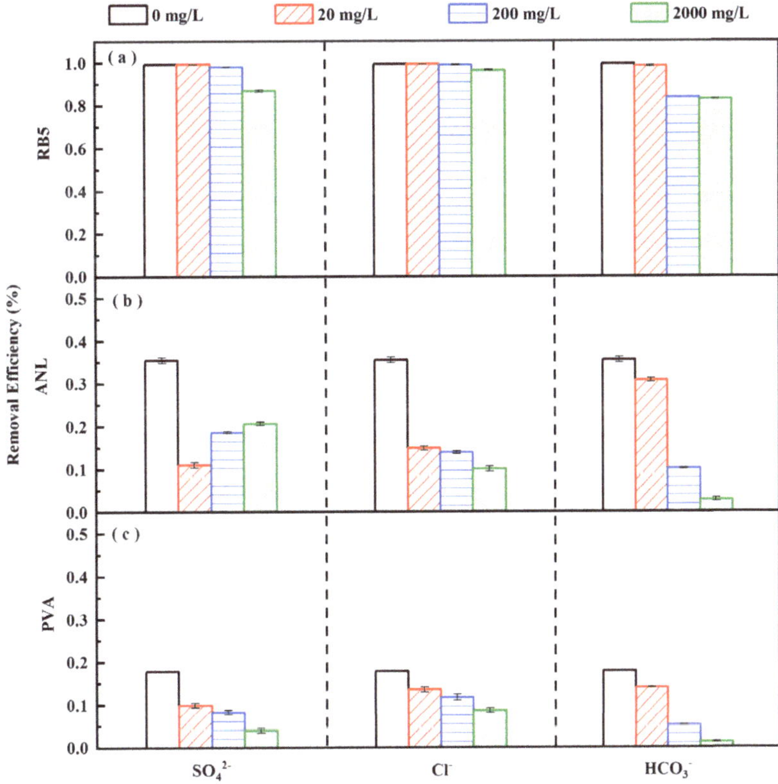

Figure 3. Effect of coexisting SO_4^{2-}, Cl^-, and HCO_3^- on the degradation of (**a**) RB5, (**b**) ANL, and (c) PVA in $Fe^{2+}/PAA/H_2O_2$ system. Conditions: [PAA] = 15 mg/L, $[RB5]_0$ = 20 mg/L, $[PVA]_0$ or $[ANL]_0$ = 10 mg/L, pH = 3.0, T = 21 ± 1 °C, [anion] = 0, 20, 200, and 2000 mg/L.

3.3.2. Effects of Background HA and Real Water Matrix

The influence of HA on the degradation of these three pollutants was investigated. Figure 4 showed that HA at a low concentration of 1 mg C/L had a minor effect on their removals, while notable inhibition at 5 and 10 mg C/L. In addition, the inhibition effect followed an order of RB5 > ANL > PVA, indicating a probably higher scavenging capacity of HA on peroxyl radicals compared with ·OH [36]. As for the effect of the water matrix, the removals of RB5, ANL, and PVA decreased from 99%, 35%, and 18% to 56%, 15%, and 7%, respectively. The more pronounced effect on PVA was probably because ·OH, with a non-selective oxidation property, tended to be consumed by background organics in practical wastewater compared to other ROSs.

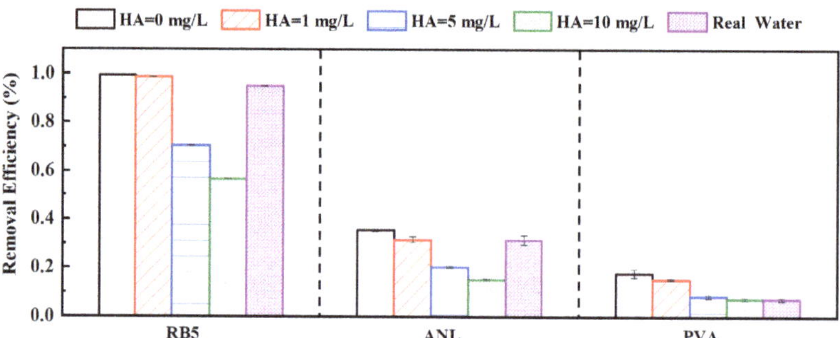

Figure 4. Effect of coexisting HA and real water matrix on the degradation of RB5, PVA and ANL in $Fe^{2+}/PAA/H_2O_2$ system. Conditions: [PAA] = 15 mg/L, $[RB5]_0$ = 20 mg/L, $[PVA]_0$ or $[ANL]_0$ = 10 mg/L, pH = 3.0, T = 21 ± 1 °C.

3.4. *Acute Toxicity Evaluation*

A bioluminescent bacteria test was used to evaluate the acute toxicity alteration induced by different AOPs. Figure 5 showed that, after $Fe^{2+}/PAA/H_2O_2$ treatment, the toxicity of RB5 and PVA decreased while that of ANL increased. For all these pollutants, the $Fe^{2+}/PAA/H_2O_2$ treated systems possessed lower toxicity compared with Fe^{2+}/H_2O_2 treated ones, implying the eco-friendly advantage of the former. In order to further reduce the toxicity of the effluent, the combination of $Fe^{2+}/PAA/H_2O_2$ and subsequent adsorption treatment might be a potential approach.

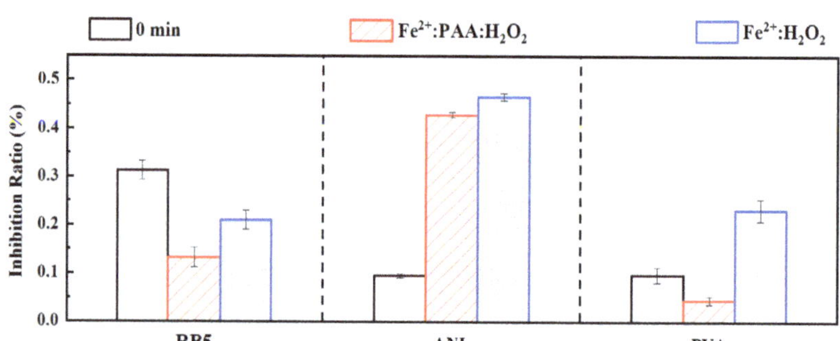

Figure 5. Acute toxicity alteration towards bioluminescent bacteria induced by different AOPs. Conditions: [PAA] = 15 mg/L, pH = 3.0, T = 21 ± 1 °C. RB5: ($[RB5]_0$ = 20 mg/L, Fe^{2+}: PAA:H_2O_2 = 1.1 mg/L:15 mg/L:5 mg/L, Fe^{2+}: H_2O_2 = 1.1 mg/L:5 mg/L); PVA or ANL: ($[PVA]_0$ or $[ANL]_0$ = 10 mg/L, Fe^{2+}: PAA:H_2O_2 = 2.2 mg/L:15 mg/L:5 mg/L, Fe^{2+}: H_2O_2 = 2.2 mg/L:5 mg/L).

3.5. Intermediate Products and Proposed Pathways

Owing to the highest degradation efficiency by $Fe^{2+}/PAA/H_2O_2$ among these three pollutants, RB5 was selected as the typical pollutant to identify the degradation products and possible pathways. Nine intermediates were identified (Figure S2) and their structures were proposed in Table 2. Accordingly, Figure 6 exhibited three possible degradation pathways of RB5 ($C_{26}H_{21}N_5Na_4O_{19}S_6$, m/z 991) to finally form $C_{10}H_{11}N_3O_3$ (m/z 221), $C_{17}H_{14}N_3NaO_7S_2$ (m/z 459), and $C_{10}H_{10}N_2O$ (m/z 174), respectively. Therein, the main mechanisms probably included hydroxylation, dehydrogenation, and demethylation, which was consistent with the previously reported oxidation mechanism related to peroxyl radicals [50,51].

Table 2. Details and proposed molecular structure of detected degradation intermediates during $Fe^{2+}/PAA/H_2O_2$ oxidation of RB5.

No.	Retention Time (min)	Chemical Formula	Molecular Mass	Experimental Mass (m/z)	Proposed Structure
(1)	5.72	$C_{18}H_{16}N_3NaO_{13}S_4$	633.58	634.15	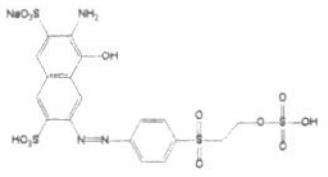
(2)	5.72	$C_{22}H_{15}N_5Na_2O_{10}S_3$	651.56	652.13	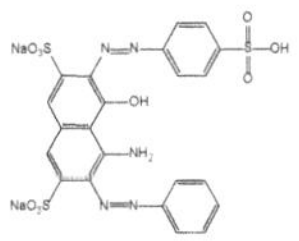
(3)	5.27	$C_{16}H_{11}N_2NaO_7S_2$	430.39	431.09	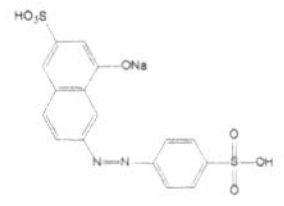
(4)	5.27	$C_{17}H_{14}N_3NaO_7S_2$	459.43	460.28	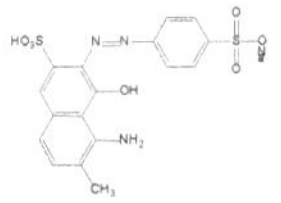
(5)	9.94	$C_{26}H_{25}N_3O_8S_2$	571.62	572.41	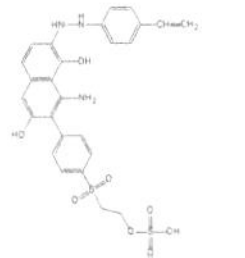
(6)	3.71	$C_{26}H_{25}N_3O_7S_2$	555.62	556.39	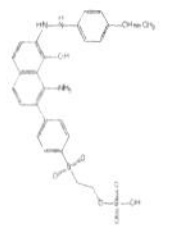
(7)	4.29	$C_{16}H_{15}N_3O$	265.31	267.00	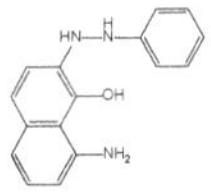
(8)	4.29	$C_{10}H_{11}N_3O_3$	221.21	223.07	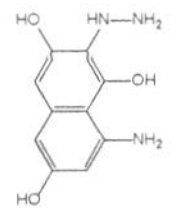

Table 2. *Cont.*

No.	Retention Time (min)	Chemical Formula	Molecular Mass	Experimental Mass (*m*/*z*)	Proposed Structure
(9)	3.71	$C_{10}H_9NO_3S$	223.25	224.07	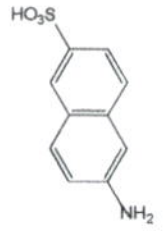
(10)	4.63	$C_{10}H_{10}N_2O$	174.20	175.11	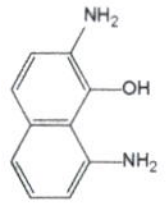

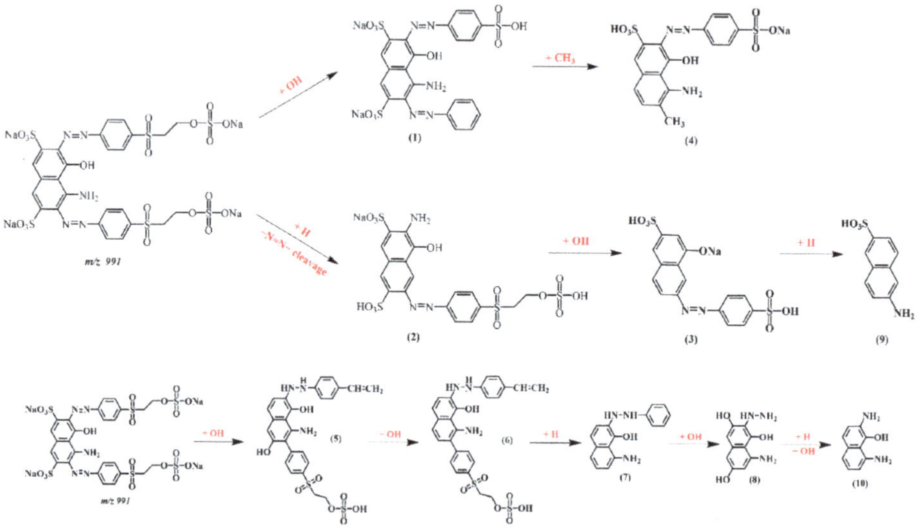

Figure 6. Proposed degradation pathways of RB5 in $Fe^{2+}/PAA/H_2O_2$ system.

For ANL, six main intermediates were identified with their proposed structures (Table S1) and possible degradation pathways (Figure S3) [52,53]. ANL was firstly attacked by ·OH to form nitrobenzene (*m*/*z* 123), N-phenylacetamide (*m*/*z* 135), and azoxybenzene (*m*/*z* 198), all of which could be further degraded to CO_2 and H_2O.

4. Conclusions

In this work, the degradation of RB5, ANL, and PVA by $Fe^{2+}/PAA/H_2O_2$ was investigated compared with Fe^{2+}/H_2O_2. Therein, $Fe^{2+}/PAA/H_2O_2$ and Fe^{2+}/H_2O_2 were relatively suitable for the degradation of RB5 (94%) and PVA (25%), respectively, while exhibiting similar removal efficiency on ANL. In addition, $Fe^{2+}/PAA/H_2O_2$ was more pH-dependent compared with Fe^{2+}/H_2O_2. Quenching test results indicated that PVA degradation mainly depended on ·OH, ANL mainly on both ·OH and other ROSs (peroxyl radicals and Fe(IV)), while RB5 was mainly on other ROSs. Both HCO_3^- (20–2000 mg/L) and HA (5–10 mg C/L) showed great inhibition in their removals. Among these pollutants, practical effluent showed the greatest inhibition on PVA removal. Toxicity test results demonstrated that, for all these pollutants, $Fe^{2+}/PAA/H_2O_2$ treated systems had lower toxicity compared with Fe^{2+}/H_2O_2 treated ones. Three pahways of RB5 degradation were proposed with the possible mechanisms including hydroxylation, dehydrogenation, and demethylation. This work may provide guidance to assess the suitability of $Fe^{2+}/PAA/H_2O_2$ to efficiently remove typical pollutants in textile wastewater.

Supplementary Materials: The following supporting information can be downloaded at: https://www.mdpi.com/article/10.3390/catal12070684/s1.

Author Contributions: Conceptualization, S.S., Q.W., N.G. and Y.Z.; Data curation, J.Y. and Y.Z.; Funding acquisition, Y.Z.; Investigation, J.Y.; Methodology, Q.W.; Project administration, Y.Z.; Resources, S.S.; Supervision, S.S.; Validation, J.Y.; Writing—original draft, J.Y. and S.S.; Writing—review & editing, Q.W., N.G. and Y.Z. All authors have read and agreed to the published version of the manuscript.

Funding: This work was supported by the Fundamental Research Funds for Center Universities (21D111311), Shanghai Sailing Program (20YF1401200), the National Natural Science Foundation of China (No. 52000023), and the Shanghai Committee of Science and Technology (No. 19DZ1204400).

Data Availability Statement: All data supporting this study are available in the supplementary information accompanying this paper.

Conflicts of Interest: The authors declare no conflict of interest.

References

1. Yukseler, H.; Uzal, N.; Sahinkaya, E.; Kitis, M.; Dilek, F.B.; Yetis, U. Analysis of the best available techniques for wastewaters from a denim manufacturing textile mill. *J. Environ. Manag.* **2017**, *203*, 1118–1125. [CrossRef]
2. Chen, H.; Yu, X.; Wang, X.; He, Y.; Zhang, C.; Xue, G.; Liu, Z.; Lao, H.; Song, H.; Chen, W.; et al. Dyeing and finishing wastewater treatment in China: State of the art and perspective. *J. Clean. Prod.* **2021**, *326*, 129353. [CrossRef]
3. Xue, G.; Wang, Q.; Qian, Y.; Gao, P.; Su, Y.; Liu, Z.; Chen, H.; Li, X.; Chen, J. Simultaneous removal of aniline, antimony and chromium by ZVI coupled with H_2O_2: Implication for textile wastewater treatment. *J. Hazard. Mater.* **2019**, *368*, 840–848. [CrossRef] [PubMed]
4. Khatri, A.; Peerzada, M.H.; Mohsin, M.; White, M. A review on developments in dyeing cotton fabrics with reactive dyes for reducing effluent pollution. *J. Clean. Prod.* **2015**, *87*, 50–57. [CrossRef]
5. Kausar, A.; Iqbal, M.; Javed, A.; Aftab, K.; Nazli, Z.; Bhatti, H.N.; Nouren, S. Dyes adsorption using clay and modified clay: A review. *J. Mol. Liq.* **2018**, *256*, 395–407. [CrossRef]
6. Martínez, C.M.; Celis, L.B.; Cervantes, F.J. Immobilized humic substances as redox mediator for the simultaneous removal of phenol and reactive red 2 in a UASB reactor. *Appl. Microbiol. Biotechnol.* **2013**, *97*, 9897–9905. [CrossRef]
7. Hou, L.; Wu, Q.; Gu, Q.; Zhou, Q.; Zhang, J. Community structure analysis and biodegradation potential of aniline-degrading bacteria in biofilters. *Curr. Microbiol.* **2018**, *75*, 918–924. [CrossRef]
8. Discharge Standards of Water Pollutants for Dyeing and Finishing of Textile Industry. 2013. Available online: https://english.mee.gov.cn/Resources/standards/water_environment/Discharge_standard/201301/t20130107_244749.shtml (accessed on 26 March 2022).
9. Zhai, S.; Zheng, Q.; Ge, M. Nanosized mesoporous iron manganese bimetal oxides anchored on natural kaolinite as highly efficient hydrogen peroxide catalyst for polyvinyl alcohol degradation. *J. Mol. Liq.* **2021**, *337*, 116611. [CrossRef]
10. Chou, W.; Wang, C.; Huang, K. Investigation of process parameters for the removal of polyvinyl alcohol from aqueous solution by iron electrocoagulation. *Desalination* **2010**, *251*, 12–19. [CrossRef]
11. Sun, W.; Chen, L.; Wang, J. Degradation of PVA (polyvinyl alcohol) in wastewater by advanced oxidation processes. *J. Adv. Oxid. Technol.* **2017**, *20*, 20170018. [CrossRef]
12. Eslami, A.; Moradi, M.; Ghanbari, F.; Mehdipour, F. Decolorization and COD removal from real textile wastewater by chemical and electrochemical Fenton processes: A comparative study. *J. Environ. Health Sci.* **2013**, *11*, 31. [CrossRef] [PubMed]
13. Cui, J.Q.; Wang, X.J.; Lin, X.L. Treatment of textile wastewater using facultative contact reactor-biological contact oxidation and ozone biological aerated filter. *Adv. Mater. Res.* **2013**, *777*, 318–325. [CrossRef]
14. Aydin, M.I.; Yuzer, B.; Öngen, A.; Ökten, H.E.; Selcuk, H. Comparison of ozonation and coagulation decolorization methods in real textile wastewater. *Desalination Water Treat.* **2018**, *103*, 5–64. [CrossRef]
15. Cardoso, J.C.; Bessegato, G.G.; Zanoni, M.V.B. Efficiency comparison of ozonation, photolysis, photocatalysis and photoelectrocatalysis methods in real textile wastewater decolorization. *Water Res.* **2016**, *98*, 39–46. [CrossRef] [PubMed]
16. Dulta, K.; Aeli, G.K.; Chauhan, P.; Jasrotia, R.; Chauhan, P.K.; Ighalo, J.O. Multifunctional CuO nanoparticles with enhanced photocatalytic dye degradation and antibacterial activity. *Sustain. Environ. Res.* **2022**, *32*, 2. [CrossRef]
17. Jasrotia, R.; Verma, A.; Verma, R.; Kumar, S.; Ahmed, J.; Krishan, B.; Kumari, S.; Tamboli, A.M.; Sharma, S.; Kalia, S. Nickel ions modified CoMg nanophotocatalysts for solar light-driven degradation of antimicrobial pharmaceutical effluents. *J. Water Process Eng.* **2022**, *47*, 102785. [CrossRef]

18. Jasrotia, R.; Prakash, J.; Kumar, G.; Verma, R.; Kumari, S.; Kumar, S.; Singh, V.P.; Nadda, A.K.; Kalia, S. Robust and sustainable Mg1-xCexNiyFe2-yO4 magnetic nanophotocatalysts with improved photocatalytic performance towards photodegradation of crystal violet and rhodamine B pollutants. *Chemosphere* **2022**, *294*, 133706. [CrossRef]
19. Kour, S.; Jasrotia, R.; Puri, P.; Verma, A.; Sharma, B.; Singh, V.P.; Kumar, R.; Kalia, S. Improving photocatalytic efficiency of $MnFe_2O_4$ ferrites via doping with Zn^{2+}/La^{3+} ions: Photocatalytic dye degradation for water remediation. *Environ. Sci. Pollut. Res.* **2021**, 1–16. [CrossRef]
20. Ao, X.; Eloranta, J.; Huang, C.; Santoro, D.; Sun, W.; Lu, Z.; Li, C. Peracetic acid-based advanced oxidation processes for decontamination and disinfection of water: A review. *Water Res.* **2021**, *188*, 116479.1–116479.23. [CrossRef]
21. Carlos, T.D.; Bezerra, L.B.; Vieira, M.M.; Sarmento, R.A.; Pereira, D.H.; Cavallini, G.S. Fenton-type process using peracetic acid: Efficiency, reaction elucidations and ecotoxicity. *J. Hazard. Mater.* **2021**, *403*, 23949.1–123949.8. [CrossRef]
22. US Environmental Protection Agency. *Emerging Technologies for Wastewater Treatment and in-Plant Wet Weather Management*; US Environmental Protection Agency: Washington, DC, USA, 2013.
23. Henao, L.D.; Turolla, A.; Antonelli, M. Disinfection by-products formation and ecotoxicological effects of effluents treated with peracetic acid: A review. *Chemosphere* **2018**, *213*, 25–40. [CrossRef] [PubMed]
24. Hassaballah, A.H.; Bhatt, T.; Nyitrai, J.; Dai, N.; Sassoubre, L. Inactivation of E. coli, Enterococcus spp., somatic coliphage, and cryptosporidium parvum in wastewater by peracetic acid (PAA), sodium hypochlorite, and combined PAA-ultraviolet disinfection. *Environ. Sci.-Water Res. Technol.* **2019**, *6*, 197–209. [CrossRef]
25. Manoli, K.; Sarathy, S.; Maffettone, R.; Santoro, D. Detailed modeling and advanced control for chemical disinfection of secondary effluent wastewater by peracetic acid. *Water Res.* **2019**, *153*, 251–262. [CrossRef] [PubMed]
26. Du, P.; Liu, W.; Cao, H.; Zhao, H.; Huang, C.H. Oxidation of amino acids by peracetic acid: Reaction kinetics, pathways and theoretical calculations. *Water Res. X* **2018**, *1*, 100002. [CrossRef]
27. Kim, J.; Zhang, T.; Liu, W.; Du, P.; Dobson, J.T.; Huang, C. Advanced oxidation process with peracetic acid and Fe(II) for contaminant degradation *Environ. Sci. Technol.* **2019**, *53*, 13312–13322. [CrossRef]
28. Lin, J.; Hu, Y.; Xiao, J.; Huang, Y.; Wang, M.; Yang, H.; Zou, J.; Yuan, B.; Ma, J. Enhanced diclofenac elimination in Fe(II)/peracetic acid process by promoting Fe(III)/Fe(II) cycle with ABTS as electron shuttle. *Chem. Eng. J.* **2021**, *420*, 129692. [CrossRef]
29. De Laat, J.; Gallard, H.; Ancelin, S.; Legube, B. Comparative study of the oxidation of atrazine and acetone by H_2O_2/UV, Fe(III)/UV, Fe(III)/H_2O_2/UV and Fe(II) or Fe(III)/H_2O_2. *Chemosphere* **1999**, *39*, 2693–2706. [CrossRef]
30. Kim, J.; Du, P.; Liu, W.; Luo, C.; Zhao, H.; Huang, C. Cobalt/peracetic acid: Advanced oxidation of aromatic organic compounds by acetylperoxyl radicals. *Environ. Sci. Technol.* **2020**, *54*, 5268–5278. [CrossRef]
31. Li, R.; Manoli, K.; Kim, J.; Feng, M.; Huang, C.H.; Sharma, V.K. Peracetic acid–ruthenium(III) oxidation process for the degradation of micropollutants in water. *Environ. Sci. Technol.* **2021**, *55*, 9150–9160. [CrossRef]
32. Cai, M.; Sun, P.; Zhang, L.; Huang, C. UV/peracetic acid for degradation of pharmaceuticals and reactive species evaluation. *Environ. Sci. Technol.* **2017**, *51*, 14217–14224. [CrossRef]
33. Zhang, T.; Huang, C. Modeling the kinetics of UV/peracetic acid advanced oxidation process. *Environ. Sci. Technol.* **2020**, *54*, 7579–7590. [CrossRef] [PubMed]
34. Zhou, F.; Lu, C.; Yao, Y.; Sun, L.; Gong, F.; Li, D.; Pei, K.; Lu, W.; Chen, W. Activated carbon fibers as an effective metal-free catalyst for peracetic acid activation: Implications for the removal of organic pollutants. *Chem. Eng. J.* **2015**, *281*, 953–960. [CrossRef]
35. Ghafoori, S.; Mehrvar, M.; Chan, P.K. Photoreactor scale-up for degradation of aqueous poly(vinyl alcohol) using UV/H_2O_2 process. *Chem. Eng. J.* **2014**, *245*, 133–142. [CrossRef]
36. Chen, S.; Cai, M.; Liu, Y.; Zhang, L.; Feng, L. Effects of water matrices on the degradation of naproxen by reactive radicals in the UV/peracetic acid process. *Water Res.* **2019**, *150*, 153–161. [CrossRef]
37. Lin, C.; Hsu, S. Performance of nZVI/H_2O_2 process in degrading polyvinyl alcohol in aqueous solutions. *Sep. Purif. Technol.* **2018**, *203*, 111–116. [CrossRef]
38. HACH. *Determination of Peracetic Acid (PAA) and Hydrogen Peroxide (H2O2) in Water*; Application Note; HACH: Loveland, CO, USA, 2014.
39. Klassen, N.V.; Marchington, D.; McGowan, H.C.E. H_2O_2 determination by the I3-method and by $KMnO_4$ titration. *Anal. Chem.* **1994**, *66*, 2921–2925. [CrossRef]
40. Liu, B.; Guo, W.; Wang, H.; Si, Q.; Zhao, Q.; Luo, H.; Ren, N. Activation of peroxymonosulfate by cobalt-impregnated biochar for atrazine degradation: The pivotal roles of persistent free radicals and ecotoxicity assessment. *J. Hazard. Mater.* **2020**, *398*, 122768. [CrossRef]
41. Bhatt, D.K.; Patel, U.D. Photocatalytic degradation of reactive black 5 using Ag_3PO_4 under visible light. *J. Phys. Chem. Solids* **2021**, *149*, 109768. [CrossRef]
42. Sharma, S.; Chokshi, N.P.; Ruparelia, J.P. Comparative studies for the degradation of reactive black 5 dye employing ozone-based AOPs. *Nanotechnol. Environ. Eng.* **2021**. [CrossRef]
43. Fadaei, S.; Noorisepehr, M.; Pourzamani, H.; Salari, M.; Moradnia, M.; Darvishmotevalli, M.; Mengelizadeh, N. Heterogeneous activation of peroxymonosulfate with Fe_3O_4 magnetic nanoparticles for degradation of reactive black 5: Batch and column study. *J. Environ. Chem. Eng.* **2021**, *9*, 105414. [CrossRef]

44. Da Silva, M.P.; De Souza, A.C.A.; De Lima Ferreira, L.E.; Pereira Neto, L.M.; Nascimento, B.F.; De Araújo, C.M.B.; Fraga, T.J.M.; Da Motta Sobrinho, M.A.; Ghislandi, M.G. Photodegradation of reactive black 5 and raw textile wastewater by heterogeneous photo-Fenton reaction using amino-Fe_3O_4-functionalized graphene oxide as nanocatalyst. *Environ. Adv.* **2021**, *4*, 100064. [CrossRef]
45. Li, L.; Guo, R.; Zhang, S.; Yuan, Y. Sustainable and effective degradation of aniline by sodium percarbonate activated with UV in aqueous solution: Kinetics, mechanism and identification of reactive species. *Environ. Res.* **2022**, *207*, 112176. [CrossRef] [PubMed]
46. Shokri, A. Using $NiFe_2O_4$ as a nano photocatalyst for degradation of polyvinyl alcohol in synthetic wastewater. *Environ. Chall.* **2021**, *5*, 100332. [CrossRef]
47. Anipsitakis, G.P.; Dionysiou, D.D. Radical generation by the interaction of transition metals with common oxidants. *Environ. Sci. Technol.* **2004**, *38*, 3705–3712. [CrossRef]
48. Grosjean, D.; Williams, E.L., II. Environmental persistence of organic compounds estimated from structure-reactivity and linear free-energy relationships. Unsaturated aliphatics. *Atmos. Environ. Part A* **1992**, *26*, 1395–1405. [CrossRef]
49. Cope, V.W.; Schoen-Nan, C.; Hoffman, M.Z. Intermediates in the photochemistry of of carbonato-amine complexes of cobalt(III). Carbonate(-) radicals and the aquocarbonato complex. *J. Am. Chem. Soc.* **1973**, *95*, 3116–3121. [CrossRef]
50. Yan, T.; Ping, Q.; Zhang, A.; Wang, L.; Dou, Y.; Li, Y. Enhanced removal of oxytetracycline by UV-driven advanced oxidation with peracetic acid: Insight into the degradation intermediates and N-nitrosodimethylamine formation potential. *Chemosphere* **2021**, *274*, 129726. [CrossRef]
51. Du, P.; Wang, J.; Sun, G.; Chen, L.; Liu, W. Hydrogen atom abstraction mechanism for organic compound oxidation by acetylperoxyl radical in Co(II)/peracetic acid activation system. *Water Res.* **2022**, *212*, 118113. [CrossRef]
52. Benito, A.; Penades, A.; Lluis Lliberia, J.; Gonzalez-Olmos, R. Degradation pathways of aniline in aqueous solutions during electro-oxidation with BDD electrodes and UV/H_2O_2 treatment. *Chemosphere* **2017**, *166*, 230–237. [CrossRef]
53. Duan, X.; Chen, Y.; Liu, X.; Chang, L. Synthesis and characterization of nanometal-ordered mesoporous carbon composites as heterogeneous catalysts for electrooxidation of aniline. *Electrochim. Acta* **2017**, *251*, 270–283. [CrossRef]

 catalysts

Article

Degradation of Tetracycline Hydrochloride by a Novel CDs/g-C_3N_4/$BiPO_4$ under Visible-Light Irradiation: Reactivity and Mechanism

Wei Qian [1,2,3], Wangtong Hu [2], Zhifei Jiang [3], Yongyi Wu [2], Zihuan Li [2], Zenghui Diao [2,*] and Mingyu Li [1,*]

1 School of Environment, Jinan University, Guangzhou 510632, China; qianwei710@zhku.edu.cn
2 College of Resources and Environment, Zhongkai University of Agriculture and Engineering, Guangzhou 510225, China; wangtongh1998@sina.com (W.H.); wuyongyi2022@sina.com (Y.W.); aayl614819344@126.com (Z.L.)
3 Huafeng Bijiang Environmental Protection Technology Co., Ltd., Zhaoqing 526108, China; 13555629364@139.com
* Correspondence: zenghuid86@zhku.edu.cn (Z.D.); tlimy@jnu.edu.cn (M.L.)

Abstract: In recent years, with the large-scale use of antibiotics, the pollution of antibiotics in the environment has become increasingly serious and has attracted widespread attention. In this study, a novel CDs/g-C_3N_4/$BiPO_4$ (CDBPC) composite was successfully synthesized by a hydrothermal method for the removal of the antibiotic tetracycline hydrochloride (TC) in water. The experimental results showed that the synthesized photocatalyst was crystalline rods and cotton balls, accompanied by overlapping layered nanosheet structures, and the specific surface area was as high as 518.50 m^2/g. This photocatalyst contains g-C_3N_4 and bismuth phosphate ($BiPO_4$) phases, as well as abundant surface functional groups such as C=N, C-O, and P-O. When the optimal conditions were pH 4, CDBPC dosage of 1 g/L, and TC concentration of 10 mg/L, the degradation rate of TC reached 75.50%. Active species capture experiments showed that the main active species in this photocatalytic system were holes (h^+), hydroxyl radicals, and superoxide anion radicals. The reaction mechanism for the removal of TC by CDBPC was also proposed. The removal of TC was mainly achieved by the synergy between the adsorption of CDBPC and the oxidation of both holes and hydroxyl radicals. In this system, TC was adsorbed on the surface of CDBPC, and then the adsorbed TC was degraded into small molecular products by an attack with holes and hydroxyl radicals and finally mineralized into carbon dioxide and water. This study indicated that this novel photocatalyst CDBPC has a huge potential for antibiotic removal, which provides a new strategy for antibiotic treatment of wastewater.

Keywords: carbon dots; photocatalysis; graphitic carbon nitride; bismuth phosphate; tetracycline hydrochloride

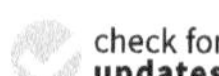

Citation: Qian, W.; Hu, W.; Jiang, Z.; Wu, Y.; Li, Z.; Diao, Z.; Li, M. Degradation of Tetracycline Hydrochloride by a Novel CDs/g-C_3N_4/$BiPO_4$ under Visible-Light Irradiation: Reactivity and Mechanism. *Catalysts* **2022**, *12*, 774. https://doi.org/10.3390/catal12070774

Academic Editors: Hao Xu and Yanbiao Liu

Received: 20 June 2022
Accepted: 11 July 2022
Published: 13 July 2022

Publisher's Note: MDPI stays neutral with regard to jurisdictional claims in published maps and institutional affiliations.

1. Introduction

According to the Chinese Academy of Sciences' antibiotic pollution map, more than 6100 tons of tetracycline (TC) antibiotics are discharged into the soil and groundwater environment every year [1]. Antibiotic concentrations in Chinese rivers generally reach 15 μg/L, far exceeding their toxicity threshold [2,3]. Tetracyclic antibiotics are mainly treated by biological, physical, and chemical oxidation methods [4]. Biological treatment methods mainly rely on microbial decomposition to breakdown pollutants. Tetracycline antibiotics are difficult to completely remove by conventional biological treatment [5,6]. Physical treatment methods include adsorption and membrane treatment technology; however, pollutants are not degraded and can easily cause secondary pollution [7–9]. Chemical oxidation methods mainly degrade various types of pollutants by oxidation and reduction reactions, including ozone oxidation [10], Fenton oxidation [11,12], persulfate oxidation [13,14] and photocatalytic technology [15,16]. Photocatalytic degradation of antibiotics with higher efficiency and broader applicability has been studied and applied by a large number of scholars at this stage for the treatment of various antibiotic contaminations [17]. In recent years, graphitic carbon nitride ($g-C_3N_4$) has been widely used in photocatalysis. $g-C_3N_4$ is a typical conjugated polymer in photocatalysts [18–20]. The conjugated polymer facilitates the transport and separation of photoinduced electrons and holes due to its unique internal conjugated π-bond structure. Additionally, its unique two-dimensional layered structure could assist the transport and migration of electrons, and with the advantages of nontoxicity and visible light response (semiconductor band gap of 2.7 eV), it is widely applied in photocatalytic degradation [21]. However, $g-C_3N_4$ suffers from a high photogenerated electron-hole recombination rate, which limits its quantum efficiency [22]. Therefore, promoting the separation of its photogenerated carriers is crucial to promoting its photocatalytic efficiency. It is an effective way to composite different materials to overcome their respective defects. Some researchers have prepared different composites, such as $g-C_3N_4/MoS_2$ [23], $g-C_3N_4$/BiOBr [24], and $g-C_3N_4/BiVO_4$ [25], which show superior photocatalytic performance. The material composite of $BiPO_4$ and $g-C_3N_4$ can compensate for each other's defects and take advantage of the high carrier separation efficiency of $BiPO_4$ and the wide range of light absorption of $g-C_3N_4$ [26]. The photocatalytic degradation rate of the $g-C_3N_4/BiPO_4$ composite photocatalyst was 4.5 times higher than that of P25 TiO_2 and 2.5 times higher than that of pure $BiPO_4$ under UV irradiation [27]. This value is 2.5 times higher than that of single mesoporous $g-C_3N_4$ under visible light irradiation [28]. Additionally, $g-C_3N_4/BiPO_4$ composites showed superior catalytic degradation performance in the degradation of TC, SMD, SD, and other pollutants [29–31].

However, the $g-C_3N_4/BiPO_4$ material could still not absorb sunlight with wavelengths longer than 420 nm. Carbon dots (CDs) have excellent optical properties and can make full use of visible light over 500 nm [32,33]. CDs take advantage of water solubility and biocompatibility. In this paper, we prepared CDs/$g-C_3N_4/BiPO_4$ composite-based precursors of $g-C_3N_4$ and inspected the performance of photocatalytic degradation TC solution, mainly in the following aspects: (a) physicochemical characterization of the CDBPC composite; (b) degradation performance of TC by the CDBPC composite under visible light; (c) degradation kinetics of TC; and (d) proposed reaction mechanism of TC degradation.

2. Results and Discussion

2.1. Characterization of CDBPC

As shown in Figure 1a, the CDBPC catalytic material presented crystal rods and spherical crystals, which agglomerated together and were accompanied by tiny spherical crystals. It was presumed that the crystal rods might be $BiPO_4$. As shown in Figure 1b, the diameter range of crystal rods varied in size, fluctuating in the range of 100–900 nm, while the diameter of spherical crystals was approximately 1 μm. The crystal rods had a smoother surface than the spherical crystals but still had tiny spherical crystals at the edges. In contrast, the surface of the spherical crystals was rougher and more porous. Figure 1c,d presents the transmission electron microscopy scans of the CDBPC composite material. Figure 1c shows that the crystal rods and spherical crystals were tightly wrapped together. Figure 1d shows that the spherical crystal structure was evenly attached to the crystal rods. Therefore, it was presumed that spherical crystals were CD crystals, which was consistent with the results of previous studies [34]. The distribution of elements on the surface of CDBPC material was analyzed by EDS mapping, and the results are shown in Figure 1e–j. The CDBPC sample was mainly composed of five elements, including C, N, O, P, Bi, etc. As shown in the SEM image (Figure 1), the rod-shaped crystals of $BiPO_4$ were mainly concentrated in the southwest corner, while $g-C_3N_4$ was mainly distributed in the northeast corner. EDS mapping scans revealed that P, O, and Bi elements were mainly concentrated in the southwest corner and N elements were mainly distributed in the northeast corner, which was consistent with the distribution of $BiPO_4$ and $g-C_3N_4$ materials. The C element is evenly distributed on the surface of the material, mainly because the C element consists of CDs and g-C_3N_4. Additionally, CDs exhibited excellent water solubility and were evenly distributed on the surface of materials.

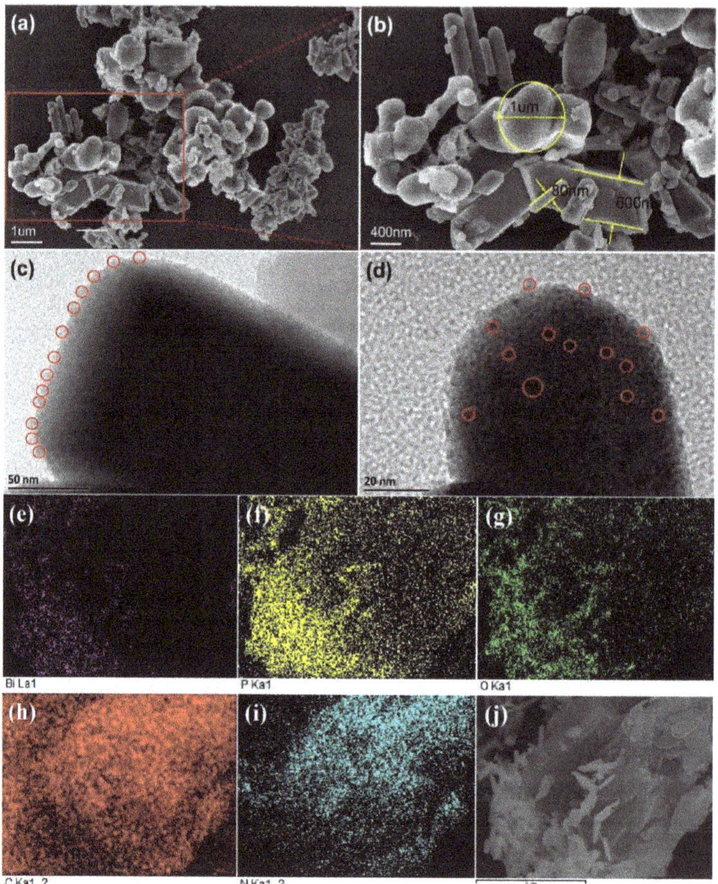

Figure 1. SEM (**a**,**b**,**j**) and TEM (**c**,**d**) images of CDBPC, EDS elemental mapping images of the corresponding area: (**e**) Bi, (**f**) P, (**g**) O, (**h**) C and (**i**) N.

The semi-quantitative mass ratio and atomic fraction analysis of each element was carried out by EDS, and the results were shown in Table 1. Accordingly, the C element accounted mass ratio of 34.84 wt%, N element mass ratio accounted for 27.98 wt%, the P element mass ratio accounted for 1.61 wt%, and the Bi mass ratio element accounted for 9.66 wt%.

Table 1. Elemental composition of CDBPC composite.

Chemical Elements	C	N	Bi	P	O
mass ratio (wt %)	34.84	37.98	15.91	1.61	9.66
atomic fraction (%)	43.26	40.45	0.69	0.78	14.83

As seen in Figure 2b, the adsorption–desorption curve of the CDBPC material was type IV, indicating the presence of a homogeneous mesoporous structure in the material, which was multilayered. Therefore, it caused the hysteresis loops. The curve belonged to the H3 type. The hysteresis loop was not closed, probably because the adsorption equilibrium was not reached when the relative pressure was close to the saturation vapor pressure [35]. This result indicated that the CDBPC materials had narrow and long pore structures. The jump in the curve of the adsorption isotherm at a relative pressure of 0.1 could illustrate the presence of microporosity in the sample material. From Figure 2a, CDBPC was a mesoporous material with a high pore volume. When the average pore size was between 25 and 5 nm, the pore volume accumulated to 0.01 cm^3/g, and the average pore size was ≥5 nm. The cumulative rate of pore volume increased when the average pore size was between 5 and 2.5 nm, reaching a maximum of 0.09 cm^3/g cumulatively at an average pore size ≥2.5 nm. The pore-specific surface area reached a maximum value of 162 m^2/g when the average pore size was ≥2.5 nm. After calculation by the BET method, the specific surface area of the CDBPC material reached 518.4978 m^2/g with an average pore diameter of 2.14 nm and a pore volume of 0.2774 cm^3/g.

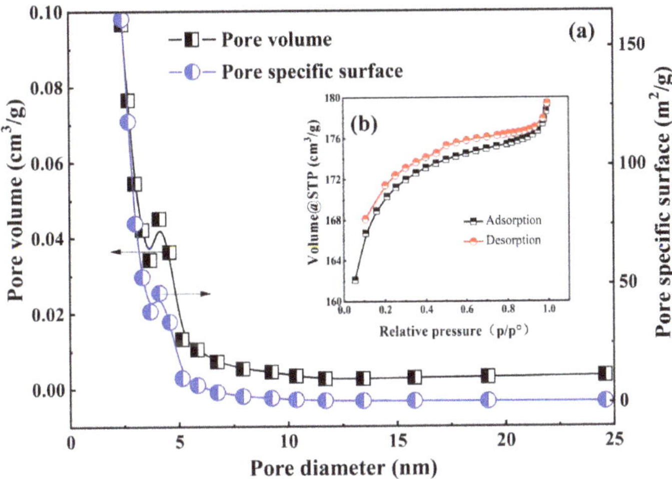

Figure 2. Pore size and volume accumulation curves (**a**) and (**b**) nitrogen adsorption–desorption isotherm of CDBPC.

As shown in Figure 3, CDBPC exhibited the peaks at 17.02°, 19.18°, 21.50°, 27.32°, 29.00°, 31.33°, 34.62°, 36.84°, 41.92°, 52.80°, 56.68°, 61.22° and 70.91°, corresponding to (101), (011), (111), (200), (120), (200), (−212), (202), (−103), (−402), (132), (233) and (−134) of $BiPO_4$ (JCPDS 80-0209), respectively. The lattice spacing of d(120) is 0.3706 and was calculated by Bragg's diffraction law, which is confirmed by the HRTEM reported in the study [10]. Our calculated value was in very good approximation to standard values. The diffraction peak of CDBPC materials at 26.1° was the characteristic peak of hexagonal CDs (JCPDS 87-1523). CDBPC had broad peaks at 13.0°, 47.62°, and 53.66°, which were characteristic of g-C_3N_4 (JCPDS 87-1524, 87-1526). The diffraction peaks of $BiPO_4$ and g-C_3N_4 were sharp and intense, indicating their highly crystalline nature. No impurity peaks were observed, confirming the high purity of the products. From the number and area of peaks detected by XRD, it was shown that $BiPO_4$, followed by g-C_3N_4 and CDs, was the main component, which was consistent with the results of previous studies [36,37].

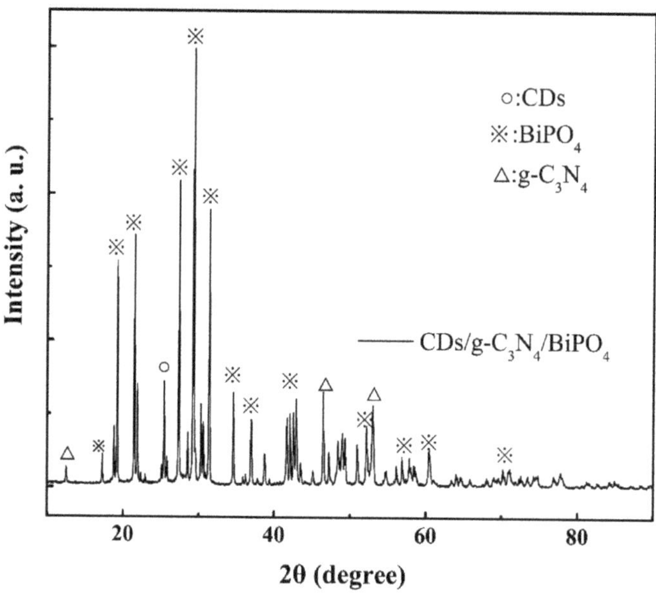

Figure 3. XRD patterns of CDBPC.

The CDBPC material was analyzed by Fourier transform infrared spectrograms, and the results are shown in Figure 4. The peaks in the wavenumber range between 680 and 480 cm^{-1} were caused by bending vibrations of $BiPO_4$, while the peaks near 560 cm^{-1} were due to O-P-O bonding vibrations. A sharp peak at approximately 1100 cm^{-1} with a small and not obvious width was due to the stretching vibration of C-O, and the adsorption peak at 1400 cm^{-1} resulted from C-OH stretching vibrations, which was consistent with previous studies [38]. The characteristic peak near 1500 cm^{-1} was attributed to the C=N stretching vibration in the carbon–nitrogen heterocycle, which was in agreement with other studies [39]. The spike near 1700 cm^{-1} with a small width belonged to the C=O stretching vibration of the carboxylic groups, which meant that carboxylic groups were present on the surface of the composite material. There was also a relatively wide peak near 3400 cm^{-1}, and such a peak was mainly caused by C-H bending; another weaker and relatively wide peak near 3500 cm^{-1} was due to the bending vibration of -OH.

As seen in Figure 5a, both CDBPC and $BiPO_4/g\text{-}C_3N_4$ had obvious absorption band edges at 280 nm and 410 nm. Compared with $BiPO_4/g\text{-}C_3N_4$, the absorption intensity of CDBPC decreased, but its absorption range increased significantly. A redshift in the absorption band edge of CDBPC implied that the addition of $g\text{-}C_3N_4$ reduced the band gap of the photocatalyst. The energy band gap of the two materials was calculated by a Tauc plot, and the results are shown in Figure 5b. The energy band gap was 2.98 eV for $BiPO_4/g\text{-}C_3N_4$, 2.68 eV for the CDBPC material, and 2.98 eV for $BiPO_4/g\text{-}C_3N_4$, indicating that the addition of CDs reduced the energy band gap and resulted in a wider light absorption range.

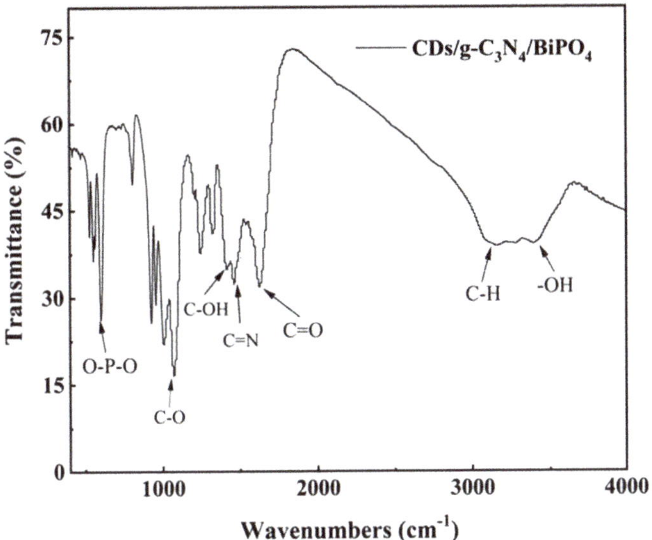

Figure 4. FT−IR spectra of the CDBPC composite.

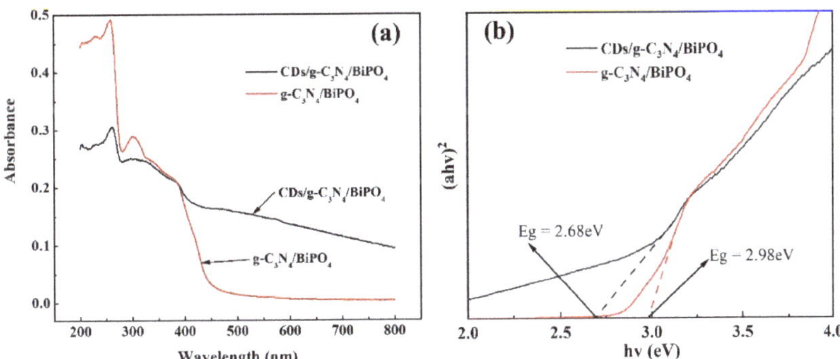

Figure 5. (**a**) UV–Vis absorption spectra, (**b**) Tauc plots curve of CDBPC.

2.2. *Effect of Preparation Conditions on CDBPC Catalytic Performance*

The effect of CD loading on the catalytic efficiency of TC degradation was investigated, with CD loading ranging from 3% to 30%, and the results are shown in Figure 6. After 60 min of light avoidance adsorption, the highest removal rate of TC was only 4%, which indicated that the adsorption performance of the CDBPC material was poor. During the photocatalytic reaction, the degradation rate of TC gradually improved with increasing time and became stable at 210 min. The catalytic degradation efficiency was 45.5%. When the CDs loading was 3 wt%, the catalytic degradation efficiency was 48.5% at 210 min, and the catalytic degradation efficiency reached 68.86% when the CDs loading was 5 wt%. Subsequently, when the CDs loading was increased to 10 wt% and 20 wt%, the degradation effect decreased slightly. When the loading of CDs was raised to 30 wt%, the removal efficiency dropped dramatically, and the final degradation rate was only 27.27%. Firstly, due to the excessive CDs covering the g-C_3N_4 and $BiPO_4$ materials, the two materials could not be excited by light to produce electron holes, and only the light energy that penetrated the CDs could have an effect [39]. Secondly, it was possible that the excessive CDs reduced the number of active sites in the CDs/g-C_3N_4/$BiPO_4$ composites, which reduced the effective collision probability between the composites and TC [40]. Thus, it was observed that increasing the loading of CDs appropriately was beneficial to improve the degradation efficiency of TC. However, excessive loading was not conducive to degradation. Under the experimental conditions of this paper, the optimal CDs loading was 5 wt%.

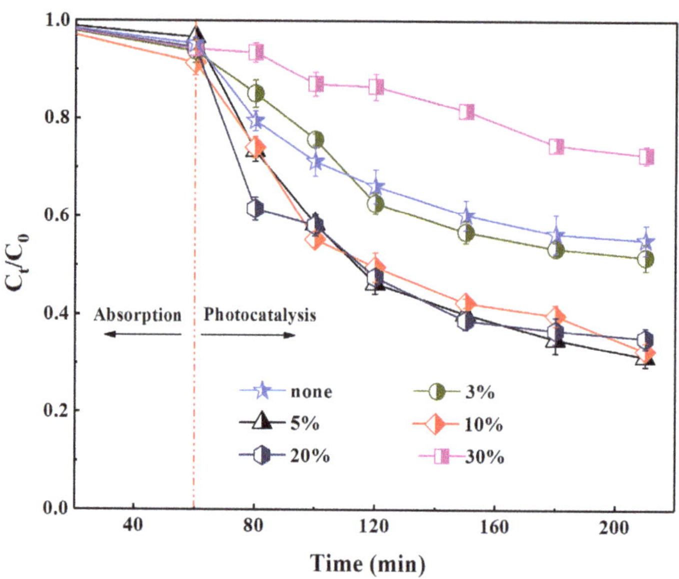

Figure 6. Degradation efficiency of TC versus time with different CD quality fractions.

2.3. *Effects of Operating Parameters on the Degradation of TC*

The pH had a strong impact on the photocatalytic degradation process. It determined the surface charge properties of the composite, affected the adsorption and degradation ability of the composite on the target pollutant, and determined the ionized state of the catalyst surface. In this paper, the effect of pH on the degradation rate of CDBPC was investigated, and the results are shown in Figure 7a. According to the figure, the pH variation had little effect on the adsorption efficiency during the light-avoidance adsorption stage, which was maintained at a low level. In the photocatalytic degradation stage, the pH was in the range of 4–14, and the degradation rate decreased gradually from 74.89% to 13.67% with increasing pH. When the pH was 2, the degradation rate of TC was 75.02%, which was close to that at pH 4. This was mainly caused by two reasons: (a) acidic conditions could inhibit the decomposition of −OH and the oxygen evolution reaction and improve the utilization of free radicals; (b) compared with the −OH oxidation potential under alkaline conditions (2.02 V), the −OH oxidation potential under acidic conditions (2.85 V) exhibited stronger oxidation performance. After comprehensive consideration, the optimal pH value was 4. In general, the acidic conditions were more favorable for the photocatalytic degradation of TC by CDBPC.

The effect of CBPC dosage on the degradation efficiency of TC was studied, and the results are shown in Figure 7b. When the doses of CDBPC were 0.1 g/L, 0.2 g/L, 0.5 g/L and 1 g/L, the degradation rates of TC were 39.6%, 49.3%, 69.5% and 73.2%, respectively. At this time, as the dosage increased, the degradation rate increased, mainly because the addition of a catalyst could enhance the chance of contact between the catalyst and solute, thereby improving the degradation efficiency. As the dosage continued to increase to 2 g/L and 3 g/L, the degradation rate of TC not only did not improve but also decreased slightly to 71.4% and 71.7%, respectively. This might be due to the high dosage of CDBPC, which caused the materials to block the light from each other and reduced the production of active species. A large number of catalysts might agglomerate to hinder the contact between the materials and the contaminants.

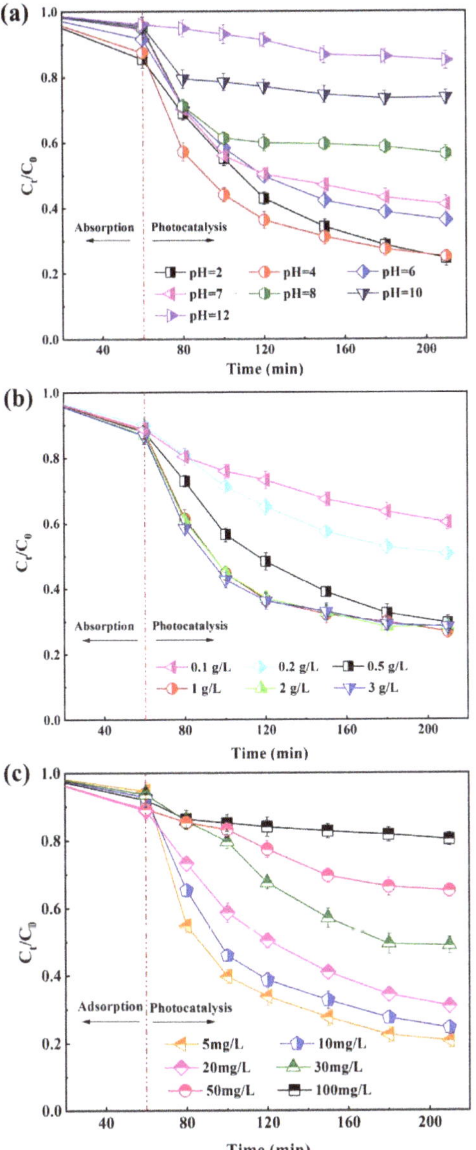

Figure 7. Degradation efficiency of TC versus time with (**a**) different initial pH values, (**b**) doses of CDBPC composite, and (**c**) initial TC concentrations.

Figure 7c shows the effect of the initial concentration of TC solution on the photocatalytic degradation rate. In general, the initial concentration of TC had little effect on the light-avoiding adsorption stage. In the photocatalytic stage, when the concentration of TC was 5 mg/L, 10 mg/L, 20 mg/L, 30 mg/L, 50 mg/L, and 100 mg/L, the degradation rates were 79.3%, 75.5%, 69.0%, 51.2%, 34.9%, and 19.7%, respectively. As the initial concentration increased, the degradation rate gradually decreased, while the amount of TC degradation increased instead. Mainly due to the high initial concentration of the TC solution, the catalyst had more contact opportunities with TC, and thus, the amount of TC degradation was greater. Similarly, because of the high initial concentration of TC solution, the degradation rate of TC was lower, which was consistent with previous studies. In summary, under the current experimental conditions, CDBPC materials provided excellent photocatalytic degradation effects for TC solutions within 20 mg/L.

2.4. Degradation Kinetics of TC

TC degradation was fitted to pseudo-first-order kinetics and pseudo-second-order kinetics models, respectively (Figure 8). The goodness of (R^2), slope (k), and half-lives ($t_{1/2}$) of the kinetic equations for the different TC initial concentrations and different initial pH values were calculated in Table 2. For different initial TC concentration conditions, the R^2 values of the pseudo-first-order kinetics model ranged from 0.9314 to 0.9937, while that of the pseudo-second-order kinetics model ranged from 0.955 to 0.9949. This suggests that TC degradation was more consistent with the pseudo-second-order kinetics model. As shown in Figure 8c,d, there was a significant difference in the degradation effect under different initial pH conditions. The fitting linear correlation coefficient was only 0.63–0.74 when the pH value was greater than 7. In the range of pH 2–7, the R^2 values for the pseudo-first-order kinetics ranged from 0.9087 to 0.972, while the R^2 values ranged from 0.9438 to 0.9985 for the pseudo-second-order kinetics, suggesting that TC degradation with different pH conditions was more consistent with the pseudo-second-order kinetics.

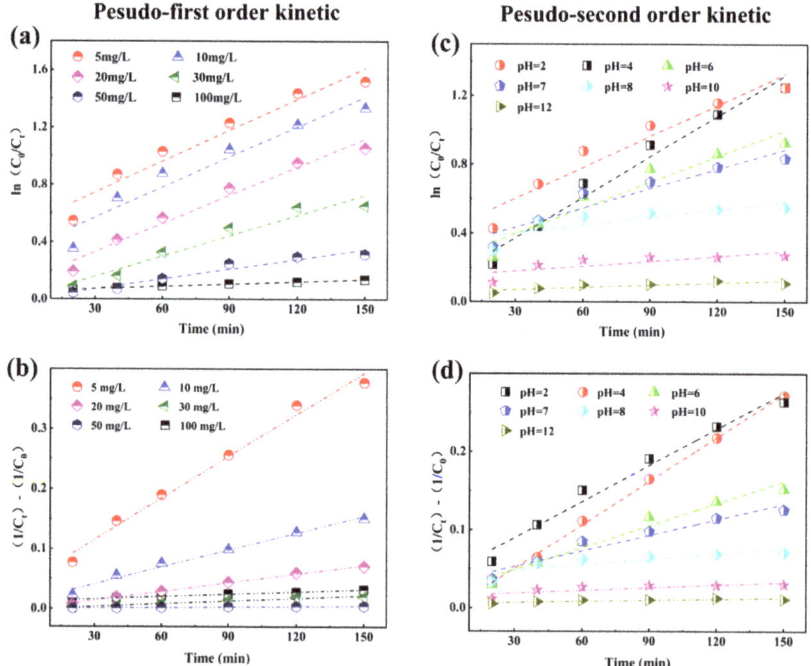

Figure 8. TC degradation of (**a**) pseudo-first-order kinetics and (**b**) pseudo-second-order kinetics with different TC concentration, (**c**) pseudo-first-order kinetics and (**d**) pseudo-second-order kinetics with different initial pH values.

As seen in Table 2, when the initial concentration of TC was 5 mg/L, the degradation rate constant was the largest (2.3×10^{-3} min^{-1}) and the $t_{1/2}$ time was the lowest (only

87 min). As the initial concentration increased, the degradation rate constant decreased continuously. When the initial concentration of TC was 100 mg/L, the degradation rate constant was 0.03×10^{-3} min^{-1}, and $t_{1/2}$ was 333.3 min. Compared with the initial concentration of TC of 5 mg/L, the degradation rate constant was only 1/76, and $t_{1/2}$ was 3.83 times longer.

Table 2. TC degradation kinetics parameters with different TC initial concentrations.

Initial Concentration (mg/L)	Pseudo-First-Order Kinetic Model			Pseudo-Second-Order Kinetic Model		
	K (min^{-1})	R^2	$t_{1/2}$ (min)	K (10^{-3} min^{-1})	R^2	$t_{1/2}$ (min)
5	0.0072	0.9459	96.3	2.3	0.9876	87.0
10	0.007	0.9314	99.0	0.9	0.9858	111.1
20	0.0065	0.9731	106.6	0.5	0.9965	100.0
30	0.0047	0.9485	147.5	0.2	0.9606	166.7
50	0.0023	0.9481	301.4	0.1	0.955	200.0
100	0.0005	0.9937	1386.3	0.03	0.9949	333.3

As shown in Table 3, a lower pH value was more favorable for the degradation of TC due to the higher OH oxidation potential. When the pH value was 2, the degradation rate was 1.5×10^{-3} min^{-1}, and $t_{1/2}$ was 67.9 min. When the pH value was 4, the highest degradation rate constant was reached, which was 1.9×10^{-3} min^{-1}, and $t_{1/2}$ was 53.2 min. As the pH value increased, the degradation rate constant gradually decreased, while $t_{1/2}$ gradually increased. When the pH value reached 12, the degradation rate constant was 0.05×10^{-3} min^{-1}, and the $t_{1/2}$ was 1904.2 min. Compared with the pH of 4, the degradation rate constant was 1/38, and the $t_{1/2}$ was 35.79 times longer.

Table 3. TC degradation kinetics parameters with different initial pH.

Initial Concentration (mg/L)	Pseudo-First-Order Kinetic Model			Pseudo-Second-Order Kinetic Model		
	K (min^{-1})	R^2	$t_{1/2}$ (min)	K (10^{-3} min^{-1})	R^2	$t_{1/2}$ (min)
5	0.0072	0.9459	96.3	2.3	0.9876	87.0
10	0.007	0.9314	99.0	0.9	0.9858	111.1
20	0.0065	0.9731	106.6	0.5	0.9965	100.0
30	0.0047	0.9485	147.5	0.2	0.9606	166.7
50	0.0023	0.9481	301.4	0.1	0.955	200.0
100	0.0005	0.9937	1386.3	0.03	0.9949	333.3

2.5. Free Radical Identification

BQ, EDTA-2Na and TBA were used as superoxide radical (O^{2-}), hole (h^+), and hydroxyl radical (OH) scavengers, respectively [41]. As shown in Figure 9, the photocatalytic reaction was almost completely inhibited by the addition of EDTA disodium, indicating that hole (h^+) was the most dominant active substance for TC degradation. The addition of BQ and TBA also had some effect on the photocatalytic performance, but the effect was not as obvious as that of the hole (h^+), indicating that $\cdot O^{2-}$ and OH were also involved in the photocatalytic reaction. The CDs/g-C_3N_4/$BiPO_4$ composite prepared in this study exhibited an excellent photocatalytic performance.

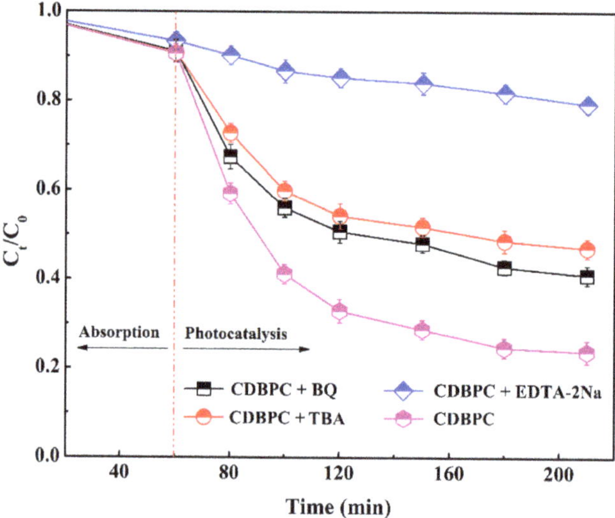

Figure 9. Effect of different quenchers on the photocatalytic degradation of TC.

2.6. Proposed Reaction Mechanism

Figure 10 displays the possible photocatalytic degradation mechanism of T composites. The semiconductor $BiPO_4$ and g-C_3N_4 were combined to form a heterogeneous structure with CDs attached to its surface and junction. The semiconductor $BiPO_4$ could directly absorb energy with wavelengths less than 320 nm in sunlight, and g-C_3N_4 could directly absorb energy with wavelengths less than 420 nm. The two materials could absorb solar energy with wavelengths greater than 500 nm by the CDs particles attached to their surfaces to enhance the utilization. After the absorption of solar energy by both materials, electrons in the VB were excited and jumped to the CB, thus generating holes in the VB. The holes (h^+) could directly degrade TC into intermediate products CO_2 and H_2O, and the holes (h^+) could also react with OH− or H_2O to produce −OH. Then, the hydroxyl radicals could interact directly with TC degradation [42–44]. The jumped electrons then combined with O^2 to produce $\cdot O^{2-}$, and superoxide radicals could catalyze TC degradation.

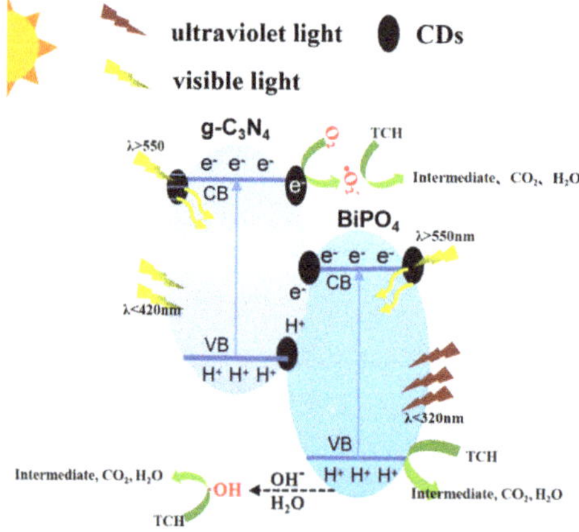

Figure 10. Schematic image of the probable photocatalytic mechanism for TC degradation over the CDBPC composite photocatalyst.

3. Materials and Methods

3.1. Materials and Chemicals

TC was purchased from Aladdin Reagent (Shanghai) Co., Ltd. (Shanghai, China) and stored at 4 °C. EDTA disodium, tert-butyl alcohol (TBA), benzoquinone (BQ) and other chemicals were obtained from Sinopharm Chemical Reagent Co., Ltd., (Shanghai, China). All chemicals used in the experiments were of analytical grade. A stock standard solution of TC (1000 mg/L) was prepared in ultrapure water and then diluted with double distilled water.

3.2. Preparation of CDs/g-C_3N_4/$BiPO_4$

Graphitic carbon nitride (g-C_3N_4) was prepared by high-temperature calcination at 550 °C employing melamine as the source material. A certain amount of citric acid and ethylenediamine monohydrate was placed in a hydrothermal reactor and reacted at 180 degrees C for 24 h. The mixed solution was subsequently freeze-dried to obtain CDs. A certain amount of g-C_3N_4 was placed in water for ultrasonic stirring, and then $Bi(NO_3)_3{\cdot}5H_2O$ and $NaH_2PO_4{\cdot}2H_2O$ were added in a molar ratio of 1:1 and synthesized in a hydrothermal reactor at 160 °C. The suspended solid was subsequently filtered, washed, and dried to obtain the g-C_3N_4/$BiPO_4$ complex. Certain amounts of CDs were codissolved with g-C_3N_4/$BiPO_4$ in ethanol and then dried at 70 °C. The materials were calcined in a muffle furnace at 300 °C to obtain 3 wt%, 5 wt%, 10 wt%, 20 wt%, and 30 wt% CDBPC composite photocatalysts.

3.3. Photocatalytic Activity of the CDBPC Composite by Beaker Experiments

All of the experiments in this work were carried out in a dedicated photocatalytic reactor, as illustrated in Figure 11. The adsorption phase was carried out in the dark, and the photocatalytic phase was carried out with a 500 W xenon lamp that simulated visible light, with the UV component isolated by a filter.

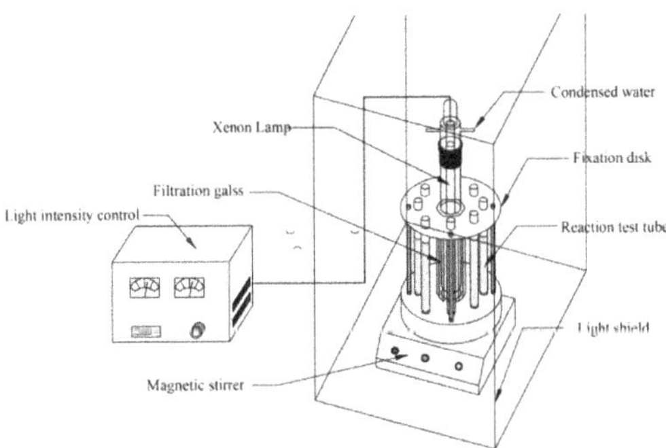

Figure 11. Schematic diagram of the photocatalysis experimental device.

To evaluate the effect of an initial pH value, experiments were conducted in the initial pH range of 2.0–12.0. 0.1 M NaOH and H_2SO_4 was used to adjust the pH values. The related description has been added in the revised manuscript accordingly. Four catalyst dosages (0.2, 0.5, 1.0, 2.0, and 3.0 g/L) were used to assess the effect of the photocatalyst dosages on TC degradation. The effect of the initial TC concentration was examined at five concentrations (5, 10, 20, 30, 50, and 100 mg/L). To identify the role of the generated reactive oxygen species such as holes (h^+), HO, and O^{2-}, scavengers such as EDTA disodium, tert-butyl alcohol, and benzoquinone (BQ) were used. All experiments except repetitive experiments were conducted three times, and the relative standard deviations were usually within 3% unless otherwise stated.

3.4. Analytical Methods

The TC concentration was analyzed by an Agilent 1100 LC system using a C18 column (5 μm, 150 × 4.6 mm, Phenomenex, Torrance, CA, USA) with a UV detector at 280 nm. The X-ray diffraction (XRD) pattern was conducted on a D/max 2200 vpc Diffractometer (Rigaku Corporation, Tokyo, Japan) with a Cu Karadiation at 30 kV and 30 mA. Scanning electron microscopy (SEM) was performed on gold-coated samples using a Quanta-400F-mode field emission scanning electron microscope (FEI, Lausanne, Switzerland). The SBET was determined with an Autosorb-iQ-MP automated gas sorption analyzer (Quantachrome Instruments, Boynton Beach, FL, USA) by nitrogen adsorption/desorption isotherms and calculated by the Brunauer–Emmett–Teller (BET) method.

3.5. Reaction Kinetics

The photocatalytic degradation of TC can be described by the pseudo-first-order kinetic and pseudo-second-order kinetic models, as shown in Equations (1) and (2), respectively [45].

$$In(C_0/C_t) = k \times t \quad (1)$$

$$C_t^{-1} - C_0^{-1} = k \times t \quad (2)$$

The half-lives ($t_{1/2}$) of the first-order kinetic and pseudo-second-order kinetic models were calculated via Equations (3) and (4):

$$t_{1/2} = In(2) \times k^{-1} \quad (3)$$

$$t_{1/2} = C_0^{-1} \times k^{-1} \quad (4)$$

where C_0 is the TC concentration at the initial time, C_t is the TC concentration at any time "t", and k is the degradation rate constant.

4. Conclusions

This paper constructed a novel CDs/g-C_3N_4/$BiPO_4$ composite photocatalyst, which could overcome the respective defects of g-C_3N_4 and $BiPO_4$. The introduction of CDs could enhance the visible light utilization range and solubility and improve catalytic efficiency. The CDs/g-C_3N_4/$BiPO_4$ composite catalytic material exhibited excellent catalytic activity. During the visible light photocatalytic reaction, CDs/g-C_3N_4/$BiPO_4$ showed the strongest catalytic activity when the loading of CDs was 5 wt%, and the TC degradation rate reached 68.85%. The CDs/g-C_3N_4/$BiPO_4$ composite had a wider adsorption range with an energy band gap of 2.68 eV, and the main active species were holes (h^+), followed by OH and O^{2-}. Finally, the probable photocatalytic mechanism for TC degradation over the CDBPC composite photocatalyst was shown.

Author Contributions: Conceptualization, Z.D. and M.L.; methodology, Z.L.; software, Z.J.; validation, Y.W. and Z.L.; formal analysis, W.Q.; writing—original draft preparation, W.Q. and W.H.; writing—review and editing, Z.D.; supervision, M.L. All authors have read and agreed to the published version of the manuscript.

Funding: This work was funded by the Project of Educational Commission of Guangdong Province of China (No. 2019KTSCX067), College Students' Innovation and Entrepreneurship Competition (No. 202211347036 and No. S202111347080), Guangdong Special Commissioners in Agricultural Science and Technology (No. KTP20190016), Special project in key areas of Guangdong Province Ordinary Universities (Nos. 2020ZDZX1003 and 2021ZDJS007), Key Realm R&D Program of Guangdong Province (Nos. 2020B1111350002 and 2020B0202080002) and the National Natural Science Foundation of China (No.21407155).

Data Availability Statement: Not applicable.

Conflicts of Interest: The authors declare no conflict of interest.

References

1. Zhang, Q.Q.; Ying, G.G.; Pan, C.G.; Liu, Y.S.; Zhao, J.L. Comprehensive evaluation of antibiotics emission and fate in the river basins of China: Source analysis, multimedia modeling, and linkage to bacterial resistance. *Environ. Sci. Technol.* **2015**, *49*, 6772. [CrossRef] [PubMed]
2. Lu, P.; Fang, Y.; Barvor, J.B.; Neth, N.L.K.; Fan, N.; Li, Z.; Cheng, J. Review of Antibiotic Pollution in the Seven Watersheds in China. *Pol. J. Environ. Stud.* **2019**, *28*, 4045–4055. [CrossRef]
3. Jia, L.; Lya, C.; Lan, Z.B.; By, B.; Li, W. Antibiotics in soil and water in China—A systematic review and source analysis. *Environ. Pollut.* **2020**, *266*, 115147.
4. Kumar, M.; Jaiswal, S.; Sodhi, K.K.; Shree, P.; Singh, D.K.; Agrawal, P.K.; Shukla, P. Antibiotics bioremediation: Perspectives on its ecotoxicity and resistance. *Environ. Int.* **2019**, *124*, 448–461. [CrossRef] [PubMed]
5. Hain, E.; Adejumo, H.; Anger, B.; Orenstein, J.; Blaney, L. Advances in antimicrobial activity analysis of fluoroquinolone, macrolide, sulfonamide, and tetracycline antibiotics for environmental applications through improved bacteria selection. *J. Hazard. Mater.* **2021**, *415*, 125686. [CrossRef] [PubMed]
6. Danner, M.C.; Robertson, A.; Behrends, V.; Reiss, J. Antibiotic pollution in surface fresh waters: Occurrence and effects. *Sci. Total Environ.* **2019**, *664*, 793–804. [CrossRef]
7. Tang, J.; Fang, J.; Tam, N.F.; Yang, Y.; Dai, Y.; Zhang, J.; Shi, Y. Impact of Phytoplankton Blooms on Concentrations of Antibiotics in Sediment and Snails in a Subtropical River, China. *Environ. Sci. Technol.* **2021**, *55*, 1811–1821. [CrossRef]
8. Chao, Y.; Zhang, J.; Li, H.; Wu, P.; Li, X.; Chang, H.; Zhu, W. Synthesis of boron nitride nanosheets with N-defects for efficient tetracycline antibiotics adsorptive removal. *Chem. Eng. J.* **2020**, *387*, 124138. [CrossRef]
9. Zhou, Y.; Gao, Y.; Jiang, J.; Shen, Y.M.; Pang, S.Y.; Wang, Z.; Duan, J.; Guo, Q.; Guan, C.; Ma, J. Transformation of tetracycline antibiotics during water treatment with unactivated peroxymonosulfate. *Chem. Eng. J.* **2020**, *379*, 122378. [CrossRef]
10. Liu, L.; Xu, Q.; Owens, G.; Chen, Z. Fenton-oxidation of rifampicin via a green synthesized rGO@nFe/Pd nanocomposite. *J. Hazard. Mater.* **2020**, *402*, 123544. [CrossRef]
11. Jeong, W.G.; Kim, J.G.; Baek, K. Removal of 1,2-dichloroethane in groundwater using Fenton oxidation. *J. Hazard. Mater.* **2022**, *428*, 128253. [CrossRef]
12. Mizuno, Y.; Yahaya, A.G.; Kristof, J.; Blajan, M.G.; Murakami, E.; Shimizu, K. Ozone Catalytic Oxidation for Gaseous Dimethyl Sulfide Removal by Using Vacuum-Ultra-Violet Lamp and Impregnated Activated Carbon. *Energies* **2022**, *15*, 3314. [CrossRef]
13. Fu, C.; Yi, X.; Liu, Y.; Zhou, H. Cu^{2+} activated persulfate for sulfamethazine degradation. *Chemosphere* **2020**, *257*, 127294. [CrossRef] [PubMed]
14. Fagan, W.P.; Zhao, J.; Villamena, F.A.; Zweier, J.L.; Weavers, L.K. Synergistic, aqueous PAH degradation by ultrasonically activated persulfate depends on bulk temperature and physicochemical parameters-ScienceDirect. *Ultrason. Sonochem.* **2020**, *67*, 105172. [CrossRef] [PubMed]
15. Huang, S.; Wang, G.; Liu, J.; Du, C.; Su, Y. A novel $CuBi_2O_4$/BiOBr direct Z-scheme photocatalyst for efficient antibiotics removal: Synergy of adsorption and photocatalysis on degradation kinetics and mechanism insight. *ChemCatChem* **2020**, *12*, 4431–4445. [CrossRef]
16. Das, S.; Ahn, Y.H. Synthesis and application of CdS nanorods for LED-based photocatalytic degradation of tetracycline antibiotic. *Chemosphere* **2022**, *291*, 132870. [CrossRef]
17. Gholami, P.; Khataee, A.; Bhatnagar, A.; Vahid, B. Synthesis of N-Doped Magnetic WO3-x @Mesoporous Carbon Using a Diatom Template and Plasma Modification: Visible-Light-Driven Photocatalytic Activities. *ACS Appl. Mater. Interfaces* **2021**, *13*, 13072–13086. [CrossRef]
18. Ahmaruzzaman, M.; Mishra, S.R. Photocatalytic performance of g-C_3N_4 based nanocomposites for effective degradation/ removal of dyes from water and wastewater. *Mater. Res. Bull.* **2021**, *143*, 111417. [CrossRef]
19. Chang, X.; Wang, Y.; Zhou, X.; Song, Y.; Zhang, M. ZIF-8-derived carbon-modified g-C_3N_4 heterostructure with enhanced photocatalytic activity for dye degradation and hydrogen production. *Dalton Trans.* **2021**, *50*, 17618–17624. [CrossRef]
20. Nemiwal, M.; Zhang, T.C.; Kumar, D. Recent Progress in g-C_3N_4, TiO_2 and ZnO Based Photocatalysts for Dye Degradation: Strategies to Improve Photocatalytic Activity. *Sci. Total Environ.* **2021**, *767*, 144896. [CrossRef]
21. Zhao, L.; Jin, R.; Zhang, B. A novel Z-scheme g-C_3N_4-Pt-$BiPO_4$ photocatalyst for enhanced photocatalytic degradation of azofuchsin and ciprofloxacin. *J. Am. Ceram. Soc.* **2021**, *104*, 6319–6334. [CrossRef]
22. Yu, H.; Xu, S.; Zhang, S.; Wang, S.; He, Z. In-situ construction of core-shell structured TiB_2-TiO_2@g-C_3N_4 for efficient photocatalytic degradation. *Appl. Surf. Sci.* **2022**, *579*, 152201. [CrossRef]
23. Nouri, A.; Faraji Dizaji, B.; Kianinejad, N.; Jafari Rad, A.; Rahimi, S.; Irani, M.; Sharifian Jazi, F. Simultaneous linear release of folic acid and doxorubicin from ethyl cellulose/chitosan/g-C_3N_4/MoS_2 core-shell nanofibers and its anticancer properties. *J. Biomed. Mater. Res. Part A* **2021**, *109*, 903–914. [CrossRef] [PubMed]
24. Shi, Z.; Zhang, Y.; Shen, X.; Duoerkun, G.; Zhu, B.; Zhang, L.; Li, M.; Chen, Z. Fabrication of g-C_3N_4/BiOBr heterojunctions on carbon fibers as weaveable photocatalyst for degrading tetracycline hydrochloride under visible light. *Chem. Eng. J.* **2020**, *386*, 124010. [CrossRef]
25. Dai, D.; Wang, P.; Bao, X.; Xu, Y.; Wang, Z.; Guo, Y.H. g-C_3N_4/ITO/Co-$BiVO_4$ Z-scheme composite for solar overall water splitting. *Chem. Eng. J.* **2022**, *433*, 134476. [CrossRef]

26. Liu, N.; Lu, N.; Yu, H.; Chen, S.; Quan, X. Enhanced degradation of organic water pollutants by photocatalytic in situ activation of sulfate based on Z-scheme g-C_3N_4/$BiPO_4$. *Chem. Eng. J.* **2022**, *428*, 132116. [CrossRef]
27. Li, J.; Zhao, L.; Zhang, R.; Teng, H.H.; Padhye, L.P.; Sun, P. Transformation of tetracycline antibiotics with goethite: Mechanism, kinetic modeling and toxicity evaluation. *Water Res.* **2021**, *199*, 117196. [CrossRef]
28. Chen, Y.; Yin, R.; Zeng, L.; Guo, W.; Zhu, M. Insight into the effects of hydroxyl groups on the rates and pathways of tetracycline antibiotics degradation in the carbon black activated peroxydisulfate oxidation process. *J. Hazard. Mater.* **2021**, *412*, 125256. [CrossRef]
29. Zhao, Z.; Zhang, G.; Zhang, Y.; Dou, M.; Li, Y. Fe_3O_4 accelerates tetracycline degradation during anaerobic digestion: Synergistic role of adsorption and microbial metabolism. *Water Res.* **2020**, *185*, 116225. [CrossRef]
30. Ahamad, T.; Mu, N.; Alzaharani, Y.; Alshehri, S.M. Photocatalytic degradation of bisphenol-A with g-C_3N_4/MoS_2-PANI nanocomposite: Kinetics, main active species, intermediates and pathways. *J. Mol. Liq.* **2020**, *311*, 113339. [CrossRef]
31. Zhu, Y.; Zhao, F.; Wang, F.; Zhou, B.; Chen, H.; Yuan, R.; Liu, Y.; Chen, Y. Combined the photocatalysis and fenton-like reaction to efficiently remove sulfadiazine in water using g-C_3N_4/ag/γ-feooh: Insights into the degradation pathway from density functional theory. *Front. Chem.* **2021**, *9*, 742459. [CrossRef] [PubMed]
32. Jaleel, U.; Devi, K.; Madhushree, R.; Pinheiro, D. Statistical and experimental studies of MoS_2/g-C_3N_4/TiO_2: A ternary Z-scheme hybrid composite. *J. Mater. Sci.* **2021**, *56*, 6922–6944. [CrossRef]
33. Yang, Q.; Fang, C.; Zhao, N.; Jiang, Y.; Xu, B.; Chai, S.; Zhou, Y. Enhancing electron-hole utilization of cds based on cucurbiturils vis electrostatic interaction in visible light. *J. Solid State Chem.* **2019**, *270*, 450–457. [CrossRef]
34. Samsudin, M.F.R.; Sufian, S. Hybrid 2D/3D g-C_3N_4/$BiVO_4$ photocatalyst decorated with RGO for boosted photoelectrocatalytic hydrogen production from natural lake water and photocatalytic degradation of antibiotics-ScienceDirect. *J. Mol. Liq.* **2020**, *314*, 113530. [CrossRef]
35. Wang, Q.; Lin, Y.; Li, P.; Ma, M.; Zhang, R. An efficient Z-scheme (Cr, B) codoped g-C_3N_4/$BiVO_4$ photocatalyst for water splitting: A hybrid DFT study. *Int. J. Hydro. Energy* **2020**, *46*, 247–261. [CrossRef]
36. Zhang, X.; Zhang, X.; Yang, P.; Jiang, S.P. Transition metals decorated g-C_3N_4/n-doped carbon nanotube catalysts for water splitting: A review. *J. Electroanal. Chem.* **2021**, *895*, 115510. [CrossRef]
37. Ma, X.; Hu, J.; Hua, H.; Shijie, D.; Chuyun, H.; Xiaobo, C. New Understanding on Enhanced Photocatalytic Activity of g-C_3N_4/$BiPO_4$ Heterojunctions by Effective Interfacial Coupling. *ACS Appl. Nano Mater.* **2018**, *1*, 5507–5515. [CrossRef]
38. Xia, J.; Zhao, J.; Chen, J.; Di, J.; Ji, M.; Xu, L. Facile fabrication of g-C_3N_4/$BiPO_4$ hybrid materials via a reactable ionic liquid for the photocatalytic degradation of antibiotic ciprofloxacin. *J. Photochem. Photobiol. A Chem.* **2017**, *339*, 59–66. [CrossRef]
39. He, Y.; Li, J.; Sheng, J.; Chen, S.; Dong, F.; Sun, Y. Crystal-Structure Dependent Reaction Pathways in Photocatalytic Formaldehyde Mineralization on $BiPO_4$. *J. Hazard. Mater.* **2021**, *420*, 126633. [CrossRef]
40. Wang, Y.; Ding, K.; Xu, R.; Yu, D.; Liu, B. Fabrication of $BiVO_4$/$BiPO_4$/GO composite photocatalytic material for the visible light-driven degradation. *J. Clean. Prod.* **2020**, *247*, 119108. [CrossRef]
41. Li, Y.; Zhang, H.; Rashid, A.; Hu, A.; Xin, K.; Li, H.; Adyari, B.; Wang, Y.; Yu, C.P.; Sun, Q. Bisphenol A attenuation in natural microcosm: Contribution of ecological components and identification of transformation pathways through stable isotope tracing. *J. Hazard. Mater.* **2020**, *385*, 121584.1–121584.10. [CrossRef] [PubMed]
42. Chen, L.C.; Pan, G.T.; Yang, C.K.; Chung, T.W.; Huang, C.M. In situ DRIFT and kinetic studies of photocatalytic degradation on benzene vapor with visible-light-driven silver vanadates. *J. Hazard. Mater.* **2010**, *178*, 644–651. [CrossRef] [PubMed]
43. Asadzadeh-Khaneghah, S.; Habibi-Yangjeh, A. g-C_3N_4/carbon dot-based nanocomposites serve as efficacious photocatalysts for environmental purification and energy generation: A review. *J. Clean. Prod.* **2020**, *276*, 124319. [CrossRef]
44. Zhu, X.; Wang, Y.; Guo, Y.; Sun, C. Environmental-friendly synthesis of heterojunction photocatalysts g-C_3N_4/$BiPO_4$ with enhanced photocatalytic performance. *Appl. Surf. Sci.* **2020**, *544*, 148872. [CrossRef]
45. Olfa, B.; Lobna, J.; Wahiba, N.; Sami, S. Photocatalytic degradation of bisphenol a in the presence of ce–zno: Evolution of kinetics, toxicity and photodegradation mechanism. *Mater. Chem. Phys.* **2016**, *173*, 95–105.

Article

Performance Optimization and Toxicity Effects of the Electrochemical Oxidation of Octogen

Yishi Qian [1,2], Kai Chen [3], Guodong Chai [3], Peng Xi [2], Heyun Yang [3], Lin Xie [3], Lu Qin [3], Yishan Lin [4], Xiaoliang Li [3], Wei Yan [1,*] and Dongqi Wang [3,5,6,*]

1 Department of Environmental Science and Engineering, School of Energy and Power Engineering, Xi'an Jiaotong University, Xi'an 710049, China; qys1017@stu.xjtu.edu.cn
2 Xi'an Modern Chemistry Research Institute, Xi'an 710065, China; pengxi204@outlook.com
3 Department of Municipal and Environmental Engineering, School of Water Resources and Hydro-Electric Engineering, Xi'an University of Technology, Xi'an 710048, China; 3180673038@xtu.xaut.edu.cn (K.C.); xaut_chaigd@yahoo.com (G.C.); heyunyang.xaut@yahoo.com (H.Y.); 1170411051@xtu.xaut.edu.cn (L.X.); 2200421182@xtu.xaut.edu.cn (L.Q.); lixiaoliang@xaut.edu.cn (X.L.)
4 State Key Laboratory of Pollution Control & Resource Reuse, School of the Environment, Nanjing University, Nanjing 210023, China; yshanlisa@nju.edu.cn
5 State Key Laboratory of Eco-Hydraulics in Northwest Arid Region, Xi'an University of Technology, Xi'an 710048, China
6 Shaanxi Key Laboratory of Water Resources and Environment, Xi'an University of Technology, Xi'an 710048, China
* Correspondence: yanwei@mail.xjtu.edu.cn (W.Y.); wangdq@xaut.edu.cn (D.W.)

Abstract: Octogen (HMX) is widely used as a high explosive and constituent in plastic explosives, nuclear devices, and rocket fuel. The direct discharge of wastewater generated during HMX production threatens the environment. In this study, we used the electrochemical oxidation (EO) method with a PbO_2-based anode to treat HMX wastewater and investigated its degradation performance, mechanism, and toxicity evolution under different conditions. The results showed that HMX treated by EO could achieve a removal efficiency of 81.2% within 180 min at a current density of 70 mA/cm^2, Na_2SO_4 concentration of 0.25 mol/L, interelectrode distance of 1.0 cm, and pH of 5.0. The degradation followed pseudo-first-order kinetics ($R^2 > 0.93$). The degradation pathways of HMX in the EO system have been proposed, including cathode reduction and indirect oxidation by •OH radicals. The molecular toxicity level (expressed as the transcriptional effect level index) of HMX wastewater first increased to 1.81 and then decreased to a non-toxic level during the degradation process. Protein and oxidative stress were the dominant stress categories, possibly because of the intermediates that evolved during HMX degradation. This study provides new insights into the electrochemical degradation mechanisms and molecular-level toxicity evolution during HMX degradation. It also serves as initial evidence for the potential of the EO-enabled method as an alternative for explosive wastewater treatment with high removal performance, low cost, and low environmental impact.

Keywords: electrochemistry; octogen; wastewater; hydroxyl radical; toxicity

Citation: Qian, Y.; Chen, K.; Chai, G.; Xi, P.; Yang, H.; Xie, L.; Qin, L.; Lin, Y.; Li, X.; Yan, W.; et al. Performance Optimization and Toxicity Effects of the Electrochemical Oxidation of Octogen. *Catalysts* **2022**, *12*, 815. https://doi.org/10.3390/catal12080815

Academic Editors: Hao Xu and Yanbiao Liu

Received: 20 June 2022
Accepted: 21 July 2022
Published: 25 July 2022

1. Introduction

Octogen [octahydro-1,3,5,7-tetranitro-1,3,5,7-tetrazocine (HMX)] is a highly explosive material widely used in plastic explosives, nuclear devices, rocket fuel, and other products [1]. As a heterocyclic compound with an eight-membered ring, HMX exhibits higher stability and detonation capacity than other conventional explosives [2,4,6-trinitrotoluene, hexahydro-1,3,5-trinitro-1,3,5-triazine (RDX)]. However, HMX poses potential threats to soil microorganisms, plants, animals, and humans [2]. It damages the central nervous, renal, and hepatic systems of mice and rats when oral exposure levels exceed 200 mg/kg [3]. HMX can also enter the human body through inhalation, dermal contact, and diet, inducing adverse effects on the central nervous system [4]. The US Environmental Protection

Agency (EPA) recommends that cumulative human exposure to HMX should not exceed 400 μg/L [4]. Owing to the persistent and non-degradable nature of HMX, it may leak into the surrounding water and soil during production, storage, and application. Continuous accumulation in the environment may impact living organisms [5] and affect the functioning of ecosystems [6]. Therefore, methods for the effective removal of HMX are urgently required.

Biological and physical methods are typically used to treat HMX wastewater. Although physical methods (e.g., membrane separation and reverse osmosis) are effective and easy to perform, they are expensive and can cause secondary pollution [7,8]. Biological treatments are the most economical method by which to remove readily biodegradable organic matter from wastewater. However, high levels of persistent organic matter and toxic substances, such as HMX, greatly reduce the efficiency of biological treatments [9–11]. Zhao et al. [12] conducted in situ HMX degradation in low-temperature marine sediments under anaerobic conditions and found that 50 days were required to degrade 50% of the HMX. Kanekar et al. [13] investigated the potential of soil yeast (*Pichia sydowiorum* MCM Y-3) to treat HMX wastewater in a fixed-film bioreactor (FFBR) and found that only 28–50% of the HMX was removed (HRT of one week). Therefore, it is crucial to choose a method that can effectively degrade HMX with a low environmental impact. Advanced oxidation processes (AOPs) can improve the biodegradability of wastewater while simultaneously treating pollutants [14,15]. Among these processes, the electrochemical oxidation (EO) process, which utilizes a cathode and anode to convert electric energy into chemical energy under an external electric field, has been widely used to treat organic wastewater that is difficult to biodegrade [16]. The EO method can be divided into direct and indirect oxidation reactions [17]. During direct oxidation, pollutants are adsorbed onto the anode surface and then oxidized into small-molecule organic matter. Indirect oxidation involves the oxidation of organic pollutants through the formation of strongly oxidizing intermediates, such as hydroxyl radicals (•OH), chlorate, ozone, and hydrogen peroxide, during the electrochemical reaction [18,19]. Because the EO method can completely oxidize organic pollutants in wastewater and results in lower chemical oxygen demand (COD) and toxicity [20,21], it can be promising for explosive wastewater treatment [22]. Recent studies have revealed that RDX-containing wastewater can be effectively treated by using EO process, resulting in 39.2% COD removal and 97.5% RDX removal, as well as significantly increased biodegradability [16]. A previous study by Bonin et al., elucidated the capability of an EO system with a boron-doped diamond (BDD) electrode to treat three nitramine explosives (RDX, HMX, and 2,4,6,8,10,12-hexanitro-2,4,6,8,10,12-hexaazaisowurtzitane (CL-20)) [23]. However, the high cost of BDD anodes limits their large-scale industrial application [24]. Instead, PbO_2-based anodes have been widely used in many EO systems and are considered a valid alternative to BDD electrodes because of their good conductivity, favorable overpotential, high chemical inertness, low cost, and excellent electrocatalytic performance [25–28]. Nonetheless, the application of the PbO_2 electrode-based EO system for HMX wastewater treatment and its optimal operating conditions (e.g., pH and current density) have not been well studied.

Moreover, limited research has focused on the HMX degradation mechanisms. Analyzing the intermediates of HMX is essential to understand its degradation pathway. Under anaerobic conditions, nitro derivatives are the intermediates formed by the reduction of the nitro groups on the HMX ring [11,29], mainly forming octahydro-1-nitroso-3,5,7-trinitro-1,3,5,7-tetrazocine (1NO-HMX) or octahydro-1,5-dinitroso-3,7-dinitro-1,3,5,7-tetrazocine (2NO-HMX) [29,30]. HMX can be converted to the ring-cleavage products, methylene nitramine, and bis(hydroxymethyl)nitramine, which are subsequently converted to formaldehyde, formic acid, and nitrous oxide [11,30]. However, the electrochemical degradation pathways of HMX have not been studied extensively. In addition, previous studies regarding the toxicity of energy-containing materials (e.g., RDX and HMX) are mostly based on cellular and mammalian bioassays [31,32], and little is known about molecular-level toxicity effects. Gou et al., recently developed a quantitative toxicogenomics-based assay by using cellular stress response pathways to reveal the potential toxicity mechanisms of target toxicants at the transcriptional level [33]. Compared to conventional resource-intensive, animal-based and isolated bioassays, this high-throughput assay is more rapid and cost-effective and can capture diverse and perhaps overlapping modes of action (MOAs) resulting from trace-level chemicals or chemical mixtures. The assay has been successfully used to evaluate the toxic effects and safety of nanomaterials and drinking water [33,34] and is therefore considered suitable for the assessment of molecular toxicity evolution during HMX degradation.

Therefore, in this study, we used a Ti/PbO_2 electrode as the anode and a copper plate as the cathode to construct the EO system. The effects of different process parameters, including initial pH, current density, and electrode distance, on HMX degradation were studied. The energy consumption (EC) for HMX degradation was estimated at varying operating conditions. Performance optimization and intermediate analysis were conducted to reveal the EO reaction mechanism, providing critical data and guidance for the industrialization of HMX wastewater treatment. The toxicogenomics-based assay was performed to investigate molecular-level toxicity effects in order to assess the potential health risks of the treated effluent. The results provide new insights into the electrochemical degradation of HMX and support the development of an effective, economically feasible, and environmentally friendly method for explosive wastewater treatment.

2. Results and Discussion

2.1. Performance Optimization of HMX Degradation by Electrochemical Oxidation

2.1.1. Electrolyte Concentration

Na_2SO_4 is considered a good electrolyte owing to its stability [35]. The effect of Na_2SO_4 concentration on HMX removal was investigated at a current density of 30 mA/cm^2, HMX concentration of 20 mg/L, interelectrode distance of 1.5 cm, and pH of 7.0. As shown in Figure 1a, the HMX removal efficiency increased rapidly within 30 min; however it slowed down thereafter. When the Na_2SO_4 concentration increased from 0.05 to 0.25 mol/L, the HMX removal efficiency increased from 50.2% to 64.5%. Figure 1a and Table S1 show the linear regression of HMX removal over time during electrochemical degradation, where $\ln(C_0/C_t)$ increased linearly with time, confirming that electrochemical degradation follows pseudo-first-order kinetics.

Electrolytes, such as Na_2SO_4, can facilitate the transport of electrons and ions. Generally, the conductivity of the EO system increases with increasing electrolyte concentration, which promotes the production of reactive groups such as •OH, thus increasing the electrooxidation reaction rate [36]. In addition, $SO_4{}^{2-}$ may be oxidized to $S_2O_8{}^2$, a powerful oxidizing agent that facilitates the oxidation of organic pollutants [16]. However, excess electrolyte concentrations lead to the accumulation of excess ions on the surface of the electrodes [37]. It will prevent the effective contact of organic contaminants, or produce unstable intermediates such as $HO_2\cdot$ radicals, which may be detrimental to the removal of persistent organic compounds [38]. As shown in Figure S1a, the cell voltage decreased with the increase of electrolyte concentrations so that the EC was saved. In this study, an Na_2SO_4 concentration of 0.25 mol/L was found to provide the highest HMX removal efficiency without adverse effects.

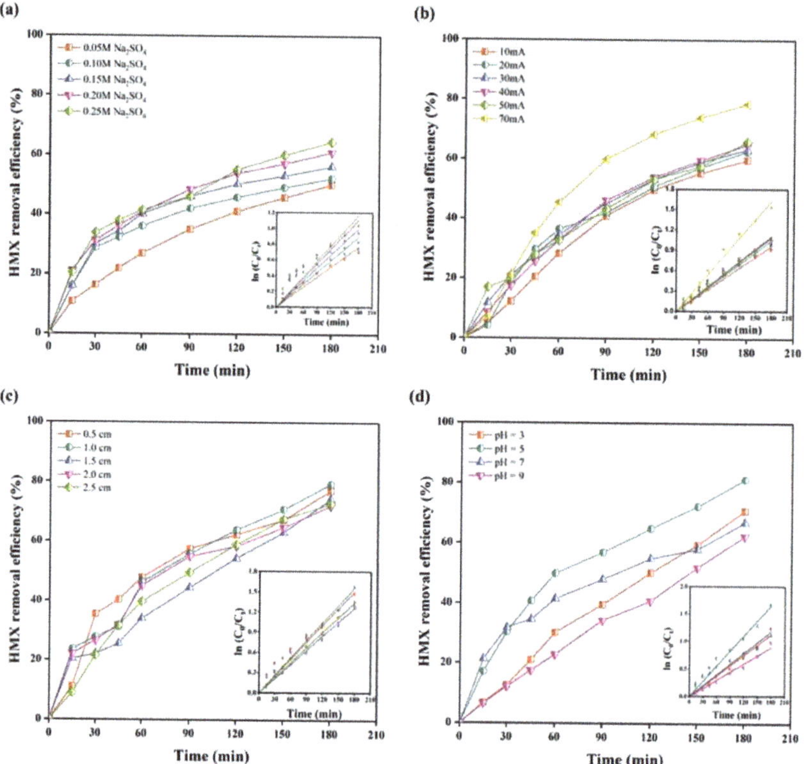

Figure 1. The effects of different operating parameters on HMX removal efficiency. (**a**) Electrolyte concentrations; (**b**) current densities; (**c**) electrode distance; (**d**) initial pH. The inset plot depicts corresponding pseudo-first-order kinetics.

2.1.2. Current Density

Several studies have concluded that increasing the current density in EO systems can improve the removal and mineralization efficiency of organic pollutants [39–41]. The effect of the current density on HMX removal was investigated at an electrolyte concentration of 0.25 mol/L, HMX concentration of 20 mg/L, interelectrode distance of 1.5 cm, and pH of 7.0. As shown in Figure 1b, HMX removal increased as the applied current density increased from 10 to 70 mA/cm^2. At a current density of 70 mA/cm^2, the HMX removal performance was significantly higher than that under other current density conditions. The HMX removal efficiency increased rapidly during the first 90 min, after which the rate of increase decreased, reaching a maximum removal efficiency of 78.6% at 180 min. Kinetic fitting of the HMX degradation curve at different current densities revealed a degradation rate constant of 5.4×10^{-2} to 9.0×10^{-2} min^{-1} (Table S1). Notably, the degradation rate of the contaminants increased slowly when the current density was increased from 10 to 50 mA/cm^2, whereas the degradation rate increased significantly when the current density was increased to 70 mA/cm^2.

Previous studies have reported that anodic oxidation is mass-transfer-controlled under low pollutants concentration or high current density conditions [42–44]. In this study, the initial HMX concentration was set constantly at 20 mg/L, resulting in the same and relatively limited mass transfer rate for all the tests. Therefore, the increase in the applied current density would not enhance the electrochemical oxidation rate of HMX significantly. Meanwhile, a continuous increase in the current density could lead to higher •OH production and electron transfer rate between the electrode surface and HMX molecules. Because

HMX is a heterocyclic nitramine whose structure is less electron-rich than the other aromatic nitramines, it is more difficult for •OH to attach to the heterocyclic compound [22,45]. Therefore, other degradation pathways (e.g., reduction on the cathode) rather than •OH radical oxidation should contribute more to HMX degradation when the current density increases to 70 mA/cm^2, resulting in a higher degradation rate. The HMX removal performance may be improved with an even higher current density. However, the EC values largely increased from 0.50 to 4.47 kWh/g (Figure S1b and Table S3) due to the increasing voltage and enhanced side reaction. It would also result in higher carbon emissions, which is not feasible for real wastewater treatment. Therefore, the current density of 70 mA/cm^2 was deemed the optimal condition for the following experiments.

2.1.3. Interelectrode Distance

The effect of interelectrode distance on HMX removal was investigated at a current density of 70 mA/cm^2, electrolyte concentration of 0.25 mol/L, HMX concentration of 20 mg/L, and pH of 7.0. Figure 1c shows that the interelectrode distance did not significantly affect the HMX removal efficiency, which was stable at approximately 75%. In general, the electric field intensity between the electrodes decreased with increasing interelectrode distance. The organic oxidation rate at the anode is faster with a shorter interelectrode distance, which may be attributed to better electrolytic performance at shorter diffusion distances [44]. When the interelectrode distance increased, the potential difference between the solution and anode decreased and weakened the driving force of the mass transfer. An increase in the mass transfer distance also reduces the concentration gradient of the solution and increases the resistance to mass transfer [46], thus leading to a decrease in the degradation performance. However, excessive reduction in the electrode distance leads to electrode breakdown or electrode short circuits [44,47]. Based on the obtained results (Figure 1c), an electrode plate spacing of 1.0 cm was considered optimal.

2.1.4. pH

The pH of the solution will affect the removal performance of organic pollutants in EO systems [48–50]. Strongly acidic (pH 3.0), acidic (pH 5.0), neutral (pH 7.0), and basic (pH 9.0) conditions were selected to assess the effect of pH on HMX removal (Figure 1d). After 180 min of EO treatment, the HMX removal efficiency reached 81.2% under acidic conditions (pH 5), whereas the efficiency was only 62.4% under alkaline conditions (pH 9). Kinetic fitting of the HMX degradation curves at different pH values revealed that the degradation rate constant was 4.9×10^{-2} to 9.2×10^{-2} min^{-1} (Table S1). It was observed that the oxidation potential of •OH was higher under acidic conditions (+2.85 V) than under basic conditions (+2.02 V) [51], suggesting that the EO system would have better pollutant treatment performance under acidic conditions. An increase in the solution pH reduces the oxygen evolution potential (OEP), facilitating the oxygen evolution reaction (OER), and thus reducing the pollutant removal efficiency [52]. Similar results were observed when the carbon felt/PbO_2 anode was used for diuron degradation, and the highest efficiency was detected at lower pH values [53,54]. However, if the acidity is too strong, the hydrogen evolution reaction (HER) will be violent, which will also reduce the treatment performance [55] and the lifetime of the Ti/PbO_2 electrode [56]. Additionally, the changes in calculated EC values (Figure S1d) suggested that the initial pH of 5.0 was more favorable for both HMX degradation and cost savings.

2.2. Possible Electrochemical Degradation Mechanism of HMX

The electrochemical degradation intermediates of HMX were analyzed by using liquid chromatography-tandem mass spectrometry (LC−MS/MS). Table S2 and Figure S2 show the MS results for HMX intermediates within 180 min of EO treatment. Based on these results, two possible HMX degradation pathways were identified (Figure 2). In pathway A, the nitro group on HMX is reduced on the Cu cathode of the EO system to produce the mononitro derivative 1NO-HMX (intermediate I, $m/z = 281$), which is similar to the previously reported microbial degradation mechanism of HMX [57]. Intermediate II (m/z = 234) contains an active imine bond (C=N) and is formed by N−denitration [58]. This intermediate reacts with water to form unstable α-hydroxy-alkylnitramine (intermediate III, m/z = 252) and then produces 4−nitro−2,4−diazabutanal (intermediate IV, m/z = 120), N_2O, and formaldehyde via ring cleavage [58]. In pathway B, the •OH generated by the EO process reacts with the carbon radical of HMX to generate hydroxyl-containing intermediate V (m/z = 313). Intermediate V would subsequently undergo ring cleavage to generate other intermediates such as methylene dinitramine and urea [59]. Both these by-products can be oxidized to form acetamide and formic acid, which continue to be oxidized, accompanied by the formation of NO_3^- and/or NH_4^+. Although some of the small-molecule substances were not determined in this study, methylene dinitramine (intermediate VI, m/z = 137) was detected, which was also found in the HMX degradation process by municipal anaerobic sludge [30].

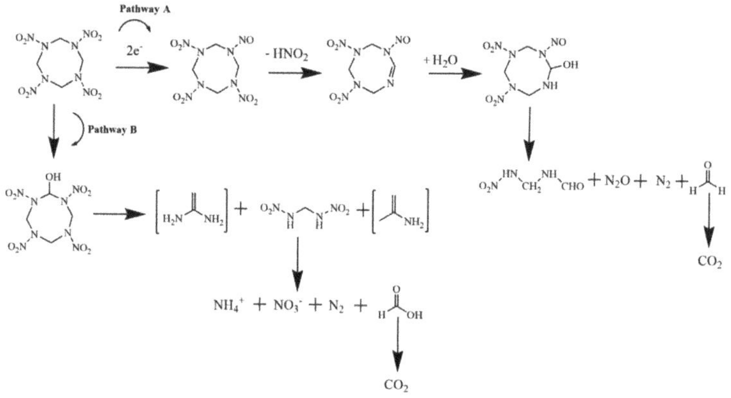

Figure 2. Possible pathways for the electrochemical degradation of HMX. Square brackets indicate undetected compounds.

To further prove the proposed pathway and evaluate the contribution of •OH to HMX degradation, a quenching experiment was conducted by adding tert-butanol (TBA) to the EO system. TBA is a strong •OH scavenger that is widely used to differentiate between direct oxidation and degradation with •OH radicals [60]. As shown in Figure S3, HMX removal efficiency decreased to 73.3% in the presence of 10 mM TBA. An increase in TBA concentration further inhibits HMX degradation. With the addition of 50 mM TBA, HMX removal efficiency was reduced to 48.6%. This demonstrates that indirect oxidation by •OH radicals (pathway B) plays an important role in HMX degradation, although cathode reduction (pathway A) or other electrocatalytic degradation mechanisms [37,61,62] will also be responsible for HMX degradation, but require further study.

2.3. *Molecular-Level Toxicity Evolution during HMX Degradation*

2.3.1. Molecular Toxicity Potency

A total of 114 reporter genes (Table S3) in five major stress categories (DNA stress, oxidative stress, protein stress, membrane stress, and general stress) were investigated to detect changes in their expression levels during HMX degradation. A slight increase in the overall transcriptional effect level index (TELI) value from 1.68 to 1.81 within 60 min was observed (Figure 3a), which could be attributed to the generation of relatively more toxic intermediates than HMX in the early stages of degradation. As these intermediates were further degraded to non-toxic end products (e.g., CO_2 and H_2O), the TELI values decreased, reaching a non-toxic level of 1.48 ($<$the threshold value of 1.5) at 180 min. According to the gene set enrichment analysis (GSEA) (Figure 3b), the main stress response pathways induced during the HMX degradation process were protein stress and oxidative stress. The

$TELI_{protein}$ and $TELI_{oxidative}$ reached a maximum value of 3.35 and 1.86 at 30 and 60 min, respectively. At 180 min, both values decreased to a minimum of 1.93 and 1.51, respectively.

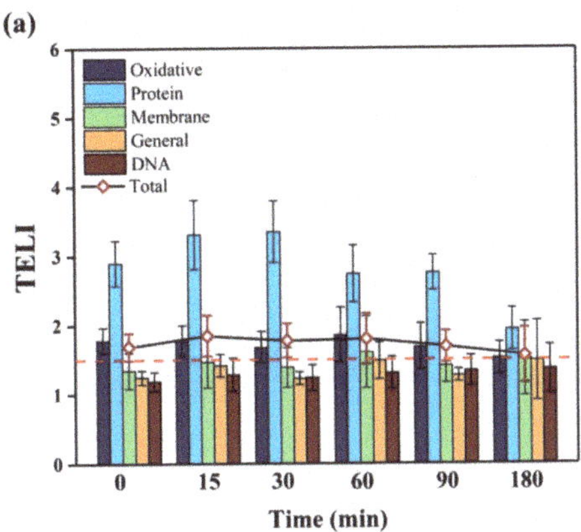

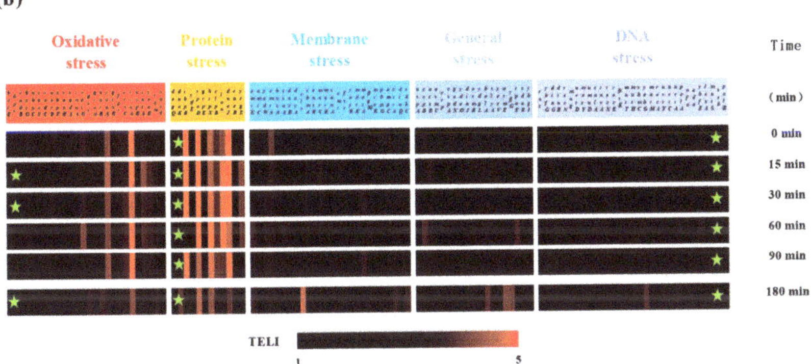

Figure 3. (**a**) The TELI-based toxicity profile changes during the electrochemical oxidation of HMX. (**a**) The TELI values for the 5 stress response categories and total TELI; (**b**) the TELI values for genes. The green star (★) indicates the significantly affected ($p < 0.05$) stress response categories revealed by the gene set enrichment analysis.

2.3.2. Molecular-Level Toxicity Effects

To further provide mechanistic insights into the evolution of molecular toxicity during HMX degradation, the differentially expressed genes ($TELI_{gene} > 1.5$) in the major stress categories (i.e., oxidative and protein stress) based on the TELI profiles and GSEA results were investigated (Figure 4). The number of genes showing altered expression during HMX degradation (Figure S4) increased from 24 at the initial stage to 35 at 60 min and then decreased to 30 at 180 min. Similarly, as shown in the hierarchical cluster (HCL) analysis diagram (Figure S5), the upregulated genes (e.g., *yea*E, *znt*A, and *tam*) and downregulated genes (e.g., *yei*G, *emr*A, *yaa*A, *fpr*, *ibp*B, *clp*B, *rpo*D, *dna*J, *trx*A, *lon*, *trx*C, *htp*X) were observed mainly from 0 to 90 min, whereas the altered gene expression was not significant at 180 min. These results suggest that, during the first 90 min, the degradation process produced more toxic intermediates, which could be further degraded and converted into less toxic end-products.

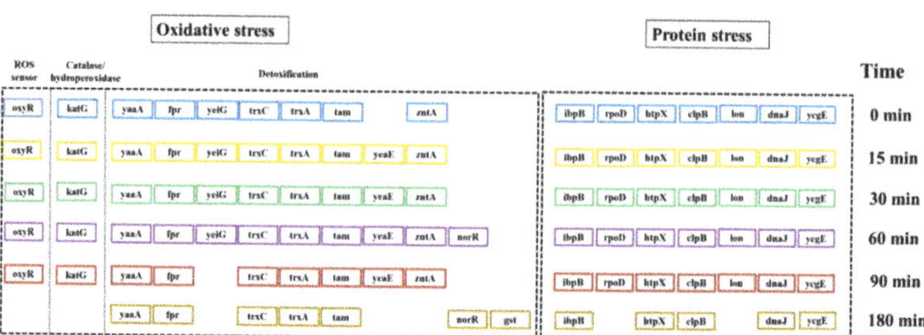

Figure 4. Major stress response pathways and biomarker genes showed altered expression ($TELI_{gene}$ > 1.5) during the electrochemical oxidation of HMX. The genes were clustered into subcategories based on their functions and involvement in various pathways (Table S1).

For oxidative stress, the differential expression of reactive oxygen species (ROS) sensors, including *oxyR*, *soxR*, and *soxS*, were induced during HMX degradation. The expression of these genes in stressful environments can scavenge oxidative free radicals and resist oxidative stress [63]. Dysregulation of *oxyR* was observed from 0 to 90 min during HMX degradation, suggesting that chemical mixtures (HMX or its degradation intermediates) led to the production of cellular ROS. At the end of the test, no differential expression of ROS sensor-related genes was observed, indicating a decreased level of ROS-induced oxidative stress. We also found that the catalase/hydroperoxidase gene *kat*G [64,65] was dysregulated from 0 to 90 min, indicating that H_2O_2 may be produced from O_2 at the cathode [66]. In addition, biomarkers involved in the detoxification pathway were dysregulated. For example, the *yaaA* gene, which is associated with cellular stress responses to peroxides, can be induced by the expression of the *oxy*R gene [67,68]. The *fpr* gene encoding a G protein-coupled receptor can be induced by exposure to H_2O_2 [69]. The number of dysregulated genes in the detoxification sub-pathway increased to 9 at 60 min and then decreased to 7 at 90–180 min, suggesting the generation of more toxic intermediate products, that could induce oxidative stress, at the beginning of the test, but that will further degrade to less or non-toxic end-products.

For the protein stress pathway, seven of the 12 genes showed altered expression during HMX degradation (Figure 4), whereas *rpo*D and *lon* were not perturbed at 180 min. Genes *rpo*D and *lon* regulate the synthesis of the RNA polymerase sigma factor and Lon proteases, respectively [70–72]. Lon proteases are required for the degradation of misfolded proteins. Therefore, the lack of dysregulation of those genes at 180 min indicates that the end products did not cause substantial protein misfolding.

3. Materials and Methods

3.1. Chemicals

HMX ($C_4H_8N_8O_8$, ≥99%) was obtained from Xi'an Modern Chemistry Research Institute (Xi'an, China). Sodium sulfate (Na_2SO_4, ≥99%), sulfuric acid (H_2SO_4, ≥99%), sodium hydroxide (NaOH pellets, ≥98%), and tert-butanol (TBA) were purchased from China National Medicines Co. Ltd. (Beijing, China). Minimal medium (M9) was purchased from Ruichu Biotechnology Co., Ltd. (Funing, China).

3.2. Experimental Design

The electrochemical reactor (8.7 × 9.2 × 12 cm, 500 mL) is shown in Figure 5. The dimensions of the Ti/PbO_2 anode and Cu cathode were 5 cm × 10 cm and 5 cm × 6 cm, respectively. The initial concentration of HMX was 20 mg/L. Because the performance of the EO system can be affected by various operating conditions [44], different influencing factors including Na_2SO_4 electrolyte concentrations (0.05, 0.1, 0.15, 0.2, 0.25 mol/L), applied current densities (10, 20, 30, 40, 50, and 70 mA/cm^2), anode-cathode distance (0.5, 1.0, 1.5, 2.0, and 2.5 cm), and initial solution pH values (3.0, 5.0, 7.0, and 9.0) were selected for factor analysis. The specifically detailed single-factor experimental parameter design is shown in Table S4. The levels of the different factors were selected based on a previous study by using an electrocatalytic reaction for RDX wastewater treatment [16]. The pH of the HMX wastewater was adjusted by using diluted H_2SO_4 (1 mol/L) and NaOH (1 mol/L). Samples were collected at different treatment time intervals and passed through a 0.45 μm filter. The supernatant was stored in vials until the HMX was measured by high-performance liquid chromatography (HPLC). All samples were analyzed at least in triplicate during the experiment to measure HMX concentrations. All test solutions were collected from the supernatant, passed through a 0.45 μm filter, and stored in vials until HPLC analysis. After determining the optimal experimental conditions, HMX wastewater was collected at different treatment times for intermediate product and toxicity analyses.

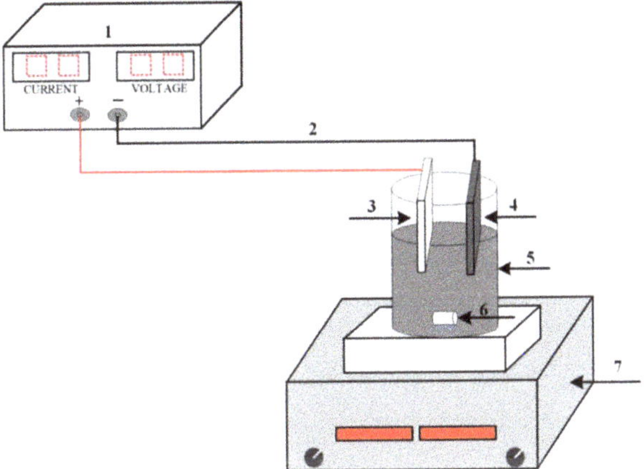

Figure 5. Electro-oxidation system for HMX wastewater treatment (1. DC power supply, 2. Copper wire, 3. Anode, 4. Cathode, 5. Electrolyte solution, 6. Rotor, 7. Magnetic stirrer).

3.3. Chemical Analysis

The concentration of HMX in the solution was determined and quantified by using HPLC (PerkinElmer, Waltham, MA, USA) and a UV detector at a flow rate of 0.7 mL/min. The detection wavelength of HMX was 236 nm with a column temperature of 35 °C. The mobile phase consisted of 40% acetonitrile and 60% water (*v*/*v*) The injection volume was 20 μL. The HMX removal efficiency was calculated as Equation (1):

$$R\ (\%) = 1 - (C_0/C_t) \times 100\% \quad (1)$$

where C_0 and C_t are the initial and residual concentrations of HMX at different times, respectively.

The HMX degradation products were further analyzed by using LC-MS/MS. The mass spectrometer (U3000, Thermo Fisher Scientific, Waltham, MA, USA) was equipped with an electrospray ionization (ESI) source and operated in the positive ESI mode.

3.4. Radical Scavenger Experiment

To investigate the role of •OH in HMX degradation, TBA scavengers at different concentrations (10, 20, and 50 mM) were added to the EO system at the beginning of the test. The HMX removal efficiency was measured after 3 h, as described in Section 3.3.

3.5. Toxicity Analysis

A toxicity analysis of HMX wastewater was performed by using a library of 114 transcriptional fusions of the green fluorescent protein (GFP), which included different promoters controlling the expression of genes involved in five known stress response pathways, namely: oxidative, DNA, protein, membrane, and general stress pathways (Table S2) in *E. coli* K12 and MG1655. A detailed toxicogenomic assay was conducted according to previous studies [32,73]. Briefly, *E. coli* cells were grown with 1 × M9 in clear-bottom black 384-well plates (Costar, Bethesda, MD, USA) for 5–6 h at 37 °C to reach early exponential growth (optical density value (OD_{600}) of ~0.2). Ten microliters of HMX samples at different degradation times were added to each well of the 384-well plate. The plate was then placed in a microplate reader (Cytation 5, Bio-Tek, Winooski, VT, USA) to simultaneously measure OD_{600} for cell growth and GFP signals (excitation at 485 nm, emission at 528 nm) at a time interval of 5 min for 2 h. All tests were performed in the dark and repeated three times.

3.6. Toxicogenomics Data Analysis

Gene expression profiling data from *E. coli* libraries were processed as previously described [30,57]. All the data was corrected for various controls, including a blank with medium control (with and without HMX samples) and promoterless bacterial controls (with and without HMX samples). The alteration in gene expression for a given gene at each time point, due to HMX sample exposure, relative to the vehicle control condition, without any HMX sample exposure, also referred to as induction factor I, was represented by $I = P_e/P_c$, where $P_e = (GFP/OD)_{experiment}$ as the normalized gene expression GFP level in the experimental condition with HMX sample exposure and $P_c = (GFP/OD)_{vehicle}$ in the vehicle control condition without any HMX sample exposure.

The quantitative molecular endpoint, the TELI, was calculated by integrating the temporal dysregulation of I values over the exposure time, as previously described [30,57]. A TELI value above the threshold of 1.5 was considered toxicity-positive [73].

3.7. Cost Estimation

To evaluate the application potential of EO systems, the energy consumption (EC) analysis was conducted based on Equation (2) [74],

$$\text{Electrical energy consumption (kwh/g)} = U_{cell}\frac{IT}{V*(C_t - C_0)}, \tag{2}$$

where U_{cell} is the average of applied voltage (V), I is the current (A), T is the time (h), V is the sample volume (L), C_0 and C_t are the initial and residual concentrations of HMX at different times.

3.8. Data Processing

Bar and scatter plots were drawn by using the Origin 2021 software (OriginLab Corporation, Northampton, MA, USA). Red, black, and green heat maps were plotted in Excel 2016 (Microsoft, Redmond, WA, USA). To assess the activity of stress categories or genes, gene set enrichment analysis (GSEA) was performed by sorting the gene list by TELI values, as described by Subramanian et al. [75]. For each stress response category, GSEA calculates the enrichment score; that is, it examines the genes sorted by TELI from highest to lowest, and gives a positive statistical value (i.e., rewarding score) if the gene belongs to the pathway of interest; otherwise, it gives a negative score (i.e., penalizing score). The significance of each pathway ($p < 0.05$) was determined by comparing its ranking score with the corresponding empirical distribution.

4. Conclusions

This study demonstrated that an EO system with a Ti/PbO_2 electrode can be employed for HMX wastewater treatment. The highest HMX removal efficiency of 81.2% was achieved within 180 min at a current density of 70 mA/cm^2, Na_2SO_4 concentration of 0.25 mol/L, interelectrode distance of 1.0 cm, and pH of 5.0. The degradation intermediates of the HMX wastewater suggest two possible electrochemical pathways: cathode reduction and indirect oxidation by •OH radicals. Protein and oxidative stress were the key stress categories in HMX wastewater. The intermediates generated during the first 90 min of degradation had relatively higher molecular toxicity levels and were then gradually converted to less toxic or non-toxic end-products. Our study provides new insights into performance optimization, degradation pathways, and molecular-level toxicity evolution during HMX degradation. More extensive studies regarding electrode material development, EO reaction mechanisms, and potential health risks of real wastewater are needed for an in-depth and comprehensive understanding of EO-enabled methods, and for developing a promising explosive wastewater treatment method with high removal performance and low environmental impact.

Supplementary Materials: The following supporting information can be downloaded at https://www.mdpi.com/article/10.3390/catal12080815/s1, Figure S1: (a) The energy consumption (EC) at different (a) electrolyte concentration, (b) current density, (c) interelectrode, and (d) initial pH; Figure S2: Ion spectra of HMX degradation intermediates in positive ion mode; Figure S3: Removal of HMX under optimal conditions in the presence of different TBA concentrations; Figure S4: The number of genes showing altered expression ($TELI_{gene}$ > 1.5) during the electrochemical oxidation of HMX; Figure S5: Hierarchical cluster (HCL) analysis diagram based on differential gene expressions (mean lnI, n = 3) of 114 selected stress genes in *E. coli* in exposure to the HMX samples at different times (Red colors indicate up-regulation, green colors indicate down-regulation); Table S1: Kinetics and energy consumption of HMX degradation under different operating conditions.; Table S2. Main degradation intermediates of HMX; Table S3: Stress gene bank and its main functions; Table S4: Single-factor experimental design.

Author Contributions: Conceptualization, Y.Q.; methodology, Y.Q. and D.W.; software, K.C.; validation, Y.Q. and K.C.; formal analysis, G.C. and P.X.; investigation, Y.Q., K.C. and G.C.; resources, P.X.; data curation, H.Y., L.Q. and L.X.; writing—original draft preparation, Y.Q.; writing—review and editing, W.Y. and D.W.; visualization, X.L. and Y.L.; supervision, W.Y. and D.W.; funding acquisition, D.W. All authors have read and agreed to the published version of the manuscript.

Funding: This research was funded by the National Natural Science Foundation of China (grant number: 52070156; 31600421). Scientific Research Program Funded by Shaanxi Provincial Education Department (No. 17JS097).

Data Availability Statement: Not applicable.

Conflicts of Interest: The authors declare no conflict of interest. The funders had no role in the study design, collection, analyses, or interpretation of the data, writing of the manuscript, or decision to publish the results.

References

1. Lyman, J.L., Liau, Y.-C.; Brand, H.V. Thermochemical functions for gas-phase, 1, 3, 5, 7-tetranitro-1, 3, 5, 7-tetraazacyclooctane (HMX), its condensed phases, and its larger reaction products. *Combust. Flame* **2002**, *130*, 185–203. [CrossRef]
2. Yang, X.; Lai, J.-l.; Li, J.; Zhang, Y.; Luo, X.-G.; Han, M.-W.; Zhu, Y.-B.; Zhao, S.-P. Biodegradation and physiological response mechanism of Bacillus aryabhattai to cyclotetramethylenete-tranitramine (HMX) contamination. *J. Environ. Manag.* **2021**, *288*, 112247. [CrossRef] [PubMed]
3. McMurry, S.T.; Jones, L.E.; Smith, P.; Cobb, G.; Anderson, T.; Lovern, M.B.; Cox, S.; Pan, X. Accumulation and effects of octahydro-1,3,5,7-tetranitro-1,3,5,7-tetrazocine (HMX) exposure in the green anole (*Anolis carolinensis*). *Ecotoxicology* **2012**, *21*, 304–314. [CrossRef]
4. USEPA. *Integrated Risk Information System (IRIS)*; Chemical Assessment Summary; USEPA: Washington, DC, USA, 2016.
5. Savard, K.; Berthelot, Y.; Auroy, A.; Spear, P.A.; Trottier, B.; Robidoux, P.Y. Effects of HMX-lead mixtures on reproduction of the earthworm Eisenia Andrei. *Arch. Environ. Contam. Toxicol.* **2007**, *53*, 351–358. [CrossRef] [PubMed]
6. Gong, P.; Hawari, J.; Thiboutot, S.; Ampleman, G.; Sunahara, G. Toxicity of octahydro-1, 3, 5, 7-tetranitro-1, 3, 5, 7-tetrazocine (HMX) to soil microbes. *Bull. Environ. Contam. Toxicol.* **2002**, *69*, 97–103. [CrossRef] [PubMed]
7. Payne, Z.M.; Lamichhane, K.M.; Babcock, R.W.; Turnbull, S.J. Pilot-scale in situ bioremediation of HMX and RDX in soil pore water in Hawaii. *Environ. Sci. Processes Impacts* **2013**, *15*, 2023–2029. [CrossRef] [PubMed]
8. Panja, S.; Sarkar, D.; Datta, R. Vetiver grass (Chrysopogon zizanioides) is capable of removing insensitive high explosives from munition industry wastewater. *Chemosphere* **2018**, *209*, 920–927. [CrossRef]
9. Adrian, N.R.; Arnett, C.M. Anaerobic biodegradation of hexahydro-1,3,5-trinitro-1,3,5-triazine (RDX) by Acetobacterium malicum strain HAAP-1 isolated from a methanogenic mixed culture. *Curr. Microbiol.* **2004**, *48*, 332–340. [CrossRef]

10. Bhatt, M.; Zhao, J.-S.; Monteil-Rivera, F.; Hawari, J. Biodegradation of cyclic nitramines by tropical marine sediment bacteria. *J. Ind. Microbiol. Biotechnol.* **2005**, *32*, 261–267. [CrossRef]
11. Singh, R.; Singh, A. Biodegradation of military explosives RDX and HMX. In *Microbial Degradation of Xenobiotics*; Springer: Berlin/Heidelberg, Germany, 2012; pp. 235–261.
12. Zhao, J.-S.; Greer, C.W.; Thiboutot, S.; Ampleman, G.; Hawari, J. Biodegradation of the nitramine explosives hexahydro-1, 3, 5-trinitro-1,3,5-triazine and octahydro-1,3,5,7-tetranitro-1,3,5,7-tetrazocine in cold marine sediment under anaerobic and oligotrophic conditions. *Can. J. Microbiol.* **2004**, *50*, 91–96. [CrossRef]
13. Kanekar, S.; Kanekar, P.; Sarnaik, S.; Gujrathi, N.; Shede, P.; Kedargol, M.; Reardon, K. Bioremediation of nitroexplosive wastewater by an yeast isolate Pichia sydowiorum MCM Y-3 in fixed film bioreactor. *J. Ind. Microbiol. Biotechnol.* **2009**, *36*, 253–260. [CrossRef] [PubMed]
14. Shang, K.; Wang, X.; Li, J.; Wang, H.; Lu, N.; Jiang, N.; Wu, Y. Synergetic degradation of Acid Orange 7 (AO7) dye by DBD plasma and persulfate. *Chem. Eng. J.* **2017**, *311*, 378–384. [CrossRef]
15. Wang, L.; Luo, Z.; Hong, Y.; Chelme-Ayala, P.; Meng, L.; Wu, Z.; El-Din, M.G. The treatment of electroplating wastewater using an integrated approach of interior microelectrolysis and Fenton combined with recycle ferrite. *Chemosphere* **2022**, *286*, 131543. [CrossRef]
16. Chen, Y.; Hong, L.; Han, W.; Wang, L.; Sun, X.; Li, J. Treatment of high explosive production wastewater containing RDX by combined electrocatalytic reaction and anoxic–oxic biodegradation. *Chem. Eng. J.* **2011**, *168*, 1256–1262. [CrossRef]
17. Moreira, F.C.; Boaventura, R.A.; Brillas, E.; Vilar, V.J. Electrochemical advanced oxidation processes: A review on their application to synthetic and real wastewaters. *Appl. Catal. B Environ.* **2017**, *202*, 217–261. [CrossRef]
18. Martinez-Huitle, C.A.; Ferro, S. Electrochemical oxidation of organic pollutants for the wastewater treatment: Direct and indirect processes. *Chem. Soc. Rev.* **2006**, *35*, 1324–1340. [CrossRef]
19. Panizza, M.; Cerisola, G. Direct and mediated anodic oxidation of organic pollutants. *Chem. Rev.* **2009**, *109*, 6541–6569. [CrossRef]
20. Oturan, M.A.; Aaron, J.-J. Advanced oxidation processes in water/wastewater treatment: Principles and applications. A review. *Crit. Rev. Environ. Sci. Technol.* **2014**, *44*, 2577–2641. [CrossRef]
21. Moreira, F.C.; Soler, J.; Fonseca, A.; Saraiva, I.; Boaventura, R.A.; Brillas, E.; Vilar, V.J. Incorporation of electrochemical advanced oxidation processes in a multistage treatment system for sanitary landfill leachate. *Water Res.* **2015**, *81*, 375–387. [CrossRef]
22. Dai Lam, T.; Van Chat, N.; Bach, V.Q.; Loi, V.D.; Van Anh, N. Simultaneous degradation of 2,4,6-trinitrophenyl-N-methylnitramine (Tetryl) and hexahydro-1,3,5-trinitro-1,3,5 triazine (RDX) in polluted wastewater using some advanced oxidation processes. *J. Ind. Eng. Chem.* **2014**, *20*, 1468–1475.
23. Bonin, P.M.; Bejan, D.; Radovic-Hrapovic, Z.; Halasz, A.; Hawari, J.; Bunce, N.J. Indirect oxidation of RDX, HMX, and CL-20 cyclic nitramines in aqueous solution at boron-doped diamond electrodes. *Environ. Chem.* **2005**, *2*, 125–129. [CrossRef]
24. Kacem, S.B.; Elaoud, S.C.; Asensio, A.M.; Panizza, M.; Clematis, D. Electrochemical and sonoelectrochemical degradation of Allura Red and Erythrosine B dyes with Ti-PbO_2 anode. *J. Electroanal. Chem.* **2021**, *889*, 115212. [CrossRef]
25. Song, S.; Zhan, L.; He, Z.; Lin, L.; Tu, J.; Zhang, Z.; Chen, J.; Xu, L. Mechanism of the anodic oxidation of 4-chloro-3-methyl phenol in aqueous solution using Ti/SnO_2–Sb/PbO_2 electrodes. *J. Hazard. Mater.* **2010**, *175*, 614–621. [CrossRef] [PubMed]
26. Suryanarayanan, V.; Nakazawa, I.; Yoshihara, S.; Shirakashi, T. The influence of electrolyte media on the deposition/dissolution of lead dioxide on boron-doped diamond electrode–A surface morphologic study. *J. Electroanal. Chem.* **2006**, *592*, 175–182. [CrossRef]
27. García-Gómez, C.; Drogui, P.; Seyhi, B.; Gortáres-Moroyoqui, P.; Buelna, G.; Estrada-Alvgarado, M.; Alvarez, L.H. Combined membrane bioreactor and electrochemical oxidation using Ti/PbO_2 anode for the removal of carbamazepine. *J. Taiwan Inst. Chem. Eng.* **2016**, *64*, 211–219. [CrossRef]
28. Wang, J.; Wang, N.; Nan, W.; Wang, C.; Chen, X.; Qi, X.; Yan, S.; Dai, S. Enhancement of electrochemical performance of $LiCoO_2$ cathode material at high cut-off voltage (4.5 V) by partial surface coating with graphene nanosheets. *Int. J. Electrochem. Sci.* **2020**, *15*, 9282–9293.
29. Zhao, J.-S.; Spain, J.; Thiboutot, S.; Ampleman, G.; Greer, C.; Hawari, J. Phylogeny of cyclic nitramine-degrading psychrophilic bacteria in marine sediment and their potential role in the natural attenuation of explosives. *FEMS Microbiol. Ecol.* **2004**, *49*, 349–357. [CrossRef]
30. Hawari, J.; Halasz, A.; Beaudet, S.; Paquet, L.; Ampleman, G.; Thiboutot, S. Biotransformation routes of octahydro-1, 3, 5, 7-tetranitro-1, 3, 5, 7-tetrazocine by municipal anaerobic sludge. *Environ. Sci. Technol.* **2001**, *35*, 70–75. [CrossRef]
31. Mukhi, S.; Patiño, R. Effects of hexahydro-1,3,5-trinitro-1,3,5-triazine (RDX) in zebrafish: General and reproductive toxicity. *Chemosphere* **2008**, *72*, 726–732. [CrossRef]
32. Nipper, M.; Carr, R.; Biedenbach, J.; Hooten, R.; Miller, K.; Saepoff, S. Development of marine toxicity data for ordnance compounds. *Arch. Environ. Contam. Toxicol.* **2001**, *41*, 308–318. [CrossRef]
33. Gou, N.; Onnis-Hayden, A.; Gu, A.Z. Mechanistic toxicity assessment of nanomaterials by whole-cell-array stress genes expression analysis. *Environ. Sci. Technol.* **2010**, *44*, 5964–5970. [CrossRef]
34. Lin, Y.; Sevillano-Rivera, M.; Jiang, T.; Li, G.; Cotto, I.; Vosloo, S.; Carpenter, C.M.; Larese-Casanova, P.; Giese, R.W.; Helbling, D.E. Impact of Hurricane Maria on drinking water quality in Puerto Rico. *Environ. Sci. Technol.* **2020**, *54*, 9495–9509. [CrossRef] [PubMed]
35. El-Desoky, H.S.; Ghoneim, M.M.; Zidan, N.M. Decolorization and degradation of Ponceau S azo-dye in aqueous solutions by the electrochemical advanced Fenton oxidation. *Desalination* **2010**, *264*, 143–150. [CrossRef]

36. Dai, Q.; Zhou, J.; Weng, M.; Luo, X.; Feng, D.; Chen, J. Electrochemical oxidation metronidazole with Co modified PbO_2 electrode: Degradation and mechanism. *Sep. Purif. Technol.* **2016**, *166*, 109–116. [CrossRef]
37. Samarghandi, M.R.; Ansari, A.; Dargahi, A.; Shabanloo, A.; Nematollahi, D.; Khazaei, M.; Nasab, H.Z.; Vaziri, Y. Enhanced electrocatalytic degradation of bisphenol A by graphite/β-PbO_2 anode in a three-dimensional electrochemical reactor. *J. Environ. Chem. Eng.* **2021**, *9*, 106072. [CrossRef]
38. Wang, Y.; Shen, Z.; Chen, X. Effects of experimental parameters on 2,4-dichlorphenol degradation over Er-chitosan-PbO_2 electrode. *J. Hazard. Mater.* **2010**, *178*, 867–874. [CrossRef] [PubMed]
39. Klidi, N.; Clematis, D.; Delucchi, M.; Gadri, A.; Ammar, S.; Panizza, M. Applicability of electrochemical methods to paper mill wastewater for reuse. Anodic oxidation with BDD and $TiRuSnO_2$ anodes. *J. Electroanal. Chem.* **2018**, *815*, 16–23. [CrossRef]
40. Zhu, X.; Hu, W.; Feng, C.; Chen, N.; Chen, H.; Kuang, P.; Deng, Y.; Ma, L. Electrochemical oxidation of aniline using Ti/RuO_2-SnO_2 and Ti/RuO_2-IrO_2 as anode. *Chemosphere* **2021**, *269*, 128734. [CrossRef]
41. Shi, H.; Wang, Q.; Ni, J.; Xu, Y.; Song, N.; Gao, M. Highly efficient removal of amoxicillin from water by three-dimensional electrode system within granular activated carbon as particle electrode. *J. Water Process Eng.* **2020**, *38*, 101656. [CrossRef]
42. Wu, H.; Xie, H.; He, G.; Guan, Y.; Zhang, Y. Effects of the pH and anions on the adsorption of tetracycline on iron-montmorillonite. *Appl. Clay Sci.* **2016**, *119*, 161–169. [CrossRef]
43. Zaky, A.M.; Chaplin, B.P. Mechanism of p-substituted phenol oxidation at a Ti4O7 reactive electrochemical membrane. *Environ. Sci. Technol.* **2014**, *48*, 5857–5867. [CrossRef]
44. Zhang, J.; Zhou, Y.; Yao, B.; Yang, J.; Zhi, D. Current progress in electrochemical anodic-oxidation of pharmaceuticals: Mechanisms, influencing factors, and new technique. *J. Hazard. Mater.* **2021**, *418*, 126313. [CrossRef] [PubMed]
45. Liou, M.-J.; Lu, M.-C.; Chen, J.-N. Oxidation of explosives by Fenton and photo-Fenton processes. *Water Res.* **2003**, *37*, 3172–3179. [CrossRef]
46. Xiang, H.; Xiao, S.; Zhang, G.; Song, Y.; Cui, P.; Shao, H.; Li, H. Treatment of simulated berberine pharmaceutical wastewater by electrochemical oxidation process. *J. Environ. Eng.* **2011**, *5*, 5.
47. Lin, H.; Niu, J.; Xu, J.; Li, Y.; Pan, Y. Electrochemical mineralization of sulfamethoxazole by Ti/SnO_2-Sb/Ce-PbO_2 anode: Kinetics, reaction pathways, and energy cost evolution. *Electrochim. Acta* **2013**, *97*, 167–174. [CrossRef]
48. Lei, L.; Dai, Q.; Zhou, M.; Zhang, X. Decolorization of cationic red X-GRL by wet air oxidation: Performance optimization and degradation mechanism. *Chemosphere* **2007**, *68*, 1135–1142. [CrossRef]
49. Radha, K.; Sridevi, V.; Kalaivani, K. Electrochemical oxidation for the treatment of textile industry wastewater. *Bioresour. Technol.* **2009**, *100*, 987–990. [CrossRef]
50. Dargahi, A.; Barzoki, H.R.; Vosoughi, M.; Mokhtari, S.A. Enhanced electrocatalytic degradation of 2, 4-Dinitrophenol (2, 4-DNP) in three-dimensional sono-electrochemical (3D/SEC) process equipped with Fe/SBA-15 nanocomposite particle electrodes: Degradation pathway and application for real wastewater. *Arab. J. Chem.* **2022**, *15*, 103801. [CrossRef]
51. Li, X.; Li, X.; Yang, W.; Chen, X.; Li, W.; Luo, B.; Wang, K. Preparation of 3D PbO_2 nanospheres@ SnO_2 nanowires/Ti electrode and its application in methyl orange degradation. *Electrochim. Acta* **2014**, *146*, 15–22. [CrossRef]
52. Neto, S.A.; De Andrade, A. Electrooxidation of glyphosate herbicide at different DSA® compositions: pH, concentration and supporting electrolyte effect. *Electrochim. Acta* **2009**, *54*, 2039–2045. [CrossRef]
53. Dargahi, A.; Vosoughi, M.; Mokhtari, S.A.; Vaziri, Y.; Alighadri, M. Electrochemical degradation of 2, 4-Dinitrotoluene (DNT) from aqueous solutions using three-dimensional electrocatalytic reactor (3DER): Degradation pathway, evaluation of toxicity and optimization using RSM-CCD. *Arab. J. Chem.* **2022**, *15*, 103648. [CrossRef]
54. Rahmani, A.; Leili, M.; Seid-Mohammadi, A.; Shabanloo, A.; Ansari, A.; Nematollahi, D.; Alizadeh, S. Improved degradation of diuron herbicide and pesticide wastewater treatment in a three-dimensional electrochemical reactor equipped with PbO_2 anodes and granular activated carbon particle electrodes. *J. Clean. Prod.* **2021**, *322*, 129094. [CrossRef]
55. Li, C.; Xiong, K.; Li, D.; Liang, J.; Guo, B. Experimental study on treatment of antibiotic pharmaceutical wastewater by electro catalytic oxidation. *Environ. Prot. Circ. Econ.* **2017**, *37*, 5.
56. Shao, D.; Wang, Z.; Zhang, C.; Li, W.; Xu, H.; Tan, G.; Yan, W. Embedding wasted hairs in Ti/PbO_2 anode for efficient and sustainable electrochemical oxidation of organic wastewater. *Chin. Chem. Lett.* **2022**, *33*, 1288–1292. [CrossRef]
57. Crocker, F.H.; Indest, K.J.; Fredrickson, H.L. Biodegradation of the cyclic nitramine explosives RDX, HMX, and CL-20. *Appl. Microbiol. Biotechnol.* **2006**, *73*, 274–290. [CrossRef] [PubMed]
58. Fournier, D.; Halasz, A.; Thiboutot, S.; Ampleman, G.; Manno, D.; Hawari, J. Biodegradation of octahydro-1,3,5,7-tetranitro-1,3,5,7-tetrazocine (HMX) by Phanerochaete chrysosporium: New insight into the degradation pathway. *Environ. Sci. Technol.* **2004**, *38*, 4130–4133. [CrossRef] [PubMed]
59. Anotai, J.; Tanvanit, P.; Garcia-Segura, S.; Lu, M.-C. Electro-assisted Fenton treatment of ammunition wastewater containing nitramine explosives. *Process Saf. Environ. Prot.* **2017**, *109*, 429–436. [CrossRef]
60. Dugandžić, A.M.; Tomašević, A.V.; Radišić, M.M.; Šekuljica, N.Ž.; Mijin, D.Ž.; Petrović, S.D. Effect of inorganic ions, photosensitisers and scavengers on the photocatalytic degradation of nicosulfuron. *J. Photochem. Photobiol. A Chem.* **2017**, *336*, 146–155. [CrossRef]
61. Sun, W.; Sun, Y.; Shah, K.J.; Chiang, P.-C.; Zheng, H. Electrocatalytic oxidation of tetracycline by Bi-Sn-Sb/γ-Al_2O_3 three-dimensional particle electrode. *J. Hazard. Mater.* **2019**, *370*, 24–32. [CrossRef]

62. Yang, J.-S.; Lai, W.W.-P.; Panchangam, S.C.; Lin, A.Y.-C. Photoelectrochemical degradation of perfluorooctanoic acid (PFOA) with GOP25/FTO anodes: Intermediates and reaction pathways. *J. Hazard. Mater.* **2020**, *391*, 122247. [CrossRef]
63. Keseler, I.M.; Bonavides-Martínez, C.; Collado-Vides, J.; Gama-Castro, S.; Gunsalus, R.P.; Johnson, D.A.; Krummenacker, M.; Nolan, L.M.; Paley, S.; Paulsen, I.T. EcoCyc: A comprehensive view of *Escherichia coli* biology. *Nucleic Acids Res.* **2009**, *37*, D464–D470. [CrossRef]
64. Seaver, L.C.; Imlay, J.A. Alkyl hydroperoxide reductase is the primary scavenger of endogenous hydrogen peroxide in *Escherichia coli*. *J. Bacteriol.* **2001**, *183*, 7173–7181. [CrossRef] [PubMed]
65. Seaver, L.C.; Imlay, J.A. Hydrogen peroxide fluxes and compartmentalization inside growing *Escherichia coli*. *J. Bacteriol.* **2001**, *183*, 7182–7189. [CrossRef] [PubMed]
66. Yeager, E. Electrocatalysts for O_2 reduction. *Electrochim. Acta* **1984**, *29*, 1527–1537. [CrossRef]
67. Zheng, M.; Wang, X.; Templeton, L.J.; Smulski, D.R.; LaRossa, R.A.; Storz, G. DNA microarray-mediated transcriptional profiling of the *Escherichia coli* response to hydrogen peroxide. *J. Bacteriol.* **2001**, *183*, 4562–4570. [CrossRef]
68. Liu, Y.; Bauer, S.C.; Imlay, J.A. The YaaA protein of the *Escherichia coli* OxyR regulon lessens hydrogen peroxide toxicity by diminishing the amount of intracellular unincorporated iron. *J. Bacteriol.* **2011**, *193*, 2186–2196. [CrossRef]
69. Manchado, M.; Michán, C.; Pueyo, C. Hydrogen peroxide activates the SoxRS regulon in vivo. *J. Bacteriol.* **2000**, *182*, 6842–6844. [CrossRef]
70. Burgess, R.R.; Travers, A.A.; Dunn, J.J.; Bautz, E.K. Factor stimulating transcription by RNA polymerase. *Nature* **1969**, *221*, 43–46. [CrossRef]
71. Rosen, R.; Biran, D.; Gur, E.; Becher, D.; Hecker, M.; Ron, E.Z. Protein aggregation in *Escherichia coli*: Role of proteases. *FEMS Microbiol. Lett.* **2002**, *207*, 9–12. [CrossRef]
72. Shineberg, B.; Zipser, D. The lon gene and degradation of β-galactosidase nonsense fragments. *J. Bacteriol.* **1973**, *116*, 1469–1471. [CrossRef]
73. Gou, N.; Gu, A.Z. A new transcriptional effect level index (TELI) for toxicogenomics-based toxicity assessment. *Environ. Sci. Technol.* **2011**, *45*, 5410–5417. [CrossRef]
74. Zeng, Q.; Zan, F.; Hao, T.; Biswal, B.K.; Lin, S.; van Loosdrecht, M.C.; Chen, G. Electrochemical pretreatment for stabilization of waste activated sludge: Simultaneously enhancing dewaterability, inactivating pathogens and mitigating hydrogen sulfide. *Water Res.* **2019**, *166*, 115035. [CrossRef] [PubMed]
75. Subramanian, A.; Tamayo, P.; Mootha, V.K.; Mukherjee, S.; Ebert, B.L.; Gillette, M.A.; Paulovich, A.; Pomeroy, S.L.; Golub, T.R.; Lander, E.S. Gene set enrichment analysis: A knowledge-based approach for interpreting genome-wide expression profiles. *Proc. Natl. Acad. Sci. USA* **2005**, *102*, 15545–15550. [CrossRef] [PubMed]

Article

Kinetics and Mechanisms of Cr(VI) Removal by nZVI: Influencing Parameters and Modification

Yizan Gao, Xiaodan Yang, Xinwei Lu, Minrui Li *, Lijun Wang and Yuru Wang *

Department of Environmental Science, School of Geography and Tourism, Shaanxi Normal University, Xi'an 710119, China
* Correspondence: author: liminrui@snnu.edu.cn (M.L.); wangyuru@snnu.edu.cn (Y.W.); Tel.: +86-029-85310524 (M.L.); +86-15891779702 (Y.W.)

Abstract: In this study, single-spherical nanoscale zero valent iron (nZVI) particles with large specific sur-face area were successfully synthesized by a simple and rapid chemical reduction method. The XRD spectra and SEM–EDS images showed that the synthesized nZVI had excellent crystal struc-ture, but oxidation products, such as γ-Fe_2O_3 and Fe_3O_4, were formed on the surface of the parti-cles. The effect of different factors on the removal of Cr(VI) by nZVI were studied, and the opti-mum experimental conditions were found. Kinetic and thermodynamic equations at different temperatures showed that the removal of Cr(VI) by nZVI was a single-layer chemical adsorption, conforming to pseudo-second-order kinetics. By applying the intraparticle diffusion model, the ad-sorption process was composed of three stages, namely rapid diffusion, chemical reduction, and in-ternal saturation. Mechanism analysis demonstrated that the removal of Cr(VI) by nZVI in-volved adsorption, reduction, precipitation and coprecipitation. Meanwhile, Cr(VI) was reduced to Cr(III) by nZVI, while $FeCr_2O_4$, $Cr_xFe_{1-x}OOH$, and $Cr_xFe_{1-x}(OH)_3$ were formed as end products. In addition, the study found that ascorbic acid, starch, and Cu modified nZVI can promote the removal efficiency of Cr(VI) in varying degrees due to the enhanced mobility of the particles. These results can provide new insights into the removal mechanisms of Cr(VI) by nZVI.

Keywords: nanoscale zero valent iron; Cr(VI); adsorption kinetics; mechanism; modification

check for updates

Citation: Gao, Y.; Yang, X.; Lu, X.; Li, M.; Wang, L.; Wang, Y. Kinetics and Mechanisms of Cr(VI) Removal by nZVI: Influencing Parameters and Modification. *Catalysts* **2022**, *12*, 999. https://doi.org/10.3390/catal12090999

Academic Editors: Hao Xu, Yanbiao Liu and Gassan Hodaifa

Received: 26 July 2022
Accepted: 30 August 2022
Published: 5 September 2022

1. Introduction

Chromium (Cr) is one of the most abundant elements in the earth, and exists primarily in two valence forms of Cr(III) and Cr(VI) in the environment [1]. Compared with Cr(III), Cr(VI) species are more soluble, mobile, and bioavailable [2,3]. However, Cr(VI) is approximately 100-fold more toxic to living organisms than Cr(III) [4]. Therefore, Cr(VI) has been listed as a priority pollutant in many countries due to its carcinogenicity, persistence, and bioaccumulation characteristics [5]. Effluents containing Cr(VI) originating from metallurgical, textile dyeing, tannery, and electroplating industries are discharged into the environment in large quantities [6], posing a serious threat to ecosystems and human health. Thus, it is of great importance to remove and detoxify Cr(VI) from industrial effluents before their discharge. In contrast, Cr(III) is an essential microelement for organisms, which can regulate the efficient metabolism of glucose, lipids, and proteins in animals and human beings [7]. Accordingly, speciation transformation from Cr(VI) into Cr(III) is usually carried out to improve the removal efficiency and reduce chromium toxicity.

In recent years, a combination of adsorption and chemical reduction technology employing environment-friendly nanoscale zero valent iron (nZVI) materials has emerged as a promising alternative for Cr(VI) removal. Indeed, nZVI has been extensively investigated for environmental remediation due to its high surface energy, strong chemical reactivity, and large surface area. As an essential part of nZVI, iron has the advantages of abundance in the environment, nontoxicity, low cost, large storage capacity, and strong reduction capacity (−0.44 V) [8]. It was found that when the particle size of iron particles decreased to nanoscale (i.e., nZVI), it had a better adsorption performance for pollutants. A comprehensive comparison of existing nZVI synthesis methods, whether top-down (lithographic grinding and precision milling) or bottom-up (borohydride chemical reduction, carbothermic reduction, ultrasonic, electrochemical, green synthesis, etc.), indicates that all have certain advantages and limitations [9]. Among these methods, the chemical reduction method is simple and feasible, and suitable for use in any laboratory; meanwhile, the product obtained is characterized by a homogeneous structure that displays a high reactivity [10] and has the potential to be widely used in practical application. Numerous studies have demonstrated that nZVI has an excellent removal efficiency for heavy metals [11,12], organic dyes [13], antibiotics [14], and other refractory organic pollutants [15]. In particular, the removal and transformation of Cr(VI) using nZVI have been extensively studied, suggesting that the removal mechanism of Cr(VI) by ZVI mainly involves adsorption, redox reaction of Fe^0 to Fe^{2+}/Fe^{3+} and Cr(VI) to Cr(III), and precipitation [16,17]. However, the reaction process of Cr(VI) adsorbed on a nanoparticle shell is still unclear, and further systematic research is needed to understand the interfacial reaction at the molecular level.

Furthermore, due to the defects of self-aggregation and rapid oxidation of nZVI, the application of nZVI was limited. To overcome these drawbacks, attempts have been made to modify nZVI and, thereby, alter the corresponding characteristics to enhance its mobility and to maintain an efficient reactivity of nZVI. The common approaches to improve the dispersion of nZVI include adhesion of nZVI on supporting materials (e.g., biochar and chitosan) by constructing skeleton structures for limiting the aggregation of iron particles [12] and the coating of a certain modifier or stabilizer (e.g., surfactant and biopolymers) on the surface of nZVI to maintain a strategic distance between nanoparticles either by electrostatic repulsion of functional groups and/or steric aversion with steric obstacles [18]. Mixing of nZVI with noble or transition metal (e.g., Ag, Pt, and Pd) to form bimetallic nanoparticle is generally employed to reduce the activation energy, increase the reactivity of nZVI particles, and enhance the reaction rate. Moreover, a hybrid modification strategy of nZVI by combining supporting materials/modifiers and bimetallic alteration to synergistically optimize the stability and reactivity of nanoparticles has also received much scientific interest. Despite the diverse approaches developed for the surface modification of nZVI and the resultant benefits, concern arises regarding the cost-effectiveness and eco-toxicity of current modifiers or supporting materials [18]. This study was also devoted to the development of easily accessible, economical, and environmentally friendly materials as possible substitutes for modified nZVI, providing as many theoretical foundations as possible for more kinds of modified nZVI.

In this study, the synthesis and characterization of nZVI and the corresponding Cr(VI) removal performance and mechanism by nZVI were investigated. The characteristics of nZVI were analyzed through different techniques (e.g., SEM–EDS, BET, XRD, and FTIR). The effect of some important parameters, including initial solution pH, contaminant concentration, adsorbent dosage, competitive anions, cations, and humic acid, on the removal of Cr(VI) by nZVI were also studied. The main mechanism of Cr(VI) removal was determined by XPS analysis, while the kinetics and thermodynamics of Cr(VI) removal were also calculated. In particular, various materials were examined as cost-effective alternatives for the surface modification of nZVI to alleviate the aggregation and oxidation of raw nZVI particles.

2. Results and Discussions

2.1. Characterization of nZVI

The morphology of nZVI was shown in Figures 1, 2 and S1. It was observed that spherical chain-like nZVI particles have been synthesized and that most of the particles were in the range of 60–200 nm. The elemental composition on the surface of nZVI was analyzed by EDS. The results showed that both Fe and O were detected on the surface of nZVI, among which the content of Fe is more than 84.25% (Table S1). The presence of O indicated that nZVI might be oxidized during the synthesis and drying processes, while the main oxides were γ-Fe_2O_3/Fe_3O_4 based on XRD analysis (Figure 2) [19]. Notably, EDS results showed that the content of O increased from 4.84% to 27.54% after the reaction of nZVI with Cr(VI) (Table S1), indicating that O was involved in the removal process of Cr(VI).

The specific surface area of the synthesized nZVI was determined as 15.19 m^2/g. Compared with the specific surface area of those nZVI particles reported by other researchers, our result was lower [9], which might due to the Van der Waals forces between particles and the ferromagnetism of iron itself leading to the agglomeration of nZVI [20,21]. Additionally, XRD patterns of the fresh nZVI revealed the crystalline structural evolution (Figure 2). An obvious diffraction peak at $2\theta = 44.68°$ (JCPDS No.06-0696) confirmed the presence of Fe^0 in the synthesized nZVI [22]. The characteristic diffraction peak of Fe^0 had a high peak height and a narrow peak width, indicating that the synthesized nZVI had a good degree of crystallization; meanwhile, the number of stray peaks was quite small, indicating that the synthesized nZVI had a high purity. In addition, some weak peaks at $2\theta = 43.5°$, 57.3°, and 62.9° representing the presence of γ-Fe_2O_3/Fe_3O_4 were observed [19]. This oxide layer could prevent oxygen from contacting nZVI inside the particles and prevent further oxidation of nZVI, which was consistent with the presence of the O element detected by EDS.

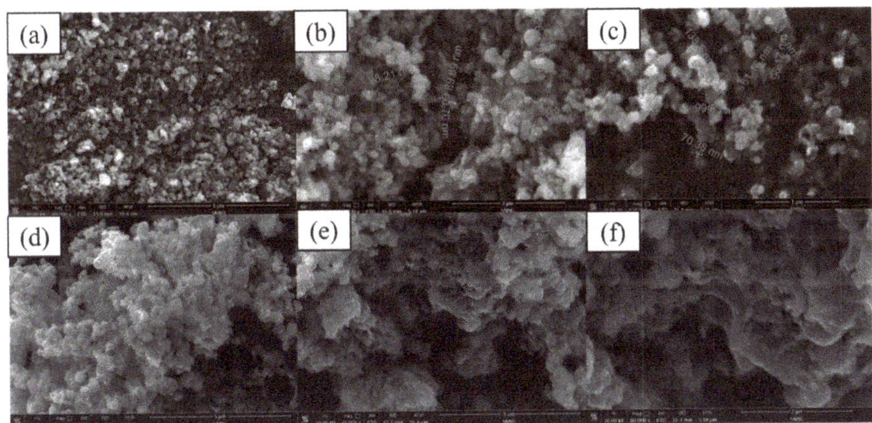

Figure 1. The SEM results of (**a–c**) pre- and (**d–f**) post-reaction of Cr(VI) with nZVI. ([Cr(VI)] = 25 mg/L, [nZVI] = 0.5 g/L, pH = 3, T = 293 K).

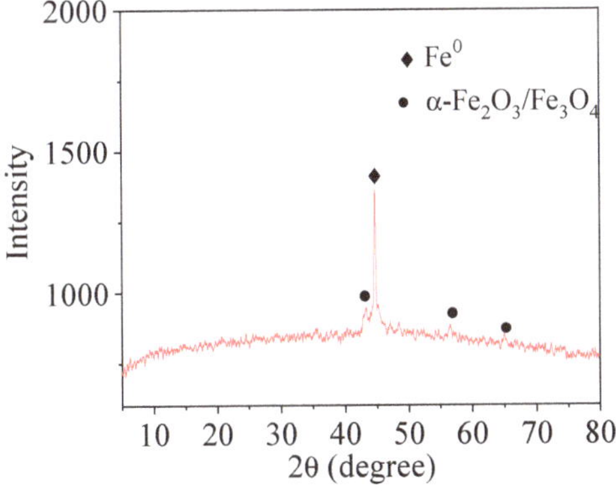

Figure 2. The XRD spectrogram of nZVI.

2.2. *Effect of Important Parameters on the Removal of Cr(VI)*

2.2.1. Effect of Initial pH

Figure 3a shows that Cr(VI) removal by nZVI strongly depended on the solution pH. The process efficiency was significantly hindered under neutral and alkaline conditions, while a promising removal of Cr(VI) was observed at acidic conditions. When the initial solution pH decreased from 11.0 to 2.0, the adsorption capacity of Cr(VI) by nZVI in 120 min remarkably increased from 7.11 to 50 mg/g (Figure S2). Additionally, Cr(VI) of 25 mg/L could be completely removed by nZVI at pH 2.0 in 5 min, while the removal efficiency was only 21.4% at pH 11.0 in 120 min. The effect of pH level on the removal of Cr(VI) by nZVI was probably attributed to the following reasons. At acidic conditions, Fe^0 reduces Cr(VI) to Cr(III) by losing electrons to form Fe^{2+}/Fe^{3+}; meanwhile the newly formed Fe^{2+} can further facilitate the reduction of Cr(VI) to Cr(III) [23]. However, higher amounts of OH^- at elevated solution pH can be easily combined with Fe^{2+}/Fe^{3+} to form $Fe(OH)_2/Fe(OH)_3$ precipitates, attaching on the surface of nZVI. This process not only consumes the effective reactant Fe^{2+}, but also blocks the reactive sites on the nZVI surface, resulting, therefore, in a substantial decrease in process efficiency. On the contrary, a large amount of H^+ in the solution at acidic conditions can accelerate the dissolution of hydroxide precipitates, which increased the effective site on the surface of nZVI and improved the specific surface area of nZVI and the adsorption/reduction capacity of Cr(VI).

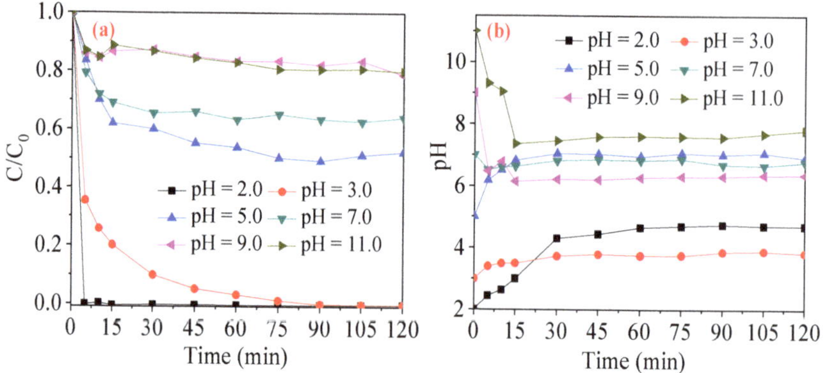

Figure 3. (**a**) Removal of Cr(VI) by nZVI as a function of time at different initial pH. (**b**) The change of solution pH with reaction time. ([Cr(VI)] = 25 mg/L, [nZVI] = 0.5 g/L, T = 293 K).

Figure 3b shows the variation of solution pH throughout the reaction. It can be found that the removal of Cr(VI) and the solution pH gradually stabilized after 30 min, probably due to the rapid adsorption and reduction of Cr(VI) on the surface of nZVI at the first 30 min. The resultant sharp decrease in Cr(VI) concentration in the solution posed a significant impact on solution pH. When the adsorption reached an equilibrium after 30 min, the solution pH also remained nearly stable.

2.2.2. Effect of Temperature

The removal kinetics of Cr(VI) by nZVI at different temperatures (293, 303, and 313 K) were shown in Figure 4. It can be seen that increasing the temperature from 293 to 303 K led to an improved removal of Cr(VI), possibly attributable to the increased collision rate between Fe^0 and Cr(VI) with the increase in temperature and, consequently, the increased removal kinetics of C(VI) by nZVI [24]. However, when the reaction temperature further increased to 313 K, the removal of Cr(VI) was retarded. This might be due to the enhanced re-release of Cr(VI) adsorbed on the particle surface to the solution at high temperature, resulting in the increase in Cr(VI) concentration in the solution.

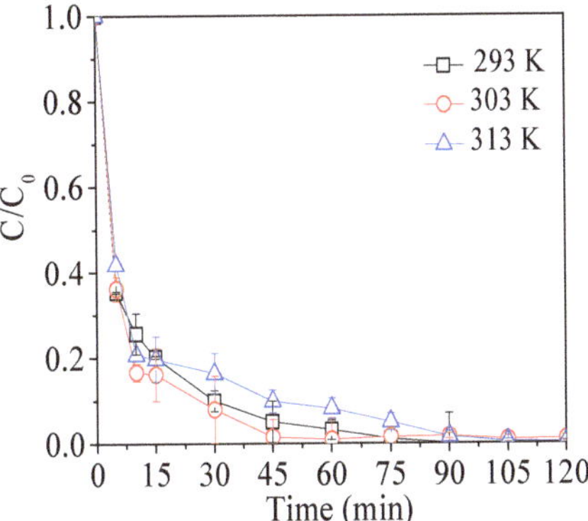

Figure 4. Removal efficiency of Cr(VI) by nZVI at different initial temperature.([Cr(VI)]= 25 mg/L, [nZVI] = 0.5 g/L, pH = 3).

2.2.3. Effects of Initial Cr(VI) Concentration and nZVI Dosage

The removal rate of Cr(VI) by nZVI at different initial Cr(VI) concentrations and nZVI dosage was illustrated in Figures 5 and S2. When the initial concentration of Cr(VI) was fixed at 25 mg/L, the increase in nZVI dosage from 0.2 to 0.5 g/L remarkably improved the removal rate of Cr(VI) and adsorption capacity of nZVI from 50% to 100% and from 29.9 to 50.0 mg/g, respectively. It can be seen from Figure 5a that, at nZVI dosage of 0.2 g/L, the reaction basically stopped after 15 min. Once nZVI of low dosage was added into the solution, a relatively large number of Cr(VI) simultaneously competed for limited adsorption sites, thereby inhibiting any further reaction. In contrast, higher nZVI dosage could provide a larger number of active sites for reaction with the target contaminant, accordingly allowing a better removal efficiency.

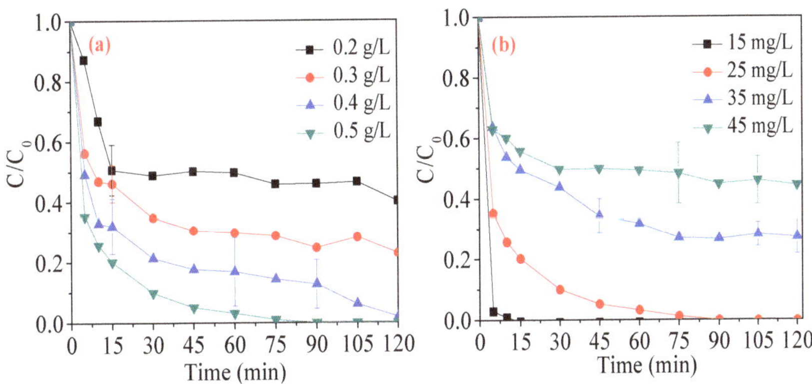

Figure 5. Removal of Cr(VI) by nZVI at different (**a**) nZVI dosage and (**b**) initial Cr(VI) concentration. Here, (**a**) [nZVI] = 0.5 g/L, pH = 3, T = 293 K; (**b**) [Cr(VI)] = 25 mg/L, pH = 3, T = 293 K.

When the applied dosage of nZVI was fixed at 0.5 g/L, removal of Cr(VI) was obviously inhibited from 100% to 54% with the increase in initial Cr(VI) concentration from 15 to 45 mg/L, corresponding to a rapid drop of nZVI adsorption capacity from 50.0 to 27.8 mg/g (see Figure S2). As more Cr(VI) approached nZVI particles, the dense passivation layers, such as Cr(III)/Fe(III)/oxide/hydroxide formed by oxidation, will be attached on the surface of nZVI, which will quickly lose the reduction ability. The higher the concentration of Cr(VI), the faster the passivation layer formed on the surface of nZVI, which significantly reduces the removal ability of nZVI [25,26], indicating that, the higher the initial concentration of Cr(VI), the lower the removal rate of Cr(VI). Therefore, the greater the dosage of nZVI and the lower the initial concentration of Cr(VI) in our study range, the better the removal of Cr(VI).

2.2.4. Effects of Co-Existing Ions

Some ions commonly present in water, including Cl^-, HCO_3^-, NO_3^-, SO_4^{2-}, Ca^{2+}, and Mg^{2+}, were selected to explore their influence on the removal of Cr(VI) by nZVI. Experimental results showed that all the studied ions inhibited the process efficiency; nevertheless, the degree of inhibition distinctively varied depending on the properties of the ions. The inhibitory effect of anions on the removal rate of Cr(VI) increased in the order of $Cl^- < NO_3^- < SO_4^{2-} < HCO_3^-$, while the inhibition of Ca^{2+} and Mg^{2+} on Cr(VI) removal was negligible.

It can be seen from Figure 6 that the removal of Cr(VI) decreased with the increase in Cl^- concentration. When the concentration of Cl^- increased from 0 to 300 mg/L, the removal rate of Cr(VI) decreased from 90.0% to 62.8% in 30 min. However, the inhibition gradually weakened with the extension of reaction time, which was similar to the results of Zhou et al. [27]. It has been reported that low concentration of Cl^- may accelerate the corrosion of the nZVI surface, while high concentration of Cl^- could facilitate the production of more iron (hydr)oxidate precipitates (e.g., β–FeOOH), which would delay the electron transfer process of nZVI to target contaminants [28,29]. Meanwhile, the existence of these oxides could block the effective sites on the nZVI surface, resulting in a lower removal of Cr(VI).

Compared with the control, the removal of Cr(VI) by nZVI in 30 min was reduced from 90% to 69.4% in the presence of 100 mg/L NO_3^- (Figure 6). However, when further increasing NO_3^- to 300 mg/L, the inhibitory effect was not significantly enhanced. The presence of NO_3^- induces two inhibitory abilities by being as an active reactant competing with Cr(VI) for nZVI and as a passivator to passivate the surface of nZVI, both of which can reduce the effective sites on the surface of particles [30]. In addition, the reactions between NO_3^- and Fe^0 according Equations (1) and (2) could consume H^+ in the solution, leading to a decrease in H^+ content and, thus, inhibiting the removal of Cr(VI) to a certain extent [31].

$$NO_3^- + 10H^+ + 4Fe^0 \rightarrow NH_4^+ + 3H_2O + 4Fe^{2+} \quad (1)$$

$$2NO_3^- + 12H^+ + 5Fe^0 \rightarrow N_2 + 6H_2O + 5Fe^{2+} \quad (2)$$

As shown in Figure 6, SO_4^{2-} had a strong inhibitory effect on the kinetics of Cr(VI) removal by nZVI and, the higher the SO_4^{2-} concentration, the stronger the inhibition. When the concentration of SO_4^{2-} increased from 0 to 300 mg/L, the removal of Cr(VI) sharply decreased from 90% to 37%. Here, SO_4^{2-} mainly acts as a competitive adsorption anion and competes for the reactive sites of nZVI with Cr(VI), resulting in limited accessibility of Cr(VI) to the surface of nZVI and, thus, reducing the removal efficiency [30]. Furthermore, SO_4^{2-} can form complexes with iron hydr(oxide) and consequently induce the production of α–FeOOH and alkaline $FeSO_4$ precipitates [32], which would further block the reactive sites of nZVI.

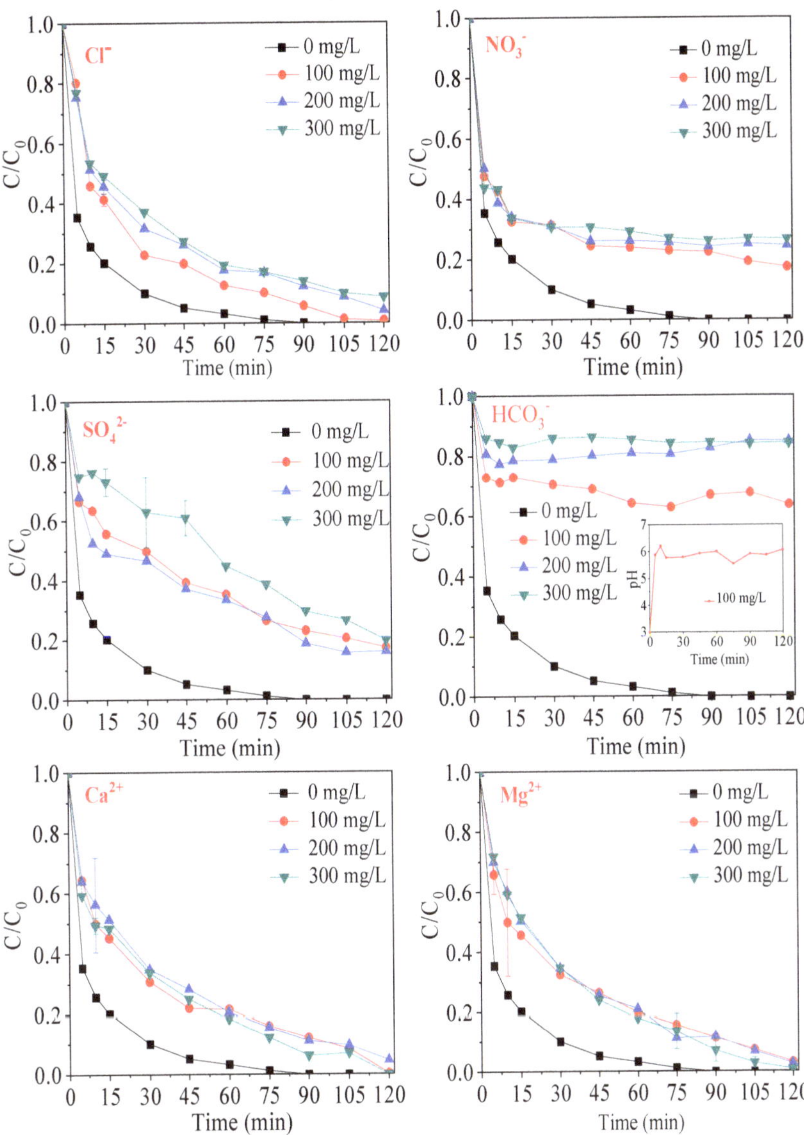

Figure 6. Effect of different anions and cations on the removal of Cr(VI) by nZVI in solution. ([Cr(VI)] = 25 mg/L, [nZVI] = 0.5 g/L, pH = 3, T = 293 K, Time = 120 min).

The addition of HCO_3^- induced the strongest inhibition on the process performance as shown in Figure 6, where the reaction rapidly stopped in 15 min in the presence of 100–300 mg/L HCO_3^-. The inhibitory effect increased with the increase in HCO_3^- concentration. Compared to the control with a Cr(VI) removal of 90%, the removal of Cr(VI) dropped to only 14% with the addition of 300 mg/L HCO_3^-. When HCO_3^- was added to the solution, a large amount of OH^- was generated according to Equation (3) This resulted in an obvious increase in solution pH from 3 to 6, favoring the formation of iron (hydr)oxides, while that of the other ions (i.e., Cl^-, NO_3^-, and SO_4^{2-}) was about 4. Therefore, this might be one of the main reasons why the inhibition of HCO_3^- on the process efficiency was the most significant. Furthermore, HCO_3^- can adhere to the surface of nZVI by forming inner-sphere surface complexes with the generated iron (hydr)oxides to further inhibit the adsorption/reduction ability of nZVI [29,31]. Moreover, HCO_3^- could react with Fe^{2+} to form $FeCO_3$ via Equation (4) or iron (oxy) hydroxyl carbonate precipitate attaching on nZVI surface, consequently blocking the reactive sites and resulting in a decreased performance [33,34].

$$HCO_3^- + H_2O \rightarrow H_2CO_3 + OH^- \quad (3)$$

$$Fe^{2+} + 2HCO_3^- \rightarrow Fe(HCO_3)_2 \rightarrow FeCO_3 + CO_2 + H_2O \quad (4)$$

Here, Ca^{2+} and Mg^{2+} were selected to examine the effect of hardness in water on the removal efficiency of the nZVI system. It can be seen from Figure 6 that Cr(VI) was completely removed in 2 h with or without Mg^{2+} and Ca^{2+}. Nevertheless, the presence of Ca^{2+} and Mg^{2+} still inhibited the reaction kinetics. The addition of 100 mg/L Ca^{2+} and Mg^{2+} decreased the removal of Cr(VI) by the nZVI in 30 min from 90% to 69.1% and 67.5%, respectively. However, the reaction kinetics was not significantly reduced with a further increase in the ionic strength of Ca^{2+} and Mg^{2+}. The inhibitory effect of Ca^{2+} and Mg^{2+} might be due to the formation of Ca^{2+} and Mg^{2+} hydroxides, which can be coated on the surface of nZVI particles.

2.2.5. Effect of Humic Acid

The effect of humic acid (HA) to simulate dissolved organic matter ubiquitously existing in aquatic systems on the process efficiency was also examined. It could be seen from Figure 7 that Cr(VI) removal kinetics remarkably decreased with the increase in HA concentration. As a highly heterogeneous mixture of macro organic molecules, HA contains a large number of functional groups that are easily adsorbed on the particle surface, blocking the effective sites of nZVI and affecting the electron transfer route between Cr(VI) and nZVI. It was reported that the –OH and –COOH groups abundantly existing in HA could occupy the effective sites on the surface of nZVI, which acted as a steric obstacle for the mass and electron transfer of target contaminant and, therefore, suppressed the process efficiency [35,36]. In addition to the strong affinity of HA to the nZVI surface, HA also exerts an influence on the nZVI system through complexation and aggregation, forming iron–humic acid complexes and colloids with Fe^0 and Fe^{3+} as coordination centers, which could alter the reactivity of nZVI and the adsorption/reduction of contaminants [37,38].

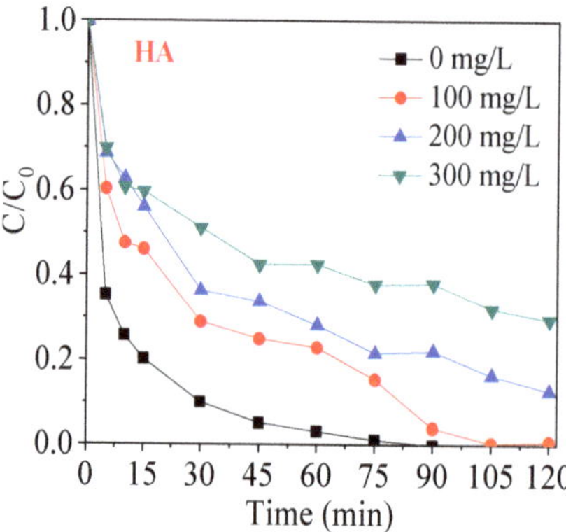

Figure 7. Effect of HA on removal of Cr(VI) by nZVI in solution. ([Cr(VI)] = 25 mg/L, [nZVI] = 0.5 g/L, pH = 3, T = 293 K, Time = 120 min).

2.3. Adsorption Kinetics and Isotherms

To understand the adsorption–reduction process and the potential rate-controlling step of Cr(VI) removal by nZVI, the first-order, second-order, pseudo-first-order, and pseudo-second-order kinetic models were utilized to interpret the uptake rate of Cr(VI) by nZVI at different temperatures (293, 303, and 313 K) (the fitting parameters are provided in Table S2). It can be ascertained that the removal of Cr(VI) by nZVI was well described by the pseudo-second-order kinetic model with high correlation coefficient ($R^2 > 0.998$) values (Figure S3). Our experimental results indicate that the valence state or electron transfer process rather than the boundary layer resistance of nZVI particles are the limiting factors affecting the chemical rate of Cr(VI) removal. Therefore, the dynamics of Cr(VI) removal by nZVI were highly related to the chemical redox reaction between nZVI and Cr(VI).

Equilibrium adsorption isotherm models are generally used to describe the relationship between the concentration of solute in solution and the amount of solute adsorbed on the adsorbent at equilibrium and to evaluate the adsorptive capacity of an adsorbent. Therefore, two widely used adsorption isotherm models (i.e., Langmuir and Freundlich isotherms [23]) were employed to fit the experimental data, respectively. It was found (see Table S3 and Figure S4) that equilibrium adsorption data of Cr(VI) removal by nZVI were more represented by the Langmuir model (i.e., $R^2 = 0.97$) than by the Freundlich model (i.e., $R^2 = 0.90$). The value of maximum adsorption capacity ($Q_m = 57$ mg/g at 20 °C) obtained from the Langmuir model matched well with the experimental value of 50 mg/g. Therefore, the removal of Cr(VI) by nZVI could be considered as a single-layer chemisorption process.

2.4. Intraparticle Diffusion

Kinetic data for the adsorption of Cr(VI) onto nZVI was modelled according to the intraparticle diffusion equation to understand the rate-determining step of this process [39]. It was found from Figure 8 and Table S4 that the adsorption process can be divided into three stages, which consisted of an initial surface diffusion stage, then internal diffusion process of particles, followed by the equilibrium dynamic process of adsorption and desorption [39]. Figure 8 shows that the three stages of adsorption are rendered as a continuous and segmented process. The surface adsorption phase was completed within 15 min of the beginning of the reaction. Due to the high specific surface area of nZVI and the large number of available reactive sites at the initial stage, Cr(VI) in the solution can be quickly adsorbed onto the surface of nZVI, leading to a rapid reduction in Cr(VI) concentration in the solution in 15 min. The internal diffusion process of nZVI was completed in the following 75 min, where Cr(VI) adsorbed on the nZVI surface reacted with Fe^0 to form Cr(III), and precipitates attached to the particle surface and then reached the adsorption equilibrium. In the first two stages, the rate-limiting step were the surface pores of the particles. For the third stage, it might be due to the saturation of adsorption, in agreement with the pseudo-second-order dynamic model analysis. As also observed by Wang et al. [40], the second stage did not pass through the origin, suggesting that both the intraparticle diffusion and chemical reactions are important rate-limiting factors in controlling the adsorption process.

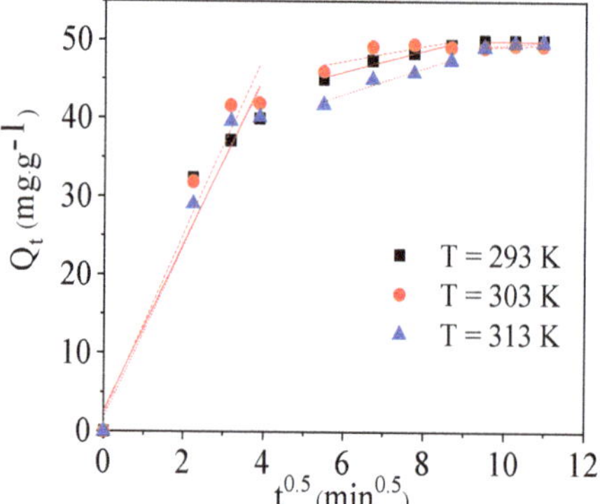

Figure 8. Internal diffusion model. ([Cr(VI)] = 25 mg/L, [nZVI] = 0.5 g/L, pH = 3, Time = 120 min).

2.5. Mechanisms of Cr(VI) Removal by nZVI

Figure 9a shows the FTIR scanning curves of nZVI before and after the reaction with Cr(VI) in the range of 400–4000 cm^{-1}. An obvious vibration band caused by O–H stretching appeared at 3300–4000 cm^{-1} [16], which may be related to the contact between nZVI and H_2O in the environment during the testing process. The peak at 1630 cm^{-1} can be considered as the characteristic bending vibration peak of O–H in H_2O adsorbed on the nZVI surface [41]. By comparing the changes of peak shapes before and after the reaction, it can be observed that a new peak appeared at 468 cm^{-1} after the reaction, which be attributed to the Fe–O of iron (hydr)oxides [42].

The XPS analysis of nZVI before and after reaction with Cr(VI) was applied to further examine the possible interaction mechanism [43,44]. Figure 9b illustrates the full-range XPS spectra of nZVI before and after the reaction with typical peaks of C1s (284.8 eV), Fe2p, O1s, and Cr2p. Figure 9c,d display the detailed XPS surveys of Fe before and after the reaction, respectively. The Fe2p$_{1/2}$ and Fe2p$_{3/2}$ characteristic peaks of Fe before the reaction appeared at 721 and 707 eV, while the characteristic peaks appeared at 722 and 709 eV after the reaction. The peak shift may be caused by the redox reaction between Fe^0 and Cr, which changed the density of the electron cloud on the nZVI surface. However, the peak of Fe^0 (706.8 eV) was not obvious in XPS spectra, which might be due to the presence of oxides on the surface of nZVI [28]. It was reported that the thickness of the oxide layer was about 2–4 nm [28,45]. Meanwhile, XPS surface analysis is subject to only several nanometer depths, making the XPS insufficient to capture the internal state of nZVI. As shown in Figure 9e, Cr2p$_{1/2}$ and Cr2p$_{3/2}$ appeared in 583 and 574 eV. However, these peaks were not observed in the same scanning range of nZVI before the reaction, indicating that Cr(VI) was reduced by nZVI to form precipitates attached to the particle surface.

The crystallinity variation of nZVI before and after the reaction with Cr(VI) was compared (Figure 9f). After the reaction, the characteristic peak of Fe^0 at 44.68° decreased, indicating that the crystalline structure of nZVI was weakened because of the chemical reaction of nZVI. A new peak was observed at 35.50°, likely due to the presence of the newly formed $FeCr_2O_4$ and iron oxides [46,47]. The enhancement of the peak at 65.0° may be attributed to the generation of Fe_3O_4 in the reaction process, resulting from the further oxidation of nZVI.

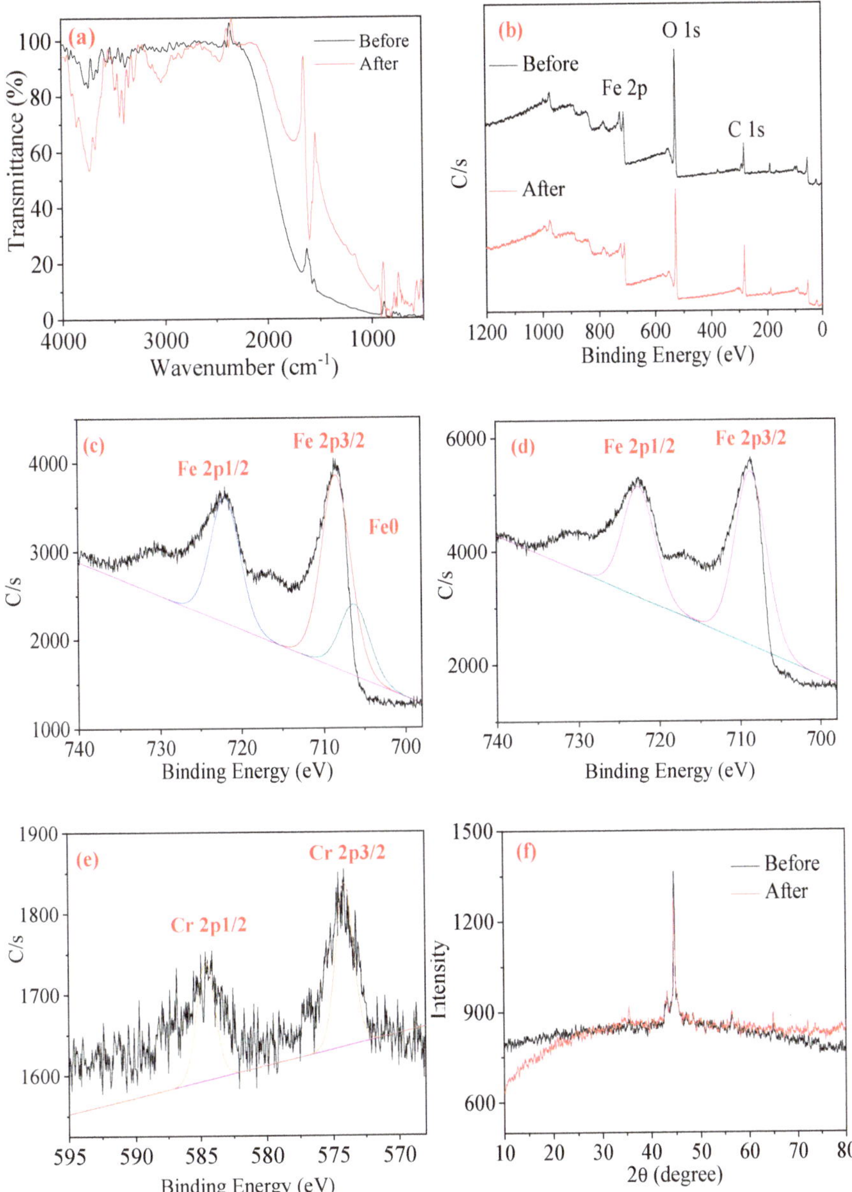

Figure 9. (**a**) The FTIR results of nZVI pre- and post- reaction with Cr(VI); (**b**) wide-scan XPS survey of nZVI before and after Cr(VI) adsorption; a magnified region in the XPS survey of nZVI before (**c**) and after (**d**) Cr(VI) adsorption; (**e**) a high-resolution Cr XPS survey of nZVI before and after Cr(VI) adsorption; (**f**) XRD analysis of nZVI before and after the reaction. ([Cr(VI)] = 25 mg/L, [nZVI] = 0.5 g/L, pH = 3, T = 293 K, Time = 120 min).

Based on the literature and the above experimental results, a reasonable mechanism of Cr(VI) removal by nZVI was proposed, as depicted in Scheme 1. It can be considered that the main steps of nZVI to remove Cr(VI) involve adsorption, reduction, precipitation, and coprecipitation [30,48]. The adsorption process was initiated through the diffusion of different forms of Cr(VI) (e.g., $HCrO_4^-$ CrO_4^{2-}, and $Cr_2O_7^{2-}$) in the solution, which were rapidly adsorbed on the surface of nZVI particles. Additionally, Fe^0 can be oxidized to Fe^{2+} at low pH according to Equations (5) and (6). Then, redox reactions, as in Equations (7)–(11), occurred via direct electron transfer from Fe^0 and dissolved Fe^{2+} to Cr(VI), leading to the generation of a higher oxidation state of iron ions (e.g., Fe^{2+} and Fe^{3+}) and Cr(III) as reduced species. Finally, $FeCr_2O_4$, $Cr_xFe_{1-x}OOH$, and $Cr_xFe_{1-x}(OH)_3$ were eventually formed through precipitation and coprecipitation via Equations (12)–(16) [30,48].

$$Fe^0 + 2H_2O \rightarrow Fe^{2+} + H_2 + 2OH^- \tag{5}$$

$$2Fe^0 + 2H_2O + O_2 \rightarrow 2Fe^{2+} + 4OH^- \tag{6}$$

$$Fe^0 + HCrO_4^- + 7H^+ \rightarrow Cr^{3+} + Fe^{3+} + 4H_2O \tag{7}$$

$$3Fe^{2+} + HCrO_4^- + 7H^+ \rightarrow Cr^{3+} + 3Fe^{3+} + 4H_2O \tag{8}$$

$$2Fe^0 + Cr_2O_7^{2-} + 14H^+ \rightarrow 2Cr^{3+} + 2Fe^{3+} + 4H_2O \tag{9}$$

$$3Fe^0 + Cr_2O_7^{2-} + 6H^+ \rightarrow FeCr_2O_4 + 2Fe^{2+} + 3H_2O \tag{10}$$

$$Fe^0 + CrO_4^{2-} + 8H^+ \rightarrow Fe^{3+} + Cr^{3+}\ 4H_2O \tag{11}$$

$$3Fe^{2+} + CrO_4^{2-} + 4OH^- + 4H_2O \rightarrow 3Fe(OH)_3 + Cr(OH)_3 \tag{12}$$

$$6Fe^{2+} + 2CrO_4^{2-} + 8OH^- \rightarrow 3Fe_2O_3 + 2Cr(OH)_3 + H_2O \tag{13}$$

$$Cr^{3+} + 3H_2O \rightarrow Cr(OH)_3 + 3H^+ \tag{14}$$

$$(1-x)Fe^{3+} + xCr^{3+} + 3H_2O \rightarrow Cr_xFe_{1-x}(OH)_3 + 3H^+ \tag{15}$$

$$(1-x)Fe^{3+} + xCr^{3+} + 2H_2O \rightarrow Cr_xFe_{1-x}OOH + 3H^+ \tag{16}$$

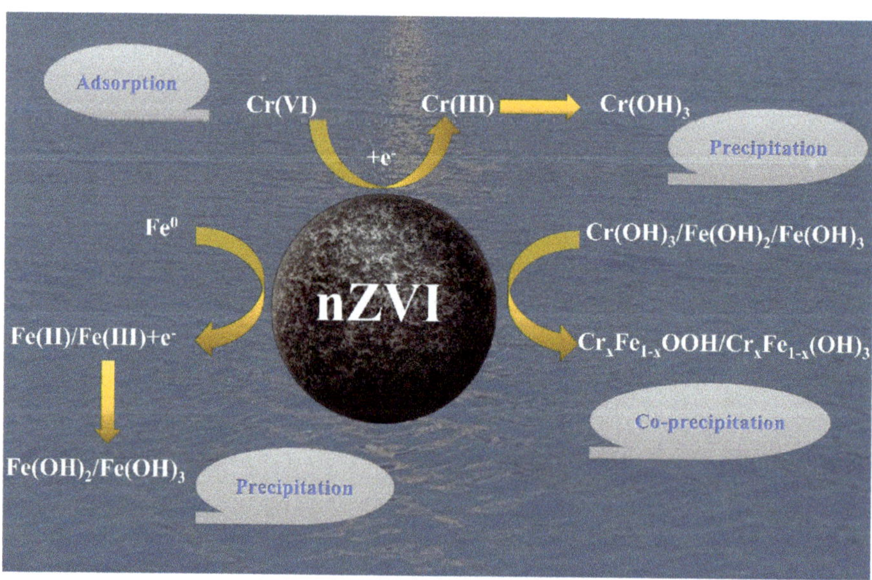

Scheme 1. A proposed mechanism for Cr(VI) reduction by nZVI.

2.6. Comparison of nZVI and Supported nZVI for Cr(VI) Removal

Raw nZVI was prone to corrosion and agglomeration due to the ferromagnetic properties of Fe. Various materials were examined as modifiers for the synthesis of modified nZVI to enhance the dispersibility and reactivity of nZVI particles. As shown in Figures 10 and S5, all six selected modifiers played either a promoting or a demoting role on the process efficiency compared with the raw nZVI. Starch, H_2A, and Cu obviously improved the removal rate of Cr(VI) from 79.9% (i.e., raw nZVI) to 100%, 90.7%, and 86%, respectively. As can be seen from the SEM images in Figure 11, the particles exhibited an obviously single spherical shape with the addition of starch and H_2A, which significantly improved the dispersity of the modified nZVI. Yang et al. [49] showed that soluble starch can act as a protecting and dispersing agent to prevent nZVI from agglomerating. The surface of the particles after adding Cu was relatively rough, which can also be considered to increase the effective sites for Cr(VI) adsorption by increasing the specific surface area. The remaining three kinds of modifiers (i.e., CMC, Zn, and Mn) with obvious inhibitory effect were agglomerated in different degrees. Therefore, it can be concluded that the aggregation degree of particles was closely related to the removal rate of Cr(VI) by nZVI. Improved dispersity of nZVI particles could facilitate the removal of target contaminants.

The promoting mechanism of H_2A and starch on the removal of Cr(VI) was similar. They were both used as surfactants to increase the surface resistance of iron particles and reduce the aggregation of iron particles. Furthermore, H_2A could form a Fe–H_2A complex on the surface of nZVI to dissolve the metal passivation layer [50]. Similar to H_2A, starch can form discrete nZVI particles with uniform shape; meanwhile, the surface functional groups make the synthesized nZVI more stable [18]. The internal structure of Fe–Cu bimetallic particles was proposed by previous researchers [51] and the shell layers of nZVI particles from inside to outside were proposed as Fe, Fe oxide, Cu, and oxide layers. Adsorption, reduction, precipitation, and coprecipitation took place in the nZVI layer, while adsorption and reduction of Cu^{2+} took place mainly in the Cu layer. Xi et al. [52] believed that Fe–Cu can form a battery system in solution, and that Cu can activate continuously and stably on the surface of nZVI to overcome the passivation of nZVI due to oxide layer. It was also considered that the current effect produced by Cu^{2+} was beneficial to electron transfer and accelerates the rapid reduction of Cr(VI). In addition, the comparison of Cr(VI) removal efficiency by nZVI and other related materials was shown in Table S5.

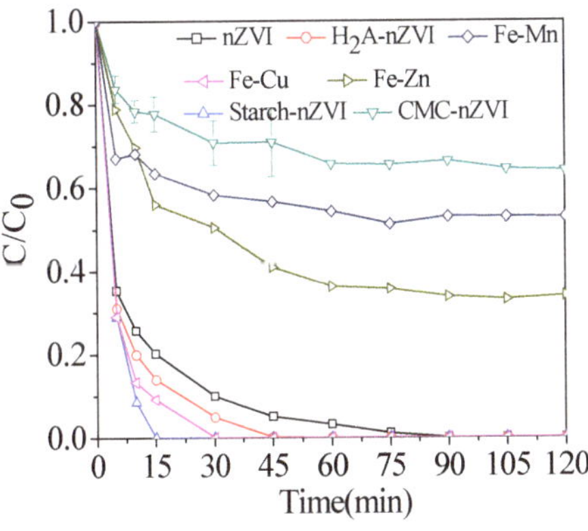

Figure 10. The concentration of Cr(VI) as a function of time with supported nZVI. ([Cr(VI)] = 25 mg/L, adsorbent = 0.5 g/L, pH = 3, T = 293 K).

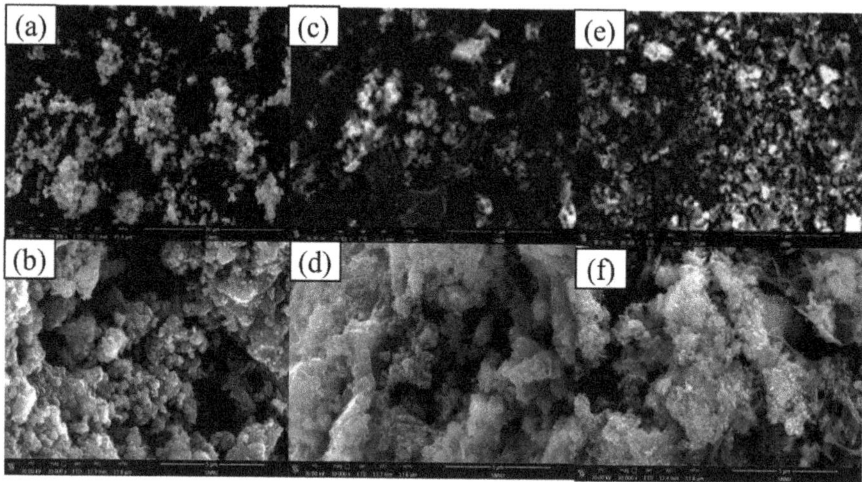

Figure 11. The SEM results of pre- and post-reaction Cr(VI) with supported nZVI. Here, (**a**,**b**) H_2A–nZVI; (**c**,**d**) starch–nZVI; (**e**,**f**) Fe–Cu.

3. Materials and Methods

3.1. Chemicals and Materials

All chemicals of analytical grade were purchased from Sinopharm Chemical Reagent Co., China and were used without further purification. Potassium borohydride (KBH_4, 97%) and ferrous sulfate heptahydrate ($FeSO_4 \cdot 7H_2O$, 99%) were used as the reductant for the synthesis of nZVI and the source of Fe^{2+}, respectively. Potassium dichromate ($K_2Cr_2O_7$, 99.8%) was employed as the source of Cr(VI). Diphenyl carbamide ($C_{13}H_{14}N_4O$, 99%) and acetone (CH_3COCH_3, 98%) were used to measure residual Cr(VI). Copper(II) sulfate pentahydrate ($CuSO_4 \cdot 5H_2O$, 99%), manganese sulfate ($MnSO_4$, 98%), zinc sulfate ($ZnSO_4$, 99.5%), carboxy methyl cellulose (CMC, 98%), starch ($(C_6H_{10}O_5)_n$, 99%), and ascorbic acid (H_2A, 99.7%) were employed for the modification of nZVI. All solutions were prepared with deionized water (ϱ = 18.25 MΩ·cm). The solution pH was adjusted with 0.1 M HCl and 0.1 M NaOH.

3.2. Preparation of nZVI and Modified nZVI

Here, nZVI was prepared by a chemical reduction method in aqueous solution using KBH_4 as reducing agent to convert Fe^{2+} to Fe^0 according to Equation (17) [53].

$$Fe^{2+} + 2BH_4^- + 6H_2O \rightarrow Fe^0 + 2B(OH)_3 + 7H_2 \quad (17)$$

First, 250 mL of 12.5 mM Fe^{2+} and 15 mM KBH_4 aqueous solution was freshly prepared with the solution pH adjusted to 3.0 and 12.0, respectively. Then, Fe^{2+} solution was transferred to a 1000 mL three-necked flask. The solution was purged with N_2 and vigorously stirred for 20 min to remove dissolved O_2. Next, 250 mL KBH_4 aqueous solution was added dropwise (1~2 drops/s) into the reactor and stirred vigorously for 1 h. When black nZVI particles appeared, the mixture was continuously stirred for 30 min. The nZVI was separated by magnet, washed with ethanol and deionized water, and then dried in a vacuum oven at 75 °C for 12 h. The obtained nZVI was used up within two days to ensure excellent reduction ability.

The preparation procedures of modified nZVI were as follows: dissolve 3.475 g $FeSO_4 \cdot 7H_2O$ (i.e., 12.5 mM) in 250 mL deionized water, and then add different dispersants (10 g/L H_2A, 0.2% starch, and 0.5% CMC) to obtain H_2A–nZVI, starch–nZVI and CMC–nZVI, while Fe–Cu, Fe–Zn, and Fe–Mn bimetals were synthesized by mixing 3.475g of 95%

Fe^{2+} and 5% Cu^{2+}, Mn^{2+}, and Zn^{2+}, respectively, with 250 mL deionized water at pH 3. The method of synthesis was the same as the previous procedures. All synthetic experiments were carried out at 293 K.

3.3. Experimental Procedures

The experiments for Cr (VI) removal by the nZVI were conducted at room temperature (T = 293 K) in 100 mL glass vials. The adsorption reaction was initiated by dosing 50 mL of 25 mg/L Cr(VI) solution and a specific amount of nZVI particles into the glass vials, and the mixture was placed in the shaking incubator (150 rpm) during the batch experiment. At pre-determined time intervals, the nZVI particles were separated from the supernatant by a strong magnet, and approximately 8 mL of the supernatant was collected with a disposable syringe, then filtered through a 0.22 μm filter for measuring the residual Cr(VI) concentrations in the solution. In addition, the separated black nZVI particles were collected and vacuum-dried for further tests. All the experiments were conducted in triplicate, whereby the average values were presented.

The influences of some main parameters on the Cr(VI) removal by nZVI particles were examined by varying initial solution pH (2–11), temperature (293–313 K), adsorbent dosage (0.2–0.5 g/L), and Cr(VI) concentration (15–45 mg/L). Furthermore, the impacts of coexisting ions (e.g., Cl^-, HCO_3^-, NO_3^-, SO_4^{2-}, Ca^{2+}, and Mg^{2+}) and humic acid ranging from 100 to 300 mg/L on the process efficiency were also evaluated. The removal efficiency of Cr(VI) by modified nZVI (e.g., H_2A–nZVI, starch–nZVI, CMC–nZVI, Fe–Cu, Fe–Zn, and Fe–Mn) under the same experimental conditions were investigated for comparison.

3.4. Characterization and Analytical Methods

The morphologies and semi-quantitative surface composition analysis of the synthesized nZVI were examined by a scanning electron microscope (SEM, MLA650F, FEI, Hillsboro, America) equipped with an energy dispersive X-ray spectrometer (EDS, XFlash6130, Bruker, Siegsdorf, Germany). The Brunauer–Emmett–Teller analysis was performed to determine the specific surface area (BET, JWGB, JW-BK112, Beijing, China). To assess the crystallinity and chemical composition of the samples, X-ray diffraction (XRD) analysis was conducted using a BrukerD8 Advance X-ray diffractometer system. The main valence state changes of Cr and Fe species on nZVI surface during the reaction process were examined using X-ray photoelectron spectroscopy (XPS, Kratos Analytical, Axis Ultra, Manchester, UK). The functional groups were identified by a Fourier-transform infrared spectrometer (FTIR, Bruker, Tensor27, Siegsdorf, Germany).

The concentration of Cr(VI) was measured by 1,5-diphenylcarbazide spectrophotometry at 540 nm using a UV–Vis Spectrophotometer [47]. The Cr(III) concentration was calculated using the difference value between Cr_{total} and Cr(VI).

4. Conclusions

Nanoscale zero valent iron was prepared by a chemical reduction method and the adsorptive and removal efficiency of Cr(VI) from aqueous solution by nZVI was explored. The XRD spectra and SEM–EDS images showed that the synthesized nZVI particles had a good crystalline spherical structure. However, oxidation products, such as γ-Fe_2O_3 and Fe_3O_4, were formed on the surface of the particles. The BET analysis indicated a specific surface area of nZVI of 15.19 m^2/g. Experimental results showed that, at pH 3 and 20 °C, the removal rate of Cr(VI) can reach 100% after 90 min of the reaction, of which the adsorption capacity was 50 mg/g, and the initial pH, Cr(VI) concentration, and nZVI dosage all affect the removal efficiency of Cr(VI). The process of Cr(VI) removal by nZVI was in accordance with the pseudo-second-order kinetic model ($R^2 > 0.99$), the Langmuir adsorption isotherm model ($R^2 > 0.97$), and the three-stage diffusion process (adsorption stage, redox stage and adsorption–analytic equilibrium stage). The order of the inhibition of coexisting ions on Cr(VI) removal from strong to weak was $HCO_3^- > NO_3^- > SO_4^{2-} > Cl^-$. The presence of humic acid also had a strong inhibitive impact on the process efficiency, while the effect of cations (i.e., Mg^{2+} and Ca^{2+}) was insignificant. Here, FTIR, XPS, and XRD were used to characterize the nZVI before and after the reaction with Cr(VI). Accordingly, a plausible mechanism for Cr(VI) removal by nZVI was proposed, including adsorption, reduction, precipitation and coprecipitation, in which the reducing products were $FeCr_2O_4$, $Cr_xFe_{1-x}OOH$, and $Cr_xFe_{1-x}(OH)_3$. In addition, the modification of nZVI by various materials has been investigated to further enhance the performance. It was found that Fe–starch, Fe–ascorbic acid, and Fe–Cu had better removal rates of Cr(VI) than pure phase nZVI, because the presence of starch and ascorbic acid could effectively reduce the agglomeration between nZVI particles. Our study suggest that nZVI is an effective and green technology for Cr(VI) removal and that it has a promising application prospects.

Supplementary Materials: The following supporting information can be downloaded at https://www.mdpi.com/article/10.3390/catal12090999/s1. Text S1; Figure S1: The EDS results of pre- (a) and post- (b) reaction of Cr(VI) with nZVI; Figure S2: Changes of adsorption capacity with pH, initial concentration of Cr(VI) and dosage of nZVI; Figure S3: Pseudo-second-order kinetic fitting curve of Cr(VI) removal by nZVI; Figure S4: Fitting results of isotherm adsorption model (a: Langmuir; b: Freundlich); Figure S5: The SEM results of pre- and post-reaction Cr(VI) with supported nZVI [(a,b) H_2A–nZVI; (c,d) Starch–nZVI; (e,f) Fe–Cu; (g,h) CMC–nZVI; (i,j) Fe–Zn; (k,l) Fe–Mn]; Table S1: The EDS results of pre- and post-reaction Cr(VI) with nZVI; Table S2: Comparison of R^2 values fitted by various kinetic models at different temperatures; Table S3: Comparison of R^2 values fitted by isothermal adsorption model; Table S4: Intraparticle diffusion coefficients and intercept values for Cr(VI) adsorption on nZVI particles at different temperatures; Table S5: Comparison of the Cr(VI) removal efficiency of nZVI and other related materials.

Author Contributions: Y.G., Conceptualization, writing—original draft; X.Y., Methodology, data curation, writing—original draft; X.L., Funding acquisition, writing—review & editing; M.L., Methodology, writing—review & editing; L.W., Funding acquisition, writing—review & editing; Y.W., Conceptualization, supervision, funding acquisition, writing—review & editing. All authors have read and agreed to the published version of the manuscript.

Funding: This research was supported by the Fundamental Research Funds for the Central Universities (GK202103145) and Natural Science Basic Research Plan of Shaanxi Province (2021JM-192).

Data Availability Statement: Not applicable.

Conflicts of Interest: The authors declare that they have no conflict of interest.

References

1. Prasad, S.; Yadav, K.K.; Kumar, S.; Gupta, N.; Cabral-Pinto, M.M.; Rezania, S.; Radwan, N.; Alam, J. Chromium contamination and effect on environmental health and its remediation: A sustainable approaches. *J. Environ. Manag.* **2021**, *285*, 112174. [CrossRef]
2. Sinha, V.; Pakshirajan, K.; Chaturvedi, R. Chromium tolerance, bioaccumulation and localization in plants: An overview. *J. Environ. Manag.* **2018**, *206*, 715–730. [CrossRef]
3. Liang, J.; Huang, X.; Yan, J.; Li, Y.; Zhao, Z.; Liu, Y.; Ye, J.; Wei, Y. A review of the formation of Cr(VI) via Cr(III) oxidation in soils and groundwater. *Sci. Total Environ.* **2021**, *774*, 145762. [CrossRef]
4. Karimi-Maleh, H.; Orooji, Y.; Ayati, A.; Qanbari, S.; Tanhaei, B.; Karimi, F.; Alizadeh, M.; Rouhi, J.; Fu, L.; Sillanpää, M. Recent advances in removal techniques of Cr(VI) toxic ion from aqueous solution: A comprehensive review. *J. Mol. Liq.* **2021**, *329*, 115062. [CrossRef]
5. Aigbe, U.O.; Osibote, O.A. A review of hexavalent chromium removal from aqueous solutions by sorption technique using nanomaterials. *J. Environ. Chem. Eng.* **2020**, *8*, 104503. [CrossRef]
6. Azeez, N.A.; Dash, S.S.; Gummadi, S.N.; Deepa, V.S. Nano-remediation of toxic heavy metal contamination: Hexavalent chromium Cr(VI). *Chemosphere* **2021**, *266*, 129204. [CrossRef]
7. Fernández, P.M.; Viñarta, S.C.; Bernal, A.R.; Cruz, E.L.; Figueroa, L.I.C. Bioremediation strategies for chromium removal: Current research, scale-up approach and future perspectives. *Chemosphere* **2018**, *208*, 139–148. [CrossRef]
8. Marvin, A.S.; Harris, F. *The Encyclopedia of Chemical Electrode Potentials*; Plenum Press: New York, NY, USA, 1982.
9. Stefaniuk, M.; Oleszczuk, P.; Ok, Y.S. Review on nano zerovalent iron (nZVI): From synthesis to environmental applications. *Chem. Eng. J.* **2016**, *287*, 618–632. [CrossRef]
10. Jamei, M.R.; Khosravi, M.R.; Anvaripour, B. Investigation of ultrasonic effect on synthesis of nano zero valent iron particles and comparison with conventional method. *Asia-Pac. J. Chem. Eng.* **2013**, *8*, 767–774. [CrossRef]

11. Li, R.; Li, Q.; Zhang, W.; Sun, X.; Li, J.; Shen, J.; Han, W. Low dose of sulfur-modified zero-valent iron for decontamination of trace Cd(II)-complexes in high-salinity wastewater. *Sci. Total Environ.* **2021**, *793*, 148579. [CrossRef]
12. Liu, K.; Li, F.; Zhao, X.; Wang, G.; Fang, L. The overlooked role of carbonaceous supports in enhancing arsenite oxidation and removal by nZVI: Surface area versus electrochemical property. *Chem. Eng. J.* **2021**, *406*, 126851. [CrossRef]
13. Pereira, C.A.A.; Nava, M.R.; Walter, J.B.; Scherer, C.E.; Dalfovo, A.D.K.; Barreto-Rodrigues, M. Application of zero valent iron (ZVI) immobilized in Ca-Alginate beads for CI Reactive Red 195 catalytic degradation in an air lift reactor operated with ozone. *J. Hazard. Mater.* **2021**, *401*, 123275. [CrossRef]
14. Xu, J.; Liu, X.; Cao, Z.; Bai, W.; Shi, Q.; Yang, Y. Fast degradation, large capacity, and high electron efficiency of chloramphenicol removal by different carbon-supported nanoscale zerovalent iron. *J. Hazard. Mater.* **2020**, *384*, 121253. [CrossRef] [PubMed]
15. Li, Q.; Chen, Z.; Wang, H.; Yang, H.; Wen, T.; Wang, S.; Hu, B.; Wang, X. Removal of organic compounds by nanoscale zero-valent iron and its composites. *Sci. Total Environ.* **2021**, *792*, 148546. [CrossRef]
16. Wang, Y.; Yu, L.; Wang, R.; Wang, Y.; Zhang, X. A novel cellulose hydrogel coating with nanoscale Fe^0 for Cr(VI) adsorption and reduction. *Sci. Total Environ.* **2020**, *726*, 138625. [CrossRef]
17. Angaru, G.K.R.; Choi, Y.-L.; Lingamdinne, L.P.; Choi, J.-S.; Kim, D.-S.; Koduru, J.R.; Yang, J.-K.; Chang, Y.-Y. Facile synthesis of economical feasible fly ash-based zeolite-supported nano zerovalent iron and nickel bimetallic composite for the potential removal of heavy metals from industrial effluents. *Chemosphere* **2021**, *267*, 128889. [CrossRef]
18. Huang, D.L.; Hu, Z.X.; Peng, Z.; Zeng, G.M.; Chen, G.M.; Zhang, C.; Cheng, M.; Wan, J.; Wang, X.; Qin, X. Cadmium immobilization in river sediment using stabilized nanoscale zero-valent iron with enhanced transport by polysaccharide coating. *J. Environ. Manag.* **2018**, *210*, 191–200. [CrossRef]
19. Hua, Y.L.; Wang, W.; Huang, X.Y.; Gu, T.; Ding, D.; Ling, L.; Zhang, W.-X. Effect of bicarbonate on aging and reactivity of nanoscale zerovalent iron (nZVI) toward uranium removal. *Chemosphere* **2018**, *201*, 603–611. [CrossRef] [PubMed]
20. Zou, Y.; Wang, X.; Khan, A.; Wang, P.; Liu, Y.; Alsaedi, A.; Hayat, T.; Wang, X. Environmental Remediation and Application of Nanoscale Zero-Valent Iron and Its Composites for the Removal of Heavy Metal Ions: A Review. *Environ. Sci. Technol.* **2016**, *50*, 7290–7304. [CrossRef]
21. Shi, D.; Zhu, G.; Zhang, X.; Zhang, X.; Li, X.; Fan, J. Ultra-small and recyclable zero-valent iron nanoclusters for rapid and highly efficient catalytic reduction of p-nitrophenol in water. *Nanoscale* **2019**, *11*, 1000–1010. [CrossRef]
22. Chen, S.; Bedia, J.; Li, H.; Ren, L.Y.; Naluswata, F.; Belver, C. Nanoscale zero-valent iron@mesoporous hydrated silica core-shell particles with enhanced dispersibility, transportability and degradation of chlorinated aliphatic hydrocarbons. *Chem. Eng. J.* **2018**, *343*, 619–628. [CrossRef]
23. Xu, H.; Gao, M.; Hu, X.; Chen, Y.; Li, Y.; Xu, X.; Zhang, R.; Yang, X.; Tang, C.; Hu, X. A novel preparation of S-nZVI and its high efficient removal of Cr(VI) in aqueous solution. *J. Hazard. Mater.* **2021**, *416*, 125924. [CrossRef]
24. Wang, Y.; Lin, N.P.; Gong, Y.S.; Wang, R.T.; Zhang, X.D. Cu-Fe embedded cross-linked 3D hydrogel for enhanced reductive removal of Cr(VI): Characterization, performance, and mechanisms. *Chemosphere* **2021**, *280*, 130663. [CrossRef]
25. Huang, X.-Y.; Ling, L.; Zhang, W.-X. Nanoencapsulation of hexavalent chromium with nanoscale zero-valent iron: High resolution chemical mapping of the passivation layer. *J. Environ. Sci.* **2018**, *67*, 4–13. [CrossRef]
26. Wen, J.; Fu, W.; Ding, S.; Zhang, Y.; Wang, W. Pyrogallic acid modified nanoscale zero-valent iron efficiently removed Cr(VI) by improving adsorption and electron selectivity. *Chem. Eng. J.* **2022**, *443*, 136510. [CrossRef]
27. Zhou, S.; Li, Y.; Chen, J.; Liu, Z.; Wang, Z.; Na, P. Enhanced Cr(VI) removal from aqueous solutions using Ni/Fe bimetallic nanoparticles: Characterization, kinetics and mechanism. *RSC Adv.* **2014**, *4*, 50699–50707. [CrossRef]
28. Qu, G.; Kou, L.; Wang, T.; Liang, D.; Hu, S. Evaluation of activated carbon fiber supported nanoscale zero-valent iron for chromium (VI) removal from groundwater in a permeable reactive column. *J. Environ. Manag.* **2017**, *201*, 378–387. [CrossRef] [PubMed]
29. Diao, Z.-H.; Yan, L.; Dong, F.-X.; Chen, Z.-L.; Guo, P.-R.; Qian, W.; Zhang, W.-X.; Liang, J.-Y.; Huang, S.-T.; Chu, W. Ultrasound-assisted catalytic reduction of Cr(VI) by an acid mine drainage based nZVI coupling with FeS_2 system from aqueous solutions: Performance and mechanism. *J. Environ. Manag.* **2021**, *278*, 111518. [CrossRef] [PubMed]
30. Zhang, S.-H.; Wu, M.-F.; Tang, T.-T.; Xing, Q.-J.; Peng, C.-Q.; Li, F.; Liu, H.; Luo, X.-B.; Zou, J.-P.; Min, X.-B.; et al. Mechanism investigation of anoxic Cr(VI) removal by nano zero-valent iron based on XPS analysis in time scale. *Chem. Eng. J.* **2018**, *335*, 945–953. [CrossRef]
31. Mangayayam, M.C.; Alonso-De-Linaje, V.; Dideriksen, K.; Tobler, D.J. Effects of common groundwater ions on the transformation and reactivity of sulfidized nanoscale zerovalent iron. *Chemosphere* **2020**, *249*, 126137. [CrossRef]
32. Yang, K.; Wang, X.; Yi, Y.; Ma, J.; Ning, P. Formulation of NZVI-supported lactic acid/PAN membrane with glutathione for enhanced dynamic Cr(VI) removal. *J. Clean. Prod.* **2022**, *363*, 132350. [CrossRef]
33. Tanboonchuy, V.; Grisdanurak, N.; Liao, C.-H. Background species effect on aqueous arsenic removal by nano zero-valent iron using fractional factorial design. *J. Hazard. Mater.* **2012**, *205–206*, 40–46. [CrossRef] [PubMed]
34. Lv, X.; Hu, Y.; Tang, J.; Sheng, T.; Jiang, G.; Xu, X. Effects of co-existing ions and natural organic matter on removal of chromium (VI) from aqueous solution by nanoscale zero valent iron (nZVI)-Fe_3O_4 nanocomposites. *Chem. Eng. J.* **2013**, *218*, 55–64. [CrossRef]
35. Tan, L.; Liang, B.; Fang, Z.; Xie, Y.; Tsang, E.P. Effect of humic acid and transition metal ions on the debromination of decabromodiphenyl by nano zero-valent iron: Kinetics and mechanisms. *J. Nanopart. Res.* **2014**, *16*, 2786. [CrossRef]

36. Diao, Z.-H.; Qian, W.; Zhang, Z.-W.; Jin, J.-C.; Chen, Z.-L.; Guo, P.-R.; Dong, F.-X.; Yan, L.; Kong, L.-J.; Chu, W. Removals of Cr(VI) and Cd(II) by a novel nanoscale zero valent iron/peroxydisulfate process and its Fenton-like oxidation of pesticide atrazine: Coexisting effect, products and mechanism. *Chem. Eng. J.* **2020**, *397*, 125382. [CrossRef]
37. Zhu, F.; Li, L.; Ren, W.; Deng, X.; Liu, T. Effect of pH, temperature, humic acid and coexisting anions on reduction of Cr(VI) in the soil leachate by nZVI/Ni bimetal material. *Environ. Pollut.* **2017**, *227*, 444–450. [CrossRef]
38. Duan, L.; Dai, Y.; Shi, L.; Wei, Y.; Xiu, Q.; Sun, S.; Zhang, X.; Zhao, S. Humic acid addition sequence and concentration affect sulfur incorporation, electron transfer, and reactivity of sulfidated nanoscale zero-valent iron. *Chemosphere* **2022**, *294*, 133826. [CrossRef]
39. Boparai, H.K.; Joseph, M.; O'Carroll, D.M. Kinetics and thermodynamics of cadmium ion removal by adsorption onto nano zerovalent iron particles. *J. Hazard. Mater.* **2011**, *186*, 458–465. [CrossRef]
40. Wang, T.; Zhang, L.; Li, C.; Yang, W.; Song, T.; Tang, C.; Meng, Y.; Dai, S.; Wang, H.; Chai, L.; et al. Synthesis of Core-Shell Magnetic Fe_3O_4@poly(m-Phenylenediamine) Particles for Chromium Reduction and Adsorption. *Environ. Sci. Technol.* **2015**, *49*, 5654–5662. [CrossRef]
41. Liu, J.; Liu, A.R.; Zhang, W.X. The influence of polyelectrolyte modification on nanoscale zero-valent iron (nZVI): Aggregation, sedimentation, and reactivity with Ni(II) in water. *Chem. Eng. J.* **2016**, *303*, 268–274. [CrossRef]
42. Zhang, W.; Qian, L.; Han, L.; Yang, L.; Ouyang, D.; Long, Y.; Wei, Z.; Dong, X.; Liang, C.; Li, J.; et al. Synergistic roles of Fe(II) on simultaneous removal of hexavalent chromium and trichloroethylene by attapulgite-supported nanoscale zero-valent iron/persulfate system. *Chem. Eng. J.* **2022**, *430*, 132841. [CrossRef]
43. Liu, T.; Wang, Z.-L.; Yan, X.; Zhang, B. Removal of mercury (II) and chromium (VI) from wastewater using a new and effective composite: Pumice—supported nanoscale zero-valent iron. *Chem. Eng. J.* **2014**, *245*, 34–40. [CrossRef]
44. Zhang, S.; Lyu, H.; Tang, J.; Song, B.; Zhen, M.; Liu, X. A novel biochar supported CMC stabilized nano zero-valent iron composite for hexavalent chromium removal from water. *Chemosphere* **2019**, *217*, 686–694. [CrossRef] [PubMed]
45. Liu, A.; Liu, J.; Han, J.; Zhang, W. Evolution of nanoscale zero-valent iron (nZVI) in water: Microscopic and spectroscopic evidence on the formation of nano- and micro-structured iron oxides. *J. Hazard. Mater.* **2017**, *322*, 129–135. [CrossRef]
46. Lv, D.; Zhou, J.; Cao, Z.; Xu, J.; Liu, Y.; Li, Y.; Yang, K.; Lou, Z.; Lou, L.; Xu, X. Mechanism and influence factors of chromium(VI) removal by sulfide-modified nanoscale zerovalent iron. *Chemosphere* **2019**, *224*, 306–315. [CrossRef]
47. Zhang, W.; Qian, L.; Ouyang, D.; Chen, Y.; Han, L.; Chen, M. Effective removal of Cr(VI) by attapulgite-supported nanoscale zero-valent iron from aqueous solution: Enhanced adsorption and crystallization. *Chemosphere* **2019**, *221*, 683–692. [CrossRef]
48. Jia, Z.; Shu, Y.; Huang, R.; Liu, J.; Liu, L. Enhanced reactivity of nZVI embedded into supermacroporous cryogels for highly efficient Cr(VI) and total Cr removal from aqueous solution. *Chemosphere* **2018**, *199*, 232–242. [CrossRef]
49. Yang, C.; Ge, C.; Li, X.; Li, L.; Wang, B.; Lin, A.; Yang, W. Does soluble starch improve the removal of Cr(VI) by nZVI loaded on biochar? *Ecotoxicol. Environ. Saf.* **2021**, *208*, 111552. [CrossRef] [PubMed]
50. Wang, X.; Du, Y.; Liu, H.; Ma, J. Ascorbic acid/Fe^0 composites as an effective persulfate activator for improving the degradation of rhodamine B. *RSC Adv.* **2018**, *8*, 12791–12798. [CrossRef] [PubMed]
51. Sepúlveda, P.; Rubio, M.A.; Baltazar, S.; Rojas-Nunez, J.; Llamazares, J.L.S.; Garcia, A.; Arancibia-Miranda, N. As(V) removal capacity of FeCu bimetallic nanoparticles in aqueous solutions: The influence of Cu content and morphologic changes in bimetallic nanoparticles. *J. Colloid Interface Sci.* **2018**, *524*, 177–187. [CrossRef] [PubMed]
52. Xi, Y.; Zou, J.; Luo, Y.; Li, J.; Li, X.; Liao, T.; Zhang, L.; Wang, C.; Lin, G. Performance and mechanism of arsenic removal in waste acid by combination of $CuSO_4$ and zero-valent iron. *Chem. Eng. J.* **2019**, *375*, 121928. [CrossRef]
53. Liu, X.; Cao, Z.; Yuan, Z.; Zhang, J.; Guo, X.; Yang, Y.; He, F.; Zhao, Y.; Xu, J. Insight into the kinetics and mechanism of removal of aqueous chlorinated nitroaromatic antibiotic chloramphenicol by nanoscale zero-valent iron. *Chem. Eng. J.* **2017**, *334*, 508–518. [CrossRef]

 catalysts

Article

Peroxymonosulfate Activation by Photoelectroactive Nanohybrid Filter towards Effective Micropollutant Decontamination

Wenchang Zhao [1], Yuling Dai [2], Wentian Zheng [2] and Yanbiao Liu [2,*]

[1] Fujian Provincial Key Laboratory of Featured Materials in Biochemical Industry, College of Chemistry and Materials, Ningde Normal University, Ningde 352100, China; wenchangzhao@foxmail.com

[2] Textile Pollution Controlling Engineering Center of Ministry of Environmental Protection, College of Environmental Science and Engineering, Donghua University, Shanghai 201620, China; 2202084@mail.dhu.edu.cn (Y.D.); 2191678@mail.dhu.edu.cn (W.Z.)

* Correspondence: yanbiaoliu@dhu.edu.cn; Tel.: +86-21-67798752

Abstract: Herein, we report and demonstrate a photoelectrochemical filtration system that enables the effective decontamination of micropollutants from water. The key to this system was a photoelectric–active nanohybrid filter consisting of a carbon nanotube (CNT) and MIL–101(Fe). Various advanced characterization techniques were employed to obtain detailed information on the microstructure, morphology, and defect states of the nanohybrid filter. The results suggest that both radical and nonradical pathways collectively contributed to the degradation of antibiotic tetracycline, a model refractory micropollutant. The underlying working mechanism was proposed based on solid experimental evidences. This study provides new insights into the effective removal of micropollutants from water by integrating state–of–the–art advanced oxidation and microfiltration techniques.

Keywords: nanohybrid filter; photoelectrochemical filtration; carbon nanotubes; MIL-101(Fe); tetracycline

Citation: Zhao, W.; Dai, Y.; Zheng, W.; Liu, Y. Peroxymonosulfate Activation by Photoelectroactive Nanohybrid Filter towards Effective Micropollutant Decontamination. *Catalysts* **2022**, *12*, 416. https://doi.org/10.3390/catal12040416

Academic Editor: Pedro B. Tavares

Received: 15 March 2022
Accepted: 4 April 2022
Published: 7 April 2022

1. Introduction

Recently, the environmental pollution associated with organic micropollutants in aquatic environments has received increasing environmental concern [1,2]. Micropollutants mainly consist of a vast and expanding category of anthropogenic substances such as pharmaceutical and personal care products (PPCP) and many other emerging compounds [3,4]. Their limited concentration and vast diversity have posed tremendous challenges for current wastewater treatment [5]. It is therefore highly desirable to develop advanced treatment approaches to remove micropollutants from water.

Peroxymonosulfate (PMS)–based advanced oxidation technology has recently emerged as a promising solution to address this issue [6]. Various heterogeneous catalysts have been widely used for PMS activation to produce highly reactive species. Among these, iron–based catalysts are of particular interest due to their exceptional high performance and cost effectiveness [7,8]. Yet, their application is significantly hindered by the challenges associated with the post-separation of powder-like catalysts from the reaction solution, potential Fe ion leaching and poor mass transport in conventional batch reactors. To immobilize the catalysts onto a carbonaceous support could avoid the agglomeration issue of powder-like materials. To this end, carbon nanotubes (CNTs) may serve as the ideal platform to host these nanoscale catalysts because of their rich surface chemistry, excellent electrical conductivity, large surface area and desirable chemical stability [9,10]. We have previously developed a photoelectrochemical filtration system that enables the effective detoxification of toxic heavy metal ions using photogenerated holes (h^+) (e.g., to transform highly toxic Sb(III) into less toxic Sb(V)) [11]. The key to this system was a nanohybrid filter consisting of electroactive CNT and photo-responsive metal-organic frameworks (MOF, MIL-88(B)). These introduced Fe ions were uniformly distributed onto oxylated functional moieties of CNT, and the exerted electric field favors the cycling of Fe^{3+}/Fe^{2+} pairs [12]. This not only avoids the post-separation of the MIL-88(B) catalysts, but also demonstrates enhanced material stability. In addition, the flow-through design outperformed the conventional batch reactor due to convection-enhanced mass transport [13].

Based on these findings, we hypothesized that such photoelectrochemical filtration design may also serve as a high-performance system towards PMS activation and micropollutant decontamination. On one hand, under irradiation, the photogenerated holes (h^+) may directly contribute to the degradation of organic micropollutant molecules upon contact or indirectly induce the generation of other aggressive reactive radicals (such as hydroxyl radicals) [14]. The applied electric field would facilitate the cycling of Fe^{3+}/Fe^{2+} within the system [15]. On the other hand, the photogenerated electrons (e^-) may cleave the O–O bond of PMS via the one-electron pathway; also, CNT itself was proven to be effective to activate PMS via nonradical pathway [16,17]. All these routes collectively contribute to the effective degradation of micropollutants in water. To do this, we firstly report on the preparation and demonstration of a MIL-101(Fe)@CNT nanohybrid filter to serve as both an electrode material and a filtration medium. Among them, the MIL-101(Fe) exhibits excellent light absorptive ability and large specific surface area properties, while the conductive CNT network structure promotes the separation of e^- and h^+. The catalytic degradation performance was tested using antibiotic tetracycline as the model micropollutant. Secondly, the effects of operational parameters and environmental factors on the efficacy of the proposed photoelectrochemical filtration system was systematically investigated. A plausible underlying working mechanism of the technology was proposed based on extensive experimental evidence. The outcomes of the present study are dedicated to providing an enhanced strategy towards the effective decontamination of micropollutants from water.

2. Results and Discussion

2.1. Characterization of the Nanohybrid Filter

Figure 1 compares the FESEM image of the conductive CNT filter in the absence and presence of MIL-101(Fe) nanoparticles. As can be seen, the pristine CNT filter showed a smooth surface with CNT intertwined with it to form a 3D porous network. The introduction of MIL-101(Fe) led to a rather rough surface, and those nanoparticles were uniformly distributed on the CNT surface. Furthermore, as the iron precursor concentration increased from 0 to 3 mM, the loading amount of MIL-101(Fe) gradually increased consequently. However, further increasing the precursor concentration to 4.5 mM leads to evident particle agglomeration. The successful loading of MIL-101(Fe) was further verified by the XRD pattern of the MIL-101(Fe)@CNT nanohybrid filter. The characteristic diffraction peaks centered at 8.5, 9.1, 18.6, 21.4 and 25.0° were observed, in good accordance with that of MIL-101(Fe) [7] (Figure S1a). The TGA curve indicated that approximately 1.1 mg of MIL-101(Fe) was loaded onto an effective filtration area of 7.1 cm^2 (Figure S1b). In addition, the atomic ratio of C, O and Fe on the composite filter surface was 86.92%, 11.80% and 1.31%, as determined by the energy-dispersive spectra analysis (Figure S2). All this evidence collectively indicated the successful preparation of the MIL-101(Fe)@CNT nanohybrid filter. Since MIL-101(Fe) is a well-known photocatalyst with several intriguing attributes, the cyclic voltammetry curve of the MIL-101(Fe)@CNT nanohybrid filter in 50 mM Na_2SO_4 electrolyte solution was examined (with/without UV light irradiation). The saturated photocurrent intensity was 2.05 times higher than that in the absence of light irradiation, suggesting a significant contribution from MIL-101(Fe) (Figure S3a). Electrochemical impedance spectroscopy analysis showed a charge transport resistance of 141.7 Ω for the MIL-101(Fe)@CNT nanohybrid filter, much lower than that of MIL-101(Fe) (1333.1 Ω, Figure S3b). This indicates that the combination of CNT and MIL-101(Fe) favors the separation of photogenerated electron-hole pairs and the rapid transport of photogenerated electrons throughout the conductive networks [11].

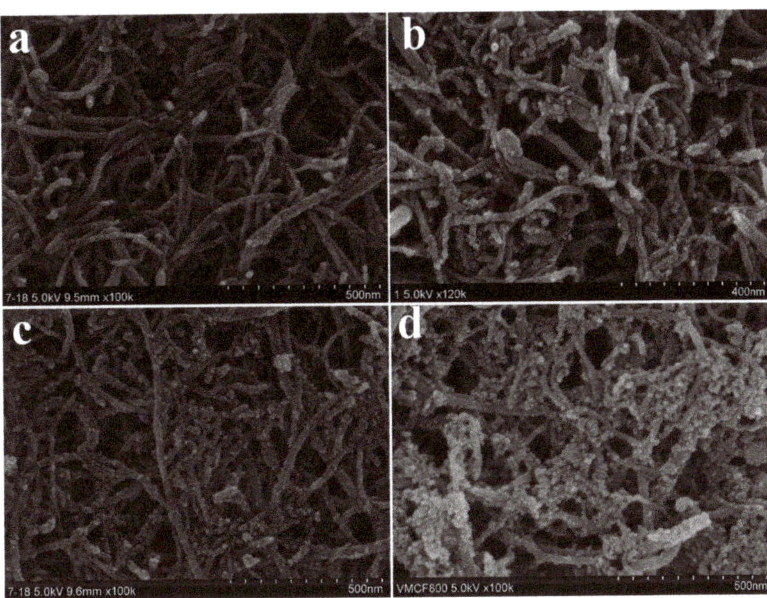

Figure 1. Characterization: comparison of the FESEM image of (**a**) CNT filter and (**b**–**d**) the presence of varying amounts of MIL-101(Fe).

2.2. *Photoelectrochemical Degradation of Tetracycline*

After successful preparation of the MIL-101(Fe)@CNT nanohybrid filter, we further examined its efficacy towards the photoelectrical activation of PMS to degrade antibiotic tetracycline in a continuous-flow configuration. As shown in Figure 2, only 13.9% TC removal within 20 min was achieved in the presence of PMS alone due to its limited oxidative ability [18].

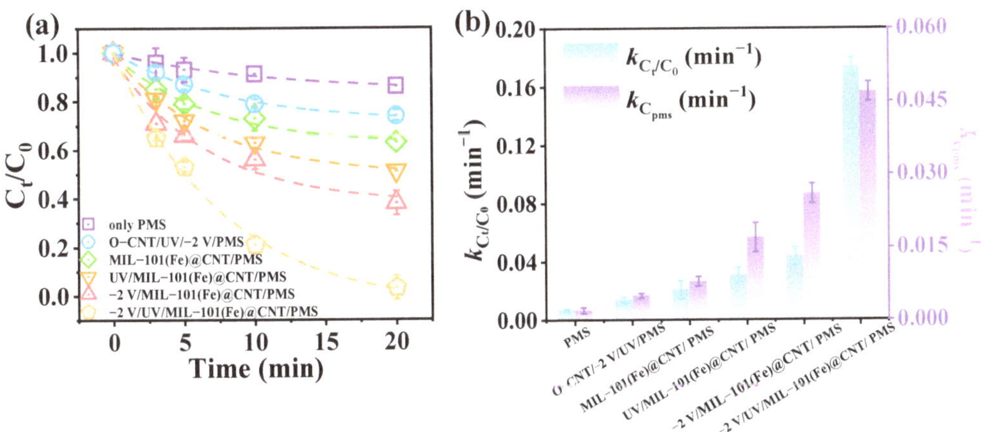

Figure 2. Photoelectrochemical degradation of TC in different systems: (**a**) TC degradation and (**b**) corresponding rate constants and rate constants of residual PMS concentrations without TC addition in different systems. Experimental conditions: MIL-101(Fe)@CNT = 3 mM, $[TC]_0$ = 20 mg L^{-1}, applied voltage = −2 V, illumination voltage = 3.8 V, pH = 6.0.

Since the catalytic reaction was conducted after the sorption saturation of the CNT filter, any contributions from physical adsorption of the filter can be excluded. A 25.9% TC removal was observed using a CNT-alone filter when exerting UV irradiation and electric field. This indicates that the PMS activation by UV and electric field is not very effective. Further combining the CNT filter with the MIL-101(Fe) catalyst, i.e., the MIL-101(Fe)@CNT nanohybrid filter, only leads to a TC removal efficiency of 36.8% without applying UV irradiation and electric field, indicating that the catalytic activity of MIL-101(Fe) towards PMS activation is limited as well, possibly due to the short residence time within the filter [19,20]. Moreover, the composite (128.1 mg L^{-1}) has a relatively high specific area compared to CNT (99.6 mg L^{-1}), which can be ascribed to the introduction of MIL-101(Fe) and the formation of a densely covered three-dimensional network. The TC removal efficiency further increased to 49.2% or 62.1%, respectively, by applying UV irradiation or electric field to the MIL-101(Fe)@CNT nanohybrid filter with corresponding pseudo-first-order rate constants (k) of 0.031 min^{-1} and 0.044 min^{-1} (Table S1). It is of note that an evident photo-electric synergistic effect was identified. When simultaneously applying UV irradiation and electric field, a complete TC removal was obtained using the nanohybrid filter within the same reaction period. The k of 0.174 min^{-1} was 5.6 times and 3.95 times higher than that only applying UV irradiation and electric field, respectively. In addition, PMS self-decomposition rate using the MIL-101(Fe)@CNT nanohybrid filter in different catalytic systems was also determined. As shown in Figure S4, the residual PMS in the photoelectrical catalytic, photocatalytic, electrocatalytic and PMS-alone systems was 0.35 mM, 0.56 mM, 0.71 mM and 0.85 mM, respectively, with a corresponding decay rate constant of 0.047 min^{-1}, 0.026 min^{-1}, 0.017 min^{-1} and 0.008 min^{-1}. As a comparison, the residual PMS and decay rate constants using a CNT-alone filter in a photocatalytic system were 0.90 mM and 0.014 min^{-1}, respectively. All these encouraging results quantitatively exemplified the advantages of combing UV irradiation, electric field, MIL-101(Fe) and CNT towards effective cleavage of the O–O bond of PMS and efficient TC removal [21].

2.3. Mechanism Insights

EPR and quenching experiments were conducted to examine the responsible reactive species within the proposed photoelectrochemical filtration system. As shown in Figure 3a and b, in the PMS-alone system, only very weak signals associated with the quartet DMPO–$HO^{\bullet}$ adduct (1:2:2:1) and the triplet TEMP–$^1O_2$ adduct (1:1:1) were identified [22,23]. This suggests that reactive oxide species (ROS) are hardly generated from the self-decomposition of PMS. In other cases, characteristic peaks of DMPO–$HO^{\bullet}$ and TEMP–$^1O_2$ adducts were detected, and the strongest signals were obtained in the photoelectrochemical filtration system. The characteristic signals of the DMPO–$SO_4^{\bullet-}$ adduct was only detected in the presence of the MIL-101(Fe)@CNT nanohybrid filter. This phenomenon could be explained by the unique Fe^{3+}/Fe^{2+} interconversion property of MIL-101(Fe) [24]. For example, Bi et al. [25] prepared graphitic carbon-nitride-functionalized MIL-101(Fe) that removed 100% of tetracycline hydrochloride within 40 min, which was attributed to the generation of $SO_4^{\bullet-}$ and $HO^{\bullet}$ active species assisted by visible light coupled with the composite.

Furthermore, quenching experiments were further employed to identify the dominant reactive species involved in the photoelectrochemical filtration system. Methanol can selectively quench with both $SO_4^{\bullet-}$ ($k = 2.5 \times 10^7$ M^{-1} s^{-1}) and $HO^{\bullet}$ ($k = 9.7 \times 10^8$ M^{-1} s^{-1}), whereas TBA is more selective for $HO^{\bullet}$ ($k = 3.8 \times 10^8$ M^{-1} s^{-1}) than $SO_4^{\bullet-}$ ($k = 4.0 \times 10^5$ M^{-1} s^{-1}) and L-histidine was used to quench the $^1O_2$ [6,19]. As displayed in Figure 3c, the TC removal efficiency decreased from 100% ($k = 0.165$ min^{-1}) to 46.7% ($k = 0.028$ min^{-1}) with the increase in spiked methanol from 0 to 100 mM (Table S2). Further increasing the methanol concentration failed to further inhibit the TC removal, suggesting a limited amount of the radical species. A similar phenomenon was also observed by spiking L-histidine as the $^1O_2$ scavenger (Figure 3d). These observations inferred the collective contribution from both radical and nonradical pathways towards TC removal. Based on the quenching analysis, the specific contribution from $SO_4^{\bullet-}$, $HO^{\bullet}$ and $^1O_2$ accounts for 16.6%, 36.7% and 46.7%, respectively (Figure S5a). In addition, to further confirm the role of $^1O_2$, the TC degradation experiment was repeated in the deuterated solvent, since the lifetime of $^1O_2$ in D_2O (20–32 μs) is one order of magnitude longer than that in H_2O (2 μs) [19]. As shown in Figure S5b, at a TC concentration of 40 mg L^{-1}, the degradation efficiency in D_2O (79.2%, k = 0.071 min^{-1}) was much higher than in H_2O (57.1%, k = 0.037 min^{-1}), again suggesting that $^1O_2$ was the dominant reactive species in the proposed system.

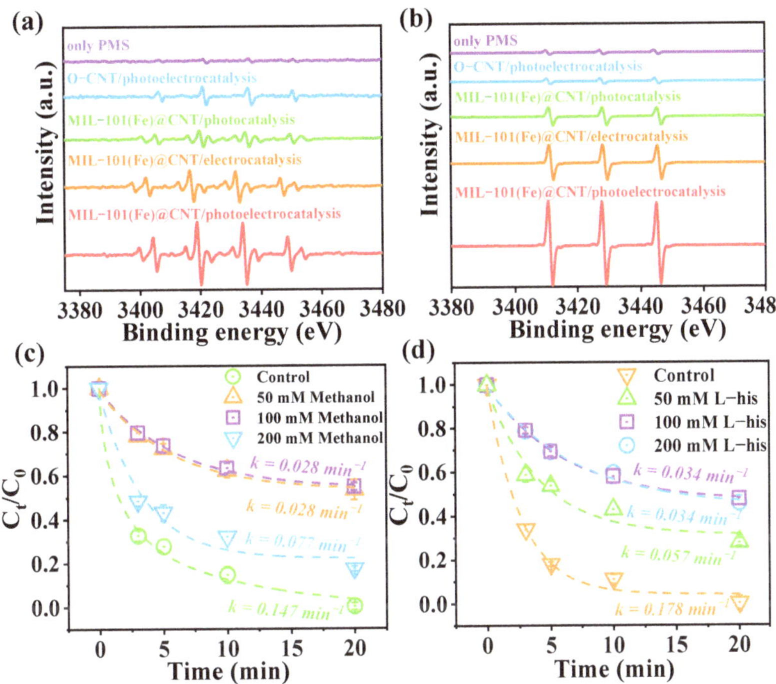

Figure 3. Identification of active species: EPR spectra for the presence of (**a**) DMPO and (**b**) TEMP, (**c**,**d**) and effect of different capture agents on TC removal. Experimental conditions: MIL-101(Fe)@CNT = 3 mM, $[TC]_0$ = 20 mg L^{-1}, applied voltage = −2 V, illumination voltage = 3.8 V, pH = 6.0.

A plausible working mechanism of the proposed photoelectrical filtration system was proposed as follows: Firstly, under UV irradiation, photogenerated electrons (e^-) and photogenerated holes (h^+) were generated at the solid/liquid interface of the MIL-101(Fe)@CNT nanohybrid filter (Equation (1)) [11]. Thus, Fe^{3+}–MOF reacts with e^- to form Fe^{2+}–MOF (Equation (2)), and the highly aggressive h^+ (E^0 =2.67 V vs. NHE) reacts with TC molecules to achieve degradation or reacts with H_2O to generate $HO^\bullet$ (E^0 =2.8 V vs. NHE) [11,19]. The applied external electric field facilitates the effective separation of e^- and h^+ and decreases their recombination rate. Meanwhile, PMS can also generate active species with strong oxidation under the action of UV light, such as $SO_4^{\bullet-}$ (E^0 = 2.5–3.1 V vs. SHE) and $HO^\bullet$ (Equation (3)). The applied electric field can not only promote the cycling of Fe^{3+}/Fe^{2+}, but also promote the decomposition of PMS to generate reactive radicals (Equation (4)) [26,27]. Subsequently, these in situformed Fe^{2+} species can activate the cleavage of the PMS O–O bond to generate $SO_4^{\bullet-}$ (Equation (5)). In addition, $SO_4^{\bullet-}$ can be further transformed into $HO^\bullet$ radicals under certain conditions (Equation (6)) [28] and participate in the oxidative degradation of TC. On the other hand, the nonradical route also plays an essential role in TC degradation. The generation of $^1O_2$ can either be generated by the reaction of SO_5^- with FeMOF

(Equations (7) and (8)) [22] or by receiving electrons from Fe^{2+} through dissolved O_2 to generate O_2^- and then $^1O_2$ (Equations (9) and (10)).

Moreover, it has been reported that the carbonyl groups (C=O) of CNT may also facilitate the PMS decomposition to generate $^1O_2$ [16]. The XPS results (Figure S6a and Table S3) suggest that the C=O content of the nanohybrid filter decreases by 9% after the catalytic reaction, while the C–O content increases by 11%. This indicates that these electronrich C=O groups may be involved in the cleavage of the PMS O–O bonds by sacrificing electrons. It has been reported that glutaraldehyde surface modification significantly increased the surface density of carbonyl groups on carbon nanotubes and promoted peroxydisulfate generation of $^1O_2$ [29]. The XPS spectra also suggested a decreased Fe^{2+}/Fe^{3+} ratio (by 10%) on a used nanohybrid filter after the photoelectrocatalytic reaction, confirming that the redox conversion of Fe^{3+}/Fe^{2+} acts as an electron transfer mediator during the PMS activation process (Figure S6b and Table S4). Finally, the effective TC degradation was further verified by a TOC mineralization of 51.7% over a 20 min reaction (Equation (11)).

$$\equiv\text{Fe-MOF} \rightarrow e^- + h^+ \quad (1)$$

$$\equiv\text{Fe}^{3+}\text{-MOF} + e^- \rightarrow \equiv\text{Fe}^{2+}\text{-MOF} \quad (2)$$

$$HSO_5^- \xrightarrow{h\nu} SO_4^{\bullet-} + HO^\bullet \quad (3)$$

$$HSO_5^- + e^- \rightarrow SO_4^{2-} + HO^\bullet \quad (4)$$

$$\equiv\text{Fe}^{2+}\text{-MOF} + HSO_5^- \rightarrow \equiv\text{Fe}^{3+}\text{-MOF} + SO_4^{\bullet-} + OH^- \quad (5)$$

$$SO_4^{\bullet-} + OH^- \rightarrow SO_4^{2-} + HO^\bullet \quad (6)$$

$$\equiv\text{Fe}^{3+}\text{-MOF} + HSO_5^- \rightarrow \equiv\text{Fe}^{2+}\text{-MOF} + SO_5^- + H^+ \quad (7)$$

$$\equiv\text{Fe}^{2+}\text{-MOF} + SO_5^- \rightarrow \equiv\text{Fe}^{3+}\text{-MOF} + SO_4^{2-} + 0.5\,^1O_2 \quad (8)$$

$$\equiv\text{Fe}^{2+}\text{-MOF} + O_2 \rightarrow \equiv\text{Fe}^{3+}\text{-MOF} + O_2^- \quad (9)$$

$$\equiv\text{Fe}^{3+}\text{-MOF} + O_2^- + e^- \rightarrow \equiv\text{Fe}^{2+}\text{-MOF} + ^1O_2 \quad (10)$$

$$HO^\bullet/SO_4^{\bullet-}/^1O_2/h^+ + TC \rightarrow CO_2 + H_2O \quad (11)$$

2.4. *Operational Parameters Optimization*

2.4.1. Impacts of MIL-101(Fe) Loading and Applied Voltage

As displayed in Figure 4a, as the precursor concentration increased from 1.5 to 3 mM, the TC degradation efficiency and the corresponding k increased from 68.1% to 97.5% and 0.053 min^{-1} to 0.179 min^{-1}, respectively (Table S5). This suggested that a higher MIL-101(Fe) loading favors the acceleration of the TC degradation kinetics by providing abundant reactive sites for the catalytic reaction. However, further increasing the precursor concentration to 4.5 mM deteriorated the TC removal kinetics, owing to the agglomeration of the as-formed MIL-101(Fe) particles as well as the inevitable burying of the surface-active sites on the nanohybrid filter (Figure 1d) [11]. An optimal loading amount of 3 mM was then used in subsequent investigations.

The applied voltage is critical for the cycling of Fe^{3+}/Fe^{2+} pairs of the Fe–MOF [30]. The TC degradation efficiency was 58.3% (k = 0.039 min^{-1}), 78.2% (k = 0.066 min^{-1}) and 98.3% (k = 0.159 min^{-1}) at applied voltages of −0.5 V, −1.5 V and −2 V, respectively (Figure 4b). The above results indicated that a more negative potential is favorable to boost the TC degradation efficiency [12]. Nevertheless, further decreasing the applied voltage to −2.5 V only led to a decreased TC removal efficiency of 83.3% (k = 0.082 min^{-1}), possibly due to the occurrence of other side reactions (e.g., hydrogen generation reaction) that decrease the current efficiencies [31].

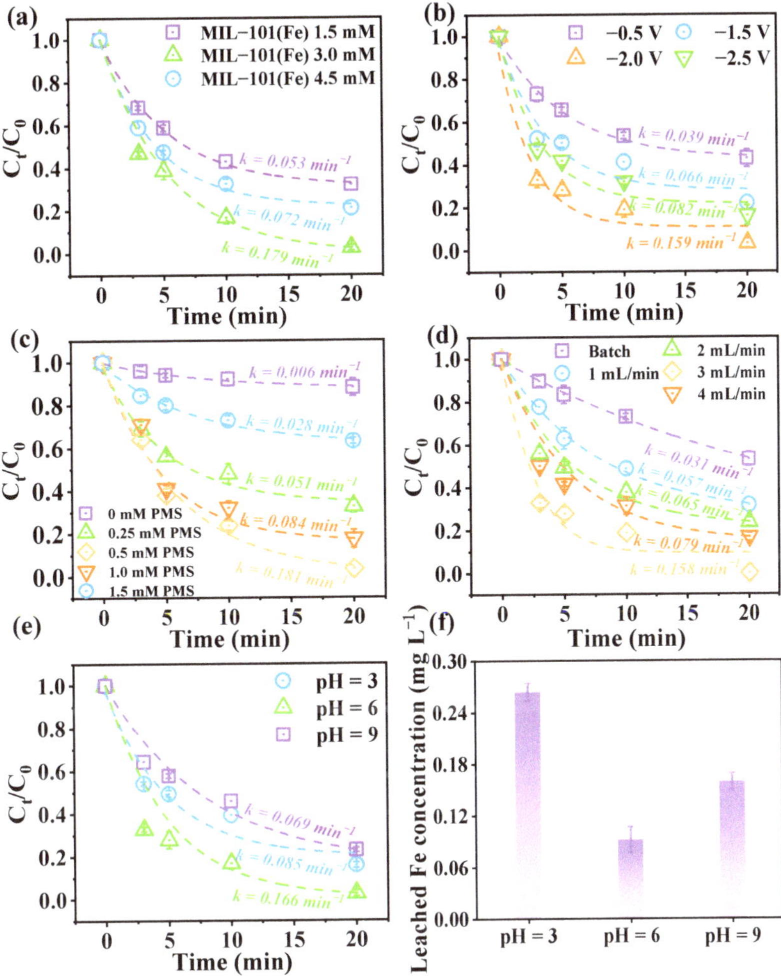

Figure 4. Impact of operational parameters on the TC degradation kinetics: effects of (**a**) MIL-101(Fe) loading, (**b**) applied voltage, (**c**) PMS concentration, (**d**) flow rate and (**e**) pH on the TC degradation and (**f**) concentration of leachable Fe at different pHs. Experimental conditions: MIL-101(Fe)@CNT = 3 mM, $[TC]_0$ = 20 mg L^{-1}, applied voltage = −2 V, illumination voltage = 3.8 V, pH = 6.0.

2.4.2. Impacts of PMS Concentration and Flow Rate

In the photoelectrochemical filtration system, PMS acts not only as an electrolyte, but also as a precursor for the reactive species. We, hence, evaluated the effect of PMS dosage on the degradation of TC. As shown in Figure 4c, the TC removal efficiency was only 11.9% in the absence of PMS. The TC removal efficiency (67.5% to 97.1%) and k value (0.028 min^{-1} to 0.181 min^{-1}) were obviously enhanced with increasing PMS concentration from 0.25 to 1.0 mM, which was attributed to more active species being generated. However, further increasing the PMS concentration to 1.5 mM contributed negatively to the TC degradation kinetics (83.1%, k = 0.084 min^{-1}). This could be associated with active species derived from the catalytic reaction being quenched at higher concentrations of PMS [26].

The effects of flow rate on TC degradation efficiency are shown in Figure 4d. The TC degradation efficiencies were 69.4%, 76.6%, 98.2%, 83.3% and 46.9%, and corresponding k

values were 0.057 min^{-1}, 0.066 min^{-1}, 0.158 min^{-1}, 0.079 min^{-1} and 0.031 min^{-1} at flow rates of 1 mL min^{-1}, 2 mL min^{-1}, 3 mL min^{-1}, 4 mL min^{-1} and batch mode, respectively. Such performance enhancement was attributed to the convection−enhanced mass transport, whereas the batch system relied on diffusion [12]. In addition, the TC degradation efficiency decreased by 14.9% at a higher flow rate in the circulation mode (4 mL min^{-1}), mainly because of the reduced residence time and contact between TC molecules and reactive species at the higher flow rates [13].

2.4.3. Impact of Solution pH

The solution pH poses an effect on the surface charge of the nanohybrid filter as well as the speciation of PMS. As displayed in Figure 4e, the highest removal efficiency of TC (98.8%, k = 0.166 min^{-1}) was obtained under a neutral pH of 6.0. Under acidic pH conditions, an 85.3% TC can still be obtained since the nanohybrid filter (pH_{zpc} = 3.5, Figure S7) became positively charged, which attracts the negatively charged HSO_5^- towards the filter surface to complete the catalytic reactions, while the evident performance decay under alkaline condition (77.6%, k = 0.069 min^{-1}) could be explained by the conversion of HSO_5^- to the weakly reactive SO_5^{2-} and the electrostatic repulsion effect (Equation (12)) [19]. In addition, the leachable Fe during the catalytic reaction was determined as a function of solution pH. As depicted in Figure 4f, the leached iron concentrations were 0.26 mg L^{-1}, 0.09 mg L^{-1} and 0.16 mg L^{-1} at pH values of 3, 6 and 9, respectively. These results indicated that the nanohybrid filter possesses relatively high pH tolerance.

$$HSO_5^- + OH^- \rightarrow SO_5^{2-} + H_2O \tag{12}$$

2.5. *System Stability Evaluation*

Several ubiquitous anions (e.g., Cl^-, NO_3^-, and HCO_3^-) may present in natural waters to negatively impact the system efficacy [32,33]. Thus, the impacts of inorganic constituents on the degradation of TC were investigated. As expected, the presence of Cl^- and NO_3^- posed a negligible influence on the system efficacy with >90% TC degradation (Figure S8a). However, HCO_3^- posed an evident inhibition on the TC degradation efficiency (72.8%) due to its buffering capability to maintain the reaction solution under basic [19]. Moreover, the efficacy of the nanohybrid filter towards the degradation of other refractory organic micropollutants was also evaluated. As displayed in Figure S8b, under similar operation conditions, the photoelectrochemical filtration system still showed excellent degradation performance toward methylene blue (99.8%), Congo red (99.7%) and *p*-nitrophenol (85.6%).

The efficacy of the nanohybrid filter was further examined by spiking TC into tap water, lake water and municipal WWTP effluent, which was considered to be much more complex compared with the ultrapure water used (Figure 5a and Table S6) [12,34].

The results showed that TC degradation efficiencies of 94.3%, 85.9% and 75.6% with corresponding k values of 0.137 min^{-1}, 0.090 min^{-1} and 0.058 min^{-1}, respectively, could be achieved in spiked tap water, lake water and municipal WWTP effluent. Moreover, the stability of the nanohybrid filter was evaluated in consecutive cycles toward TC degradation. The TC degradation efficiencies slightly dropped by 6.1%, 8.9%, 12.2% and 14.2% over five consecutive running cycles, which indicated a decent stability of the system (Figure 5b). These encouraging results indicate that the proposed technology with promising practical application prospects may provide a viable solution for water decontamination.

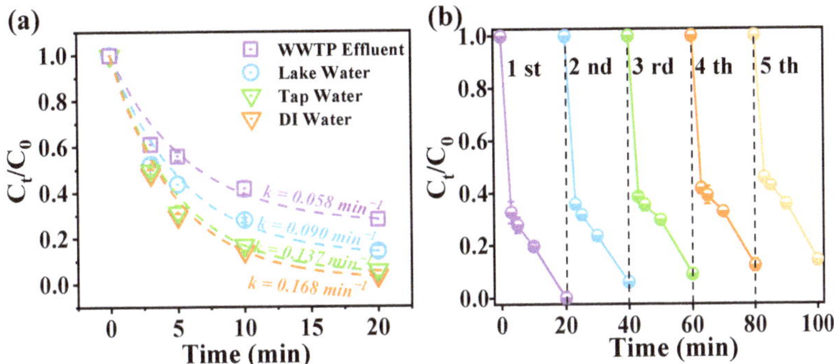

Figure 5. System stability evaluation: (**a**) The degradation efficiency of TC in different water matrixes. (**b**) Degradation efficiency of TC in different cycle times. Experimental conditions: MIL-101(Fe)@CNT = 3 mM, $[TC]_0$ = 20 mg L^{-1}, applied voltage = −2 V, illumination voltage = 3.8 V, pH = 6.0.

3. Materials and Methods

3.1. Chemicals and Materials

Potassium peroxymonosulfate (PMS, $HKSO_5 \cdot 0.5HKSO_4 \cdot 0.5K_2SO_4$, ≥98.0%), dimethylformamide (DMF, C_3H_7NO, ≥99.5%), iron chloride hexahydrate ($FeCl_3 \cdot 6H_2O$, ≥99.5%), 1,4–benzenedicarboxylic acid (BDC, $C_8H_6O_4$, ≥99.9%), tetracycline hydrochloride (TC, $C_{22}H_{25}ClN_2O_8$, ≥99.5%), Congo red ($C_{32}H_{22}N_6Na_2O_6S_2$, ≥98.0%), methylene blue ($C_{16}H_{18}ClN_3S$, ≥99.9%), *p*-nitrophenol ($C_6H_5NO_3$, ≥98.0%), ethanol (C_2H_5OH, ≥96.0%), methanol (CH_3OH, ≥98.0%), hydrochloric acid (HCl, ≥36–38%), sodium hydroxide (NaOH, ≥96.0%), sodium nitrate ($NaNO_3$, ≥99.9%), sodium sulfate (Na_2SO_4, ≥99.0%), sodium chloride (NaCl, ≥99.5%) and sodium bicarbonate ($NaHCO_3$, ≥99.5%) were obtained from Sinopharm Chemical Reagent Co., Ltd. (Shanghai, China). 2,2,6,6-tetramethyl-4-piperidinol (TEMP, $C_9H_{20}N_2$, ≥96.0%), 5,5-dimethyl-1-pyrroline-n-oxide (DMPO, $C_6H_{11}NO$, ≥97.0%), furfuryl alcohol (FFA, $C_5H_6O_2$, ≥98.0%), L-histidine ($C_6H_9N_3O_2$, ≥99.5%), tert-butyl alcohol (TBA, $C_4H_{10}O$, ≥98.0%) and deuterium oxide (D_2O, ≥99.0%) were purchased from Sigma-Aldrich (St. Louis, MO, USA). Acidified multi-walled carbon nanotubes (O–CNT) were provided by Times Nano Co., Ltd. (Chengdu, China). All aqueous solutions were prepared by using the ultrapure water produced from a Milli-Q Direct 8 purification system (Millipore, Billerica, MA, USA).

3.2. Synthesis of the Nanohybrid Filter

An MIL-101(Fe)@CNT hybrid filter was synthesized by using a reported solvothermal protocol with modifications [18]. Briefly, in a typical process, 24.3 mg (3 mM) $FeCl_3 \cdot 6H_2O$ and 25 mg CNT were first separately dissolved into 15 mL DMF, followed by sonication for 40 min. Afterwards, 7.5 mg BDC was mixed into 15 mL DMF and then added dropwise into the above solution. The as-obtained mixture was transferred to a 50 mL Teflon-lined autoclave and heated at 110 °C for 24 h. The product was loaded onto a polytetrafluoroethylene membrane by vacuum filtration and further purified with DMF and ethanol, and then dried at 60 °C under vacuum overnight to obtain the MIL-101(Fe)@CNT nanohybrid filter.

3.3. Characterization

The crystallinity and morphology of the filter samples were characterized, respectively, on a Rigaku D/max-2550 PC thin-film X-ray diffractometer (XRD, Rigaku, Japan) and a field emission scanning electron microscope (FESEM, Hitachi S-4800, Hitachi, Tokyo, Japan). The X-ray photoelectron spectroscopy (XPS) analysis was performed under high vacuum (1×10^{-9} Torr) using a Thermo Fisher Scientific ESCALAB 250Xi (Thermo Fisher Scientific, Waltham, MA, USA) spectrometer. The photochemical and electrochemical activity of the prepared filters were characterized on a CHI660E electrochemical workstation (Shanghai

Chenhua Co., Ltd., Shanghai, China) in a three-electrode system (i.e., MIL-101(Fe)@CNT working anode, Pt counter electrode and Ag/AgCl reference electrode). The preparation of MIL-101(Fe) electrode was referred to in a previous report [11]. Zeta potential was determined on a JS94H micro electrophoresis instrument (Shanghai Zhongchen, China). Reactive species were determined by using a Bruker EMX nano Bench-Top electron paramagnetic resonance (EPR) spectrometer (Bruker, Karlsruhe, Germany).

3.4. TC Degradation Experiments

A commercial filtration casing with photoelectrochemical modifications was employed for organic degradation experiments (Figure S9) [11,13]. All TC degradation experiments were performed after the filter adsorption saturation was reached to eliminate the contribution of physical adsorption. Two operational modes, i.e., recirculated filtration and batch, were conducted for comparison. For the batch mode, the MIL-101(Fe)@CNT nanohybrid filter and titanium sheet were used as a cathode and anode, respectively, and suspended in a beaker containing 50 mL of 20 mg L^{-1} TC and 1 mM PMS at 2 V. UV LED illumination was introduced by eight NSPU510CS UV LED lamps (NICHIA, Japan) installed on the top of the filtration apparatus. For the recirculation mode, the solution was passed through the filtration system and returned. The PMS concentration (0.25 to 1.5 mM), applied voltage (0.5 to 2 V), MIL-101(Fe) dose (1.5 to 4.5 mM), flow rate (1 to 4 mL min^{-1}) and solution pH (3 to 9) were evaluated and optimized.

The stability of the nanohybrid filter was assessed by performing five consecutive cycles under optimal conditions. A new cycle was initiated after washing the exhausted filter with NaOH solution (5 mM 100 mL) and abundant water solution until the effluent became neutral. The effluent (1 mL) was sampled at predetermined time intervals and immediately mixed with 300 μL methanol to quench any remaining radical species. Subsequently, the TC concentration was quantified by high-performance liquid chromatography (HPLC) [13]. The PMS residual concentration was measured by potassium iodide method [27]. To examine the impact of common coexisting anions on the TC degradation kinetics, 5 mM of nitrate, carbonate or chloride were spiked into TC solution before the recirculated filtration. The practical application potential of the system was verified using the TC-spiked tap water, lake water and municipal wastewater treatment plant (WWTP) effluent as well as an array of other refractory organic contaminants. Detailed characteristics of different water matrices are available in Table S6. The concentration of Congo red, methylene blue and *p*-nitrophenol were determined via a UV-2600 Shimadzu ultraviolet-visible spectrophotometry (Japan) at wavelengths of 497 nm, 665 nm and 317 nm, respectively.

4. Conclusions

In conclusion, we reported and demonstrated a photoelectrochemical filtration system that enables the effective decontamination of antibiotic tetracycline from water. The key to this technology was a nanohybrid filter consisting of electrically conductive CNT and photo-responsive MIL-101(Fe). Various advanced characterizations collectively provided detailed compositional and morphological information on the filter. Results suggested that both radical and nonradical pathways contributed to the effective tetracycline degradation. Moreover, the plug-flow configuration facilitated a convection-enhanced mass transport, further promoting the organic degradation kinetics. Such excellent system efficacy can be maintained across a wide range of solution pH as well as complex water matrices. Overall, the outcomes of this study provide a viable strategy toward water remediation by integrating the state-of-the-art photoelectrochemistry, membrane separation, nanotechnology, and advanced oxidation technologies.

Supplementary Materials: The following are available online at: https://www.mdpi.com/article/10.3390/catal12040416/s1. Figure S1: (a) XRD pattern and (b) TGA curves of 3 mM MIL-101(Fe)@CNT and O–CNT; Figure S2: Energy dispersive spectra of the MIL-101(Fe)@CNT filter. Figure S3: (a) Cyclic voltammetry curves of 3 mM MIL-101(Fe)@CNT with or without UV irradiation. (b) Nyquist diagrams of 3 mM MIL-101(Fe)@CNT and MIL-101(Fe). Environmental conditions: $[Na_2SO_4]_0$ = 50 mM; Figure S4: Residual concentration of PMS in different systems. Experimental conditions: MIL-101(Fe)@CNT = 3 mM, $[TC]_0$ = 20 mg L^{-1}, applied voltage = −2 V, illumination voltage = 3.8 V, pH = 6.0; Figure S5: Effects of (a) adding TBA and (b) reaction solvent (H_2O and D_2O) on the TC degradation. Experimental conditions: MIL-101(Fe)@CNT = 3 mM, $[TC]_0$ = 20 mg L^{-1}, applied voltage = −2 V, illumination voltage = 3.8 V, pH = 6.0; Figure S6: XPS spectra of MIL-101(Fe)@CNT: (a) O 1s and (b) Fe 2p before and after reaction; Figure S7: Zeta spectra of 3 mM MIL-101(Fe)@CNT and O–CNT; Figure S8: (a) Effect of coexisting ions on TC degradation and (b) degradation efficiency of four typical organic compounds by MIL-101(Fe)@CNT filter in circulation mode. Experimental conditions: [Congo red]$_0$ = [Methylene blue]$_0$ = 20 mg L^{-1}, [*p*-nitrophenol]$_0$ = 10 mg L^{-1}, MIL-101(Fe)@CNT = 3 mM, $[TC]_0$ = 20 mg L^{-1}, applied voltage = −2 V, illumination voltage = 3.8 V, pH = 6.0; Figure S9: Schematic illustration of photoelectric reactor device; Table S1: Corresponding *k* value to TC degradation in different systems according to the pseudo first order kinetic model; Table S2: Corresponding *k* value to TC degradation in different quenchers according to the pseudo first order kinetic model; Table S3: XPS results of O1s; Table S4: XPS results of Fe 2p; Table S5: Corresponding *k* value to TC degradation of different operational parameters according to the pseudo first order kinetic model; Table S6: Characteristics of different water samples and corresponding *k* value to TC degradation.

Author Contributions: W.Z. (Wenchang Zhao) conducted data curation and wrote the original draft. Y.D. performed the experiments. W.Z. (Wentian Zheng) performed formal analysis. Y.L. supervised the study, provided resources, and reviewed and edited the paper. All authors have read and agreed to the published version of the manuscript.

Funding: This work was supported by the Natural Science Foundation of China (No. 52170068).

Data Availability Statement: All relevant data are available from the corresponding author on request.

Conflicts of Interest: The authors declare no conflict of interest.

References

1. Cristian, M.; Antonio, V.; José Luis, V.; Leire, R. Hybrid organic-inorganic membranes for photocatalytic water remediation. *Catalysts* **2022**, *12*, 180. [CrossRef]
2. Vadivel, D.; Sturini, M.; Speltini, A.; Dondi, D. Tungsten catalysts for visible light driven ofloxacin photocatalytic degradation and hydrogen production. *Catalysts* **2022**, *12*, 310. [CrossRef]
3. Azar, F.; Ivana, J.; Nivetha, S.; Leslie, B.; Robert, L.; Norman, Z.; Mark, S.; Maricor, A. Effect of background water matrices on pharmaceutical and personal care product removal by UV-LED/TiO_2. *Catalysts* **2021**, *11*, 576. [CrossRef]
4. Xu, Y.; Yu, X.; Xu, B.; Peng, D.; Guo, X. Sorption of pharmaceuticals and personal care products on soil and soil components: Influencing factors and mechanisms. *Sci. Total Environ.* **2020**, *753*, 141891. [CrossRef] [PubMed]
5. Zhang, C.; Chen, X.; Ho, S. Wastewater treatment nexus: Carbon nanomaterials towards potential aquatic ecotoxicity. *J. Hazard. Mater.* **2021**, *417*, 125959. [CrossRef] [PubMed]
6. Zheng, W.T.; Liu, Y.B.; Liu, W.; Ji, H.D.; Li, F.; Shen, C.S.; Fang, X.F.; Li, X.; Duan, X.G. A novel electrocatalytic filtration system with carbon nanotube supported nanoscale zerovalent copper toward ultrafast oxidation of organic pollutants. *Water Res.* **2021**, *194*, 116961. [CrossRef] [PubMed]
7. Zhou, C.; Zhu, L.; Deng, L.; Zhang, H.; Zeng, H.; Shi, Z. Efficient activation of peroxymonosulfate on CuS@MIL-101(Fe) spheres featured with abundant sulfur vacancies for coumarin degradation: Performance and mechanisms. *Sep. Purif. Technol.* **2021**, *276*, 119404. [CrossRef]
8. Wu, D.; Xia, K.; Fang, C.; Chen, X.; Ye, Y. Rapid removal of azophloxine via catalytic degradation by a novel heterogeneous catalyst under visible light. *Catalysts* **2020**, *10*, 138. [CrossRef]
9. Liu, Y.B.; Liu, F.Q.; Ding, N.; Hu, X.H.; Shen, C.S.; Li, F.; Huang, M.H.; Wang, Z.W.; Sand, W.; Wang, C.C. Recent advances on electroactive CNT-based membranes for environmental applications: The perfect match of electrochemistry and membrane separation. *Chin. Chem. Lett.* **2020**, *31*, 2539–2548. [CrossRef]
10. Ren, Y.L.; Liu, Y.B.; Liu, F.Q.; Li, F.; Shen, C.S.; Wu, Z.C. Extremely efficient electro-Fenton-like Sb(III) detoxification using nanoscale Ti-Ce binary oxide: An effective design to boost catalytic activity via non-radical pathway. *Chin. Chem. Lett.* **2021**, *32*, 2519–2523. [CrossRef]

11. Li, M.H.; Liu, Y.B.; Shen, C.S.; Li, F.; Wang, C.C.; Huang, M.H.; Yang, B.; Wang, Z.W.; Yang, J.M.; Sand, W. One-step Sb(III) decontamination using a bifunctional photoelectrochemical filter. *J. Hazard. Mater.* **2020**, *389*, 121840. [CrossRef] [PubMed]
12. Liu, Y.B.; Gao, G.D.; Vecitis, C.D. Prospects of an electroactive carbon nanotube membrane toward environmental applications. *Acc. Chem. Res.* **2020**, *53*, 2892–2902. [CrossRef] [PubMed]
13. Guo, D.L.; Liu, Y.B.; Ji, H.D.; Wang, C.C.; Chen, B.; Shen, C.S.; Li, F.; Wang, Y.X.; Lu, P.; Liu, W. Silicate-enhanced heterogeneous flow-through electro-Fenton system using iron oxides under nanoconfinement. *Environ. Sci. Technol.* **2021**, *55*, 4045–4053. [CrossRef] [PubMed]
14. Debashis, R.; Sudarsan, N.; Sirshendu, D. Mechanistic investigation of photocatalytic degradation of Bisphenol-A using MIL-88A(Fe)/MoS_2 Z-scheme heterojunction composite assisted peroxymonosulfate activation. *Chem. Eng. J.* **2021**, *428*, 131028. [CrossRef]
15. Zhang, S.; Zhao, Y.; Yang, K.; Liu, W.; Xu, Y.; Liang, P.; Zhang, X.; Huang, X. Versatile zero valent iron applied in anaerobic membrane reactor for treating municipal wastewater: Performances and mechanisms. *Chem. Eng. J.* **2020**, *382*, 123000. [CrossRef]
16. Liu, B.; Song, W.; Wu, H.; Liu, Z.; Teng, Y.; Sun, Y.; Xu, Y.; Zheng, H. Degradation of norfloxacin with peroxymonosulfate activated by nanoconfinement Co_3O_4@CNT nanocomposite. *Chem. Eng. J.* **2020**, *398*, 125498. [CrossRef]
17. Guo, D.L.; You, S.J.; Li, F.; Liu, Y.B. Engineering carbon nanocatalysts towards efficient degradation of emerging organic contaminants via persulfate activation: A review. *Chin. Chem. Lett.* **2022**, *33*, 1–10. [CrossRef]
18. Xiao, Z.; Feng, X.; Shi, H.; Zhou, B.; Wang, W.; Ren, N. Why the cooperation of radical and non-radical pathways in PMS system leads to a higher efficiency than a single pathway in tetracycline degradation. *J. Hazard. Mater.* **2021**, *424*, 127247. [CrossRef]
19. Jin, L.M.; You, S.J.; Duan, X.G.; Yao, Y.; Yang, J.M.; Liu, Y.B. Peroxymonosulfate activation by Fe_3O_4-MnO_2/CNT nanohybrid electroactive filter towards ultrafast micropollutants decontamination: Performance and mechanism. *J. Hazard. Mater.* **2022**, *423*, 127111. [CrossRef]
20. Wang, C.; Du, J.; Liang, Z.; Liang, J.; Zhao, Z.; Cui, F.; Shi, W. High efficiency oxidation of fluoroquinolones by the synergistic activation of peroxymonosulfate via vacuum ultraviolet and ferrous iron. *J. Hazard. Mater.* **2021**, *422*, 126884. [CrossRef]
21. Li, N.; Tang, S.; Rao, Y.; Qi, J.; Zhang, Q.; Yuan, D. Peroxymonosulfate enhanced antibiotic removal and synchronous electricity generation in a photocatalytic fuel cell. *Electrochim. Acta* **2018**, *298*, 56–69. [CrossRef]
22. Xie, X.; Cao, J.; Xiang, Y.; Xie, R.; Suo, Z.; Ao, Z.; Yang, X.; Huang, H. Accelerated iron cycle inducing molecular oxygen activation for deep oxidation of aromatic VOCs in MoS_2 co-catalytic Fe^{3+}/PMS system. *Appl. Catal. B Environ.* **2022**, *309*, 121235. [CrossRef]
23. Yang, Q.; Yan, Y.; Yang, X.; Liao, G.; He, J.; Wang, D. The effect of complexation with metal ions on tetracycline degradation by Fe^{2+}/Fe^{3+} and Ru^{3+} activated peroxymonosulfate. *Chem. Eng. J.* **2021**, *429*, 132178. [CrossRef]
24. Jiang, Y.; Wang, Z.; Huang, J.; Yan, F.; Du, Y.; He, C.; Liu, Y.; Yao, G.; Lai, B. A singlet oxygen dominated process through photocatalysis of CuS-modified MIL-101(Fe) assisted by peroxymonosulfate for efficient water disinfection. *Chem. Eng. J.* **2022**, *439*, 135788. [CrossRef]
25. Bi, H.; Liu, C.; Li, J.; Tan, J. Insights into the visible-light-driving MIL-101 (Fe)/g-C_3N_4 materials-activated persulfate system for efficient hydrochloride water purification. *J. Solid State Chem.* **2022**, *306*, 122741. [CrossRef]
26. Zheng, W.T.; You, S.J.; Yao, Y.; Jin, L.M.; Liu, Y.B. Development of atomic hydrogen-mediated electrocatalytic filtration system for peroxymonosulfate activation towards ultrafast degradation of emerging organic contaminants. *Appl. Catal. B Environ.* **2021**, *298*, 120593. [CrossRef]
27. Zuo, S.; Xia, D.; Guan, Z.; Yang, F.; Zhang, B.; Xu, H.; Huang, M.; Guo, X.; Li, D. The polarized electric field on Fe_2O_3/g-C_3N_4 for efficient peroxymonosulfate activation: A synergy of $^1O_2$, electron transfer and pollutant oxidation. *Sep. Purif. Technol.* **2021**, *269*, 118717. [CrossRef]
28. Li, J.; Liu, Q.; Gou, G.; Kang, S.; Tan, X.; Tan, B.; Li, L.; Li, N.; Liu, C.; Lai, B. New insight into the mechanism of peroxymonosufate activation by Fe_3S_4: Radical and non-radical oxidation. *Sep. Purif. Technol.* **2022**, *286*, 120471. [CrossRef]
29. Cheng, X.; Guo, H.G.; Zhang, Y.L.; Wu, X.; Liu, Y. Non-photochemical production of singlet oxygen via activation of persulfate by carbon nanotubes. *Water Res.* **2017**, *113*, 80–88. [CrossRef]
30. Zhang, Y.; Chu, W. G–C_3N_4 induced acceleration of Fe^{3+}/Fe^{2+} cycles for enhancing metronidazole degradation in Fe^{3+}/peroxymonosulfate process under visible light. *Chemosphere* **2022**, *293*, 133611. [CrossRef]
31. Dai, Y.L.; Yao, Y.; Li, M.H.; Fang, X.F.; Shen, C.S.; Li, F.; Liu, Y.B. Carbon nanotube filter functionalized with MIL-101(Fe) for enhanced flow-through electro-Fenton. *Environ. Res.* **2022**, *204*, 112117. [CrossRef] [PubMed]
32. Chen, X.; Han, Y.; Gao, P.; Li, H. New insight into the mechanism of electro-assisted pyrite minerals activation of peroxymonosulfate: Synergistic effects, activation sites and electron transfer. *Sep. Purif. Technol.* **2021**, *274*, 118817. [CrossRef]
33. Wang, W.; Wu, Q.; Huang, N.; Wang, T.; Hu, H. Synergistic effect between UV and chlorine (UV/chlorine) on the degradation of carbamazepine: Influence factors and radical species. *Water Res.* **2016**, *98*, 190–198. [CrossRef] [PubMed]
34. Zhong, S.; Yang, B.; Xiong, Q.; Cai, W.; Lan, Z.; Ying, G. Hydrolytic transformation mechanism of tetracycline antibiotics: Reaction kinetics, products identification and determination in WWTPs. *Ecotoxicol. Environ. Saf.* **2021**, *229*, 113063. [CrossRef] [PubMed]

Editorial

New Insights into Novel Catalysts for Treatment of Pollutants in Wastewater

Hao Xu [1,*] and Yanbiao Liu [2]

1 Department of Environmental Science & Engineering, Xi'an Jiaotong University, Xi'an 710049, China
2 College of Environmental Science and Engineering, Textile Pollution Controlling Engineering Center of the Ministry of Ecology and Environment, Donghua University, Shanghai 201620, China; yanbiaoliu@dhu.edu.cn
* Correspondence: xuhao@xjtu.edu.cn

Water scarcity has become a worldwide problem. Wastewater treatment and reuse has become an effective way to expand water resources. Among many treatment methods, the wastewater treatment method using catalysts as media is unique. Obviously, catalysts are the core of these treatment methods, and their properties directly determine the treatment effect and cost. In recent years, with the development of science and technology, a variety of new catalysts has emerged in an endless stream, and the related fields have become the focus of current scientific research. To this end, we are organizing a Special Issue: New Insights into Novel Catalysts for Treatment of Pollutants in Wastewater, focusing on the preparation, modification, and application of novel catalysts for catalytic wastewater treatment. Below, we will introduce each of the 13 works published in this Special Issue.

To achieve sustainable, low-carbon development, it is imperative to explore water treatment technologies in a carbon-neutral model. Because of the advantages of high efficiency, low consumption, and a lack of secondary pollution, electrocatalytic oxidation technology has attracted increasing attention for tackling the challenges of organic wastewater treatment. The performance of an electrocatalytic oxidation system depends mainly on the properties of the electrode materials. Compared with the instability of graphite electrodes, the high expenditure of noble metal electrodes and boron doped diamond electrodes, and the hidden dangers of titanium-based metal oxide electrodes, a titanium sub-oxide material has been characterized as an ideal choice of anode material due to its unique crystal and electronic structure, including high conductivity, decent catalytic activity, intense physical and chemical stability, corrosion resistance, low cost, long service life, etc. Guo's paper [1] systematically reviews the electrode preparation technology of the Magnéli phase titanium sub-oxide and its research progress in the advanced electrochemical oxidation treatment of organic wastewater in recent years, with technical difficulties highlighted. Future research directions are further proposed in terms of process optimization, material modification, and application expansion. It is worth noting that Magnéli phase titanium sub-oxides have played very important roles in organic degradation. There is no doubt that titanium sub-oxides will become indispensable materials in the future.

Lead dioxide electrodes are also typical electrocatalytic anode materials. In Kang's paper [2], active granule (WC/Co_3O_4)-doping $Ti/Sb\text{-}SnO_2/PbO_2$ electrodes were successfully synthesized by composite electrodeposition. The as-prepared electrodes were systematically characterized by scanning electron microscopy (SEM), energy-dispersive X-ray spectroscopy (EDS), X-ray diffraction (XRD), X-ray photoelectron spectroscopy (XPS), electrochemical performance, zeta potential, and accelerated lifetime. It was found that the doping of active granules (WC/Co_3O_4) can reduce the average grain size and increase the number of active sites on the electrode surface. Moreover, it can improve the proportion of surface oxygen vacancies and non-stoichiometric PbO_2, resulting in outstanding conductivity, which can improve the electron transfer and catalytic activity of the electrode. Electrochemical measurements implied that $Ti/Sb\text{-}SnO_2/Co_3O_4\text{-}PbO_2$ and

Citation: Xu, H.; Liu, Y. New Insights into Novel Catalysts for Treatment of Pollutants in Wastewater. *Catalysts* **2023**, *13*, 840. https://doi.org/10.3390/catal13050840

Received: 27 April 2023
Accepted: 4 May 2023
Published: 5 May 2023

$Ti/Sb\text{-}SnO_2/WC\text{-}Co_3O_4\text{-}PbO_2$ electrodes have superior oxygen evolution reactions (OERs) relative to those of $Ti/Sb\text{-}SnO_2/PbO_2$ and $Ti/Sb\text{-}SnO_2/WC\text{-}PbO_2$ electrodes. A $Ti/Sb\text{-}SnO_2/Co_3O_4\text{-}PbO_2$ electrode is considered as the optimal modified electrode due to its long lifetime (684 h) and the remarkable stability of plating solutions. The treatment of copper wastewater suggests that composite electrodes exhibit low cell voltage and excellent extraction efficiency. Furthermore, pilot simulation tests verified that a composite electrode consumes less energy than other electrodes. Therefore, it is inferred that composite electrodes may be promising for the treatment of wastewater containing high concentrations of copper ions.

The composition of the electrode is also an important issue. Magnetic activated carbon particles (Fe_3O_4/active carbon composites) as auxiliary electrodes (AEs) were fixed onto the surface of $Ti/Sb\text{-}SnO_2$ foil by a NdFeB magnet to form a new magnetically assembled electrode (MAE) in Shao's paper [3]. Characterizations including cyclic voltammetry, Tafel analysis, and electrochemical impedance spectroscopy were carried out. The electrochemical oxidation performances of the new MAE towards different simulated wastewaters (azo dye acid red G, phenol, and lignosulfonate) were also studied. Series of the electrochemical properties of MAE were found to be varied with the loading amounts of AEs. The electrochemical area, as well as the number of active sites, increased significantly with the loading of AEs, and the charge transfer was also facilitated by these AEs. Target pollutants' removal of all simulated wastewaters were found to be enhanced when loading appropriate amounts of AEs. The accumulation of intermediate products was also determined by the loading amount of AEs. This new MAE may provide a more cost-effective and flexible method of electrochemical oxidation wastewater treatment (EOWT).

Octogen (HMX) is widely used as a high explosive and constituent in plastic explosives, nuclear devices, and rocket fuel. The direct discharge of wastewater generated during HMX production threatens the environment. In Qian's study [4], they used the electrochemical oxidation (EO) method with a PbO_2-based anode to treat HMX wastewater and investigated its degradation performance, mechanism, and toxicity evolution under different conditions. The results showed that HMX treated by EO was able to achieve a removal efficiency of 81.2% within 180 min at a current density of 70 mA/cm^2, Na_2SO_4 concentration of 0.25 mol/L, interelectrode distance of 1.0 cm, and pH of 5.0. The degradation followed pseudo-first-order kinetics ($R^2 > 0.93$). Degradation pathways of HMX in the EO system have been proposed, including cathode reduction and indirect oxidation by •OH radicals. The molecular toxicity level (expressed as the transcriptional effect level index) of HMX wastewater first increased to 1.81 and then decreased to a non-toxic level during the degradation process. Protein and oxidative stress were the dominant stress categories, possibly because of the intermediates that evolved during HMX degradation. This study provides new insights into the electrochemical degradation mechanisms and molecular-level toxicity evolution during HMX degradation. It also serves as initial evidence for the potential of the EO-enabled method as an alternative for explosive wastewater treatment with high removal performance, low cost, and low environmental impact.

Zhao [5] reported and demonstrated a photoelectrochemical filtration system that was shown to enable the effective decontamination of micropollutants from water. The key to this system was a photoelectric-active nanohybrid filter consisting of a carbon nanotube (CNT) and MIL–101(Fe). Various advanced characterization techniques were employed to obtain detailed information on the microstructure, morphology, and defect states of the nanohybrid filter. The results suggest that both radical and nonradical pathways collectively contributed to the degradation of antibiotic tetracycline, a model refractory micropollutant. The underlying working mechanism was proposed based on solid experimental evidence. This study provides new insights into the effective removal of micropollutants from water by integrating state-of-the-art advanced oxidation and microfiltration techniques.

Noble metal nanoparticle-loaded catalytic membrane reactors (CMRs) have emerged as a promising method for water decontamination. In Yan's study [6], a convenient and green strategy was proposed to prepare gold nanoparticle (Au NPs)-loaded CMRs. First, a redox-active substrate membrane (CNT-MoS_2), composed of carbon nanotubes (CNTs) and molybdenum disulfide (MoS_2), was prepared by an impregnation method. Water-diluted Au(III) precursor ($HAuCl_4$) was then spontaneously adsorbed on the CNT-MoS_2 membrane through filtration and reduced into Au(0) nanoparticles in situ, which involved a "adsorption–reduction" process between Au(III) and MoS_2. The constructed CNT-MoS_2@Au membrane demonstrated excellent catalytic activity and stability; a complete 4-nitrophenol transformation could be obtained within a hydraulic residence time of <3.0 s. In addition, thanks to the electroactivity of CNT networks, the as-designed CMR could also be applied to the electrocatalytic reduction of bromate (>90%) at an applied voltage of -1 V. More importantly, by changing the precursors, one could further obtain the other noble metal-based CMR (e.g., CNT-MoS_2@Pd) with superior (electro)catalytic activity. This study provided new insights into the rational design of high-performance CMRs for various environmental applications.

The water pollution caused by industry emissions makes effluent treatment a serious matter that needs to be settled. Heterogeneous Fenton oxidation has been recognized as an effective means to degrade pollutants in water. Attapulgite can be used as a catalyst carrier because of its distinctive spatial crystal structure and surface ion exchange. In Zhou's study [7], iron ions were transported on attapulgite particles to generate an iron-supporting attapulgite particle catalyst. BET, EDS, SEM, and XRD characterized the catalysts. The particle was used as a heterogeneous catalyst to degrade rhodamine B (RhB) dye in wastewater. The effects of H_2O_2 concentration, initial pH value, catalyst dosage, and temperature on the degradation of the dyes were studied. The results showed that the decolorization efficiency was consistently maintained after consecutive use of a granular catalyst five times, and the removal rate was more than 98%. The degradation and mineralization effect of cationic dyes by granular catalyst was better than that of anionic dyes. Hydroxyl radicals play a dominant role in RhB catalytic degradation. The dynamic change and mechanism of granular catalysts in the catalytic degradation of RhB were analyzed. In this study, the application range of attapulgite was widened. The prepared granular catalyst was inexpensive, stable, and efficient, and could be used to treat refractory organic wastewater.

In Yu's work [8], the degradation performance of $Fe^{2+}/PAA/H_2O_2$ on three typical pollutants (reactive black 5, ANL, and PVA) in textile wastewater was investigated in comparison with Fe^{2+}/H_2O_2. Therein, $Fe^{2+}/PAA/H_2O_2$ showed a high removal of RB5 (99%), mainly owing to the contribution of peroxyl radicals and/or Fe(IV). Fe^{2+}/H_2O_2 showed a relatively high removal of PVA (28%), mainly resulting from $\cdot$OH. $Fe^{2+}/PAA/H_2O_2$ and Fe^{2+}/H_2O_2 showed comparative removals of ANL. Additionally, $Fe^{2+}/PAA/H_2O_2$ was more sensitive to pH than Fe^{2+}/H_2O_2. The coexisting anions (20–2000 mg/L) showed inhibition on their removals and followed an order of $HCO_3^- > SO_4^{2-} > Cl^-$. Humic acid (5 and 10 mg C/L) showed notable inhibition on their removal following an order of reactive black 5 (RB5) > ANL > PVA. In a practical wastewater effluent, PVA removal was dramatically inhibited by 88%. Test results regarding bioluminescent bacteria suggested that the toxicity of $Fe^{2+}/PAA/H_2O_2$-treated systems was lower than that of Fe^{2+}/H_2O_2. RB5 degradation had three possible pathways with the proposed mechanisms of hydroxylation, dehydrogenation, and demethylation. The results may favor the performance evaluation of $Fe^{2+}/PAA/H_2O_2$ in the advanced treatment of textile wastewater.

In recent years, with the large-scale use of antibiotics, the pollution of antibiotics in the environment has become increasingly serious and has attracted widespread attention. In Qian's study [9], a novel CDs/g-C_3N_4/$BiPO_4$ (CDBPC) composite was successfully synthesized by a hydrothermal method for the removal of the antibiotic tetracycline hydrochloride (TC) in water. The experimental results showed that the synthesized photocatalyst was crystalline rods and cotton balls, accompanied by overlapping layered nanosheet structures, and the specific surface area was as high as 518.50 m^2/g. This photocatalyst contains

$g-C_3N_4$ and bismuth phosphate ($BiPO_4$) phases, as well as abundant surface functional groups such as C=N, C-O, and P-O. When the optimal conditions were pH 4, CDBPC dosage of 1 g/L, and TC concentration of 10 mg/L, the degradation rate of TC reached 75.50%. Active species capture experiments showed that the main active species in this photocatalytic system were holes (h+), hydroxyl radicals, and superoxide anion radicals. The reaction mechanism for the removal of TC by CDBPC was also proposed. The removal of TC was mainly achieved by the synergy between the adsorption of CDBPC and the oxidation of both holes and hydroxyl radicals. In this system, TC was adsorbed on the surface of CDBPC; then, the adsorbed TC was degraded into small molecular products by an attack with holes and hydroxyl radicals and, finally, mineralized into carbon dioxide and water. This study indicated that this novel photocatalyst, CDBPC, has significant potential for antibiotic removal, which provides a new strategy for antibiotic treatment of wastewater.

In Gan's study [10], nitrogen-doped biochar (N-PPB) and nitrogen-doped activated biochar (AN-PPB) were prepared and used for removing bisphenol A (BPA) in water through activating peroxymonosulfate. It was found from the results that N-PPB exhibited superior catalytic performance over pristine biochar, since nitrogen was able to bring about abundant active sites to the biochar structure. The non-radical singlet oxygen ($^1O_2$) was determined to be the dominant active species responsible for BPA degradation. Having a non-radical pathway in the N-PPB/PMS system, the BPA degradation was barely influenced by many external environmental factors, including solution pH value, temperature, and foreign organic and inorganic matters. Furthermore, AN-PPB had richer porosity than N-PPB, which showed even faster BPA removal efficiency than N-PPB through adsorptive/catalytic synergy. The finding of this study introduced a novel way of designing hieratical structured biochar catalysts for effective organic pollutant removal in water.

The purpose of Wang's work [11] was to optimize the catalytic performance of biochar (BC), improve the removal effect of BC composites on organic pollutants in wastewater, and promote the recycling and sustainable utilization of water resources. Firstly, the various characteristics and preparation principles of new BC are discussed. Secondly, the types of organic pollutants in wastewater and their removal principles are discussed. Finally, based on the principle of removing organic pollutants, a BC/zero valent iron (BC/ZVI) composite is designed, in which BC is mainly used for catalysis. The effect of BC/ZVI on the removal tetracycline (TC) was comprehensively evaluated. The research results revealed that the TC removal effect of pure BC is not ideal, and that of ZVI is general. The BC/ZVI composite prepared by combining the two had a better removal effect on TC, with a removal amount of about 275 mg/g. Different TC concentrations, ethylene diamine tetraacetic acid (EDTA), pH environment, tert-butanol, and calcium ions were shown to affect the TC removal effect of BC composites. The overall effect was the improvement of the TC removal amount of BC composites. This reveals that BC has a very suitable catalytic effect on ZVI, and the performance of BC composite material integrating the BC catalyst and ZVI was effectively improved; it can play a very suitable role in wastewater treatment. This exploration provides a technical reference for the effective removal of organic pollutants in wastewater and contributes to the development of water resource recycling.

In Wang's study [12], single-spherical nanoscale zero valent iron (nZVI) particles with large specific surface areas were successfully synthesized by a simple and rapid chemical reduction method. The XRD spectra and SEM–EDS images showed that the synthesized nZVI had excellent crystal structure, but oxidation products, such as γ-Fe_2O_3 and Fe_3O_4, were formed on the surface of the particles. The effects of different factors on the removal of Cr(VI) by nZVI were studied, and the optimum experimental conditions were found. Kinetic and thermodynamic equations at different temperatures showed that the removal of Cr(VI) by nZVI was a single-layer chemical adsorption, conforming to pseudo-second-order kinetics. Applying the intraparticle diffusion model, the adsorption process was composed of three stages, namely rapid diffusion, chemical reduction, and internal saturation. Analysis of the mechanism demonstrated that the removal of Cr(VI) by nZVI involved adsorption, reduction, precipitation, and coprecipitation. Meanwhile, Cr(VI) was reduced to Cr(III) by nZVI, while $FeCr_2O_4$, $CrxFe_{1\text{-}x}OOH$, and $CrxFe_{1-x}(OH)_3$ were formed as end products. In addition, the study found that ascorbic acid, starch, and Cu-modified nZVI were able to promote the removal efficiency of Cr(VI) in varying degrees due to the enhanced mobility of the particles. These results can provide new insights into the removal mechanisms of Cr(VI) by nZVI.

In Xiong's study [13], a Co-Mn/CeO_2 composite was prepared through a facile sol-gel method and used as an efficient catalyst for the ozonation of norfloxacin (NOR). The Co-Mn/CeO_2 composite was characterized via XRD, SEM, BET, and XPS analysis. The catalytic ozonation of NOR by Co-Mn/CeO_2 under different conditions was systematically investigated, including the effect of the initial solution's pH, Co-Mn/CeO_2 composite dose, O_3 dose, and NOR concentration on degradation kinetics. Only about 3.33% of the total organic carbon (TOC) and 72.17% of NOR could be removed within 150 min by single ozonation under the conditions of 60 mg/L of NOR and 200 mL/min of O_3 at pH = 7 and room temperature, whereas in the presence of 0.60 g/L of the Co-Mn/CeO2 composite under the same conditions, 87.24% NOR removal was obtained through the catalytic ozonation process. The results show that catalytic ozonation with the Co-Mn/CeO_2 composite can effectively enhance the degradation and mineralization of NOR compared to a single ozonation system alone. The catalytic performance of CeO_2 was significantly improved by modification with Mn and Co. Co-Mn/CeO_2 represents a promising way to prepare efficient catalysts for the catalytic ozonation of organic polluted water. The removal efficiency of NOR in five cycles indicates that Co-Mn/CeO_2 is stable and recyclable for catalytic ozonation in water treatment.

In the future, there will be additional new catalysts and new catalytic methods proposed by scholars, and our Special Issue will continue to focus on the latest progress in this field.

Conflicts of Interest: The authors declare no conflict of interest.

References

1. Guo, S.; Xu, Z.; Hu, W.; Tang, D.; Wang, X.; Xu, H.; Xu, X.; Long, Z.; Yan, W. Progress in Preparation and Application of Titanium Sub-Oxides Electrode in Electrocatalytic Degradation for Wastewater Treatment. *Catalysts* **2022**, *12*, 618. [CrossRef]
2. Kang, X.; Wu, J.; Wei, Z.; Jia, B.; Feng, Q.; Xu, S.; Wang, Y. Modification of Ti/Sb-SnO_2/PbO_2 Electrode by Active Granules and Its Application in Wastewater Containing Copper Ions. *Catalysts* **2023**, *13*, 515. [CrossRef]
3. Zhang, F.; Shao, D.; Yang, C.; Xu, H.; Yang, J.; Feng, L.; Wang, S.; Li, Y.; Jia, X.; Song, H. New Magnetically Assembled Electrode Consisting of Magnetic Activated Carbon Particles and Ti/Sb-SnO_2 for a More Flexible and Cost-Effective Electrochemical Oxidation Wastewater Treatment. *Catalysts* **2023**, *13*, 7. [CrossRef]
4. Qian, Y.; Chen, K.; Chai, G.; Xi, P.; Yang, H.; Xie, L.; Qin, L.; Lin, Y.; Li, X.; Yan, W.; et al. Performance Optimization and Toxicity Effects of the Electrochemical Oxidation of Octogen. *Catalysts* **2022**, *12*, 815. [CrossRef]
5. Zhao, W.; Dai, Y.; Zheng, W.; Liu, Y. Peroxymonosulfate Activation by Photoelectroactive Nanohybrid Filter towards Effective Micropollutant Decontamination. *Catalysts* **2022**, *12*, 416. [CrossRef]
6. Yan, H.; Liu, F.; Zhang, J.F.; Liu, Y. Synthesis and Environmental Applications of Noble Metal-Based Catalytic Membrane Reactors. *Catalysts* **2022**, *12*, 861. [CrossRef]
7. Zhou, P.; Dai, Z.; Lu, T.; Ru, X.; Ofori, M.A.; Yang, W.; Hou, J.; Jin, H. Degradation of Rhodamine B in Wastewater by Iron-Loaded Attapulgite Particle Heterogeneous Fenton Catalyst. *Catalysts* **2022**, *12*, 669. [CrossRef]
8. Yu, J.; Shu, S.; Wang, Q.; Gao, N.; Zhu, Y. Evaluation of Fe^{2+}/Peracetic Acid to Degrade Three Typical Refractory Pollutants of Textile Wastewater. *Catalysts* **2022**, *12*, 684. [CrossRef]
9. Qian, W.; Hu, W.; Jiang, Z.; Wu, Y.; Li, Z.; Diao, Z.; Li, M. Degradation of Tetracycline Hydrochloride by a Novel CDs/g-C_3N_4/$BiPO_4$ under Visible-Light Irradiation: Reactivity and Mechanism. *Catalysts* **2022**, *12*, 774. [CrossRef]
10. Lu, H.; Xu, G.; Gan, L. N Doped Activated Biochar from Pyrolyzing Wood Powder for Prompt BPA Removal via Peroxymonosulfate Activation. *Catalysts* **2022**, *12*, 1449. [CrossRef]
11. Wang, G.; Zong, S.; Ma, H.; Wan, B.; Tian, Q. Removal Efficiency and Performance Optimization of Organic Pollutants in Wastewater Using New Biochar Composites. *Catalysts* **2023**, *13*, 184. [CrossRef]

12. Gao, Y.; Yang, X.; Lu, X.; Li, M.; Wang, L.; Wang, Y. Kinetics and Mechanisms of Cr(VI) Removal by nZVI: Influencing Parameters and Modification. *Catalysts* **2022**, *12*, 999. [CrossRef]
13. Li, R.; Xiong, J.; Zhang, Y.; Wang, S.; Zhu, H.; Lu, L. Catalytic Ozonation of Norfloxacin Using Co-Mn/CeO_2 as a Multi-Component Composite Catalyst. *Catalysts* **2022**, *12*, 1606. [CrossRef]

MDPI
St. Alban-Anlage 66
4052 Basel
Switzerland
www.mdpi.com

Catalysts Editorial Office
E-mail: catalysts@mdpi.com
www.mdpi.com/journal/catalysts

www.ingramcontent.com/pod-product-compliance
Lightning Source LLC
LaVergne TN
LVHW071009170726
843515LV00006B/1119

* 9 7 8 3 0 3 6 5 9 8 6 6 6 *